P9-AQW-968

INTRODUCTION TO LOGIC

PATRICK SUPPES

Lucie Stern Professor of Philosophy, Emeritus
Stanford University

DOVER PUBLICATIONS, INC.
Mineola, New York

Copyright

Published in Canada by General Publishing Company, Ltd., 30 Lesmill Road, Don Mills, Toronto, Ontario.

Bibliographical Note

This Dover edition, first published in 1999, is an unabridged republication of the work originally published in New York in 1957 by Van Nostrand Reinhold Company, a division of Litton Educational Publishing, Inc.

Library of Congress Cataloging-in-Publication Data

Suppes, Patrick, 1922–
Introduction to logic / Patrick Suppes.
p. cm.
Originally published: New York : Van Nostrand Reinhold, 1957.
Includes index.
ISBN 0-486-40687-3 (pbk.)
1. Logic. I. Title.
BC108.S85 1999
160—dc21 99-13623
CIP

Manufactured in the United States of America
Dover Publications, Inc., 31 East 2nd Street, Mineola, N.Y. 11501

PREFACE

This book has been written primarily to serve as a textbook for a first course in modern logic. No background in mathematics or philosophy is supposed. My main objective has been to familiarize the reader with an exact and complete theory of logical inference and to show how it may be used in mathematics and the empirical sciences. Since several books already available have aims closely related to the one just stated, it may be well to mention the major distinguishing features of the present book.

Part I (the first eight chapters) deals with formal principles of inference and definition. Beginning with the theory of sentential inference in Chapter 2 there is continual emphasis on application of the method of interpretation to prove arguments invalid, premises consistent, or axioms of a theory independent. There is a detailed attempt (Chapter 7) to relate the formal theory of inference to the standard informal proofs common throughout mathematics. The theory of definition is presented (Chapter 8) in more detail than in any other textbook known to the author; a discussion of the method of Padoa for proving the independence of primitive concepts is included.

Part II (the last four chapters) is devoted to elementary intuitive set theory, with separate chapters on sets, relations, and functions. The treatment of ordering relations in Chapter 10 is rather extensive. Part II is nearly self-contained and can be read independently of Part I. The last chapter (Chapter 12) is concerned with the set-theoretical foundations of the axiomatic method. The idea that the best way to axiomatize a branch of mathematics is to define appropriate set-theoretical predicates is familiar to modern mathematicians and certainly does not originate with the author, but the exposition of this idea, which provides a sharp logical foundation for the axiomatic method, has been omitted from the excellent elementary textbooks on modern mathematics which have appeared in recent years.

Beginning with Chapter 4, numerous examples of axiomatically formulated theories are introduced in the discussion and exercises. These examples range from the theory of groups and the algebra of the real numbers to elementary probability theory, classical particle mechanics and the theory of measurement of sensation intensities. (The section on mechanics is the one exception to the general statement that there are no mathematical prerequisites for the reading of this book; some knowledge of the differential and integral calculus is required for the full understanding of this section, the final one in the book.) Certain of the exercises included in connection with these substantive examples are more difficult than those ordinarily put in an elementary logic text. The purpose of these exercises is to challenge the ablest students. There is, however, a very large number of additional exercises of relatively simple character which adequately illustrate all the general principles introduced. It is hoped that the material on measurement, probability, and mechanics in Chapter 12 may be useful in some Philosophy of Science courses.

The system of inference for first-order predicate logic developed in Chapters 2, 4, and 5 has been designed to correspond as closely as possible to the author's conception of the most natural techniques of informal proof. Probably the most novel feature of the system is the method of handling existential quantifiers by the use of "ambiguous names"; the central idea of this approach is related to Hilbert's ϵ symbol. Since many teachers of logic have their own preferred rules for handling inferences with quantifiers, it should be mentioned that the particular rules introduced here play a major role only in Chapters 4 and 5.

Numerous people have contributed to the gradual development of this book. I am particularly indebted to Professor Robert McNaughton for many useful criticisms and suggestions, based on his teaching experience with earlier drafts; to Professor Herman Rubin, who contributed to the formulation of the system of natural deduction presented in Chapters 2 and 4; and to Mr. Dana Scott for many helpful suggestions concerning Chapters 8 and 12. I am also indebted to Professor Moffatt Hancock of the Stanford Law School for several exercises in Chapters 2 and 4. Various teaching assistants at Stanford have aided in the preparation of exercises and made numerous useful criticisms—notably Mr. Leonard Leving, Mrs. Muriel Wood Gerlach and Mrs. Rina Ullmann.

The last set of revisions has benefited from the comments and criticisms of Professors Ernest Adams, Herman Chernoff, Benson Mates, John Myhill, David Nivison, Hartley Rogers, Jr., Leo Simons, Robert Vaught, and Mr. Richard Robinson. The extraordinarily detailed and perspicacious criticisms of Professor Vaught were especially valuable. Miss Peggy Reis and Mrs. Karol Valpreda Walsh have been of much assistance in reading proofs. In addition, Miss Reis has cheerfully and accurately typed the several preliminary editions used in courses since the summer of 1954.

Originally this book was planned jointly with Professor J. C. C. McKinsey, but he died in 1953. Professor McKinsey wrote the original drafts of chapters 9, 10, and 11, but these chapters have been revised three times and the length of chapters 10 and 11 has been more than doubled, so his handiwork is now little evident. The other nine chapters are my sole responsibility in all respects.

PATRICK SUPPES

Stanford, California
April, 1957

TABLE OF CONTENTS

INTRODUCTION

Our everyday use of language is vague, and our everyday level of thinking is often muddled. One of the main purposes of this book is to introduce you to a way of thinking that encourages carefulness and precision. There are many ways to learn how to use language and ideas precisely. Our approach shall be through a study of logic. In modern times logic has become a deep and broad subject. We shall initially concentrate on that portion of it which is concerned with the theory of correct reasoning, which is also called the theory of logical inference, the theory of proof or the theory of deduction. The principles of logical inference are universally applied in every branch of systematic knowledge. It is often said that the most important critical test of any scientific theory is its usefulness and accuracy in predicting phenomena before the phenomena are observed. Any such prediction must involve application of the principles of logical inference. For example, if we know what forces are acting on a body and we know at a given time where the body is and what its velocity is, we may use the theory of mechanics together with the rules of logical inference and certain theorems of mathematics to predict where the body will be at some later time.

For over two thousand years mathematicians have been making correct inferences of a systematic and intricate sort, and logicians and philosophers have been analyzing the character of valid arguments. It is, therefore, somewhat surprising that a fully adequate formal theory of inference has been developed only in the last three or four decades. In the long period extending from Aristotle in the fourth century B.C. to Leibniz in the seventeenth century, much of importance and significance was discovered about logic by ancient, medieval and post-medieval logicians, but the most important defect in this classical tradition was the failure to relate logic as the theory of inference to the kind of deductive reasonings that are continually used in mathematics.

Leibniz had some insight into the necessity of making this connection, but not until the latter part of the nineteenth century and the beginning of this century were systematic relations between logic and mathematics established, primarily through the work of Frege, Peano, and Russell. In spite of the scope and magnitude of their researches, only in recent years has there been formulated a completely explicit theory of inference adequate to deal with all the standard examples of deductive reasoning in mathematics and the empirical sciences. The number of people who have contributed to these recent developments is large, but perhaps most prominent have been Kurt Gödel, David Hilbert, and Alfred Tarski.

Yet it is a mistake to think that the theory of inference developed in the first part of this book has relevance exclusively to scientific contexts. The theory applies just as well to proceedings in courts of law or to philosophical analyses of the eternal verities. Indeed, it is not too much to claim that the theory of inference is pertinent to every serious human deliberation.

A correct piece of reasoning, whether in mathematics, physics or casual conversation, is valid by virtue of its logical form. Because most arguments are expressed in ordinary language with the addition of a few technical symbols particular to the discipline at hand, the logical form of the argument is not transparent. Fortunately, this logical structure may be laid bare by isolating a small number of key words and phrases like 'and', 'not', 'every' and 'some'. In order to fix upon these central expressions and to lay down explicit rules of inference depending on their occurrence, one of our first steps shall be to introduce logical symbols for them. With the aid of these symbols it is relatively easy to state and apply rules of valid inference, a task which occupies the first seven chapters.

To bring logical precision to our analysis of ideas, it is not ordinarily enough to be able to construct valid inferences; it is also essential to have some mastery of methods for defining in an exact way one concept in terms of other concepts. In any given branch of science or mathematics one of the most powerful methods for eliminating conceptual vagueness is to isolate a small number of concepts basic to the subject at hand and then to define the other concepts of the discipline in terms of the basic set. The purpose of Chapter 8 is to lay down exact rules for giving such definitions. Correct definitions, like correct inferences, will be shown to depend primarily on matters of logical form. However, certain subtle questions of existence arise in the theory of definition which have no counterpart in the theory of inference.

The first eight chapters constitute Part I, which is devoted to general principles of inference and definition. Part II, the last four chapters, is concerned with elementary set theory. Because the several respects in which set theory is intimately tied to logic will not be familiar to many readers, some explanation for the inclusion of this material will not be amiss.

Set theory, or the general theory of classes as it is sometimes called, is the basic discipline of mathematics, for with a few rare exceptions the entities which are studied and analyzed in mathematics may be regarded as certain particular sets or classes of objects. As we shall see, the objects studied in a branch of pure mathematics like the theory of groups or in a branch of mathematical physics like the theory of mechanics may be characterized as certain sets. For this reason any part of mathematics may be called a special branch of set theory. However, since this usage would identify set theory with the whole of mathematics it is customary to reserve the term 'set theory' for the general theory of classes or sets and certain topics, such as the construction of the integers and real numbers as sets, which are closely connected historically with investigations into the foundations of mathematics.

The first chapter of Part II is concerned with an intuitive account of the more important relationships among arbitrary sets. There are, for example, simple operations on sets which correspond to the arithmetical operations of addition, multiplication, and subtraction. The next chapter (Chapter 10) deals with the theory of relations, which is brought within set theory via the notion of an ordered couple of objects. Emphasis is given to ordering relations because of their importance in many branches of mathematics and science. Chapter 11 deals with functions, which from the standpoint of set theory are just relations having a special property.

While the first three chapters of Part II are concerned with general set theory, the final chapter (Chapter 12) turns to the relation between set theory and certain methodological or foundational questions in mathematics and philosophy. The central point of this chapter is to indicate how any branch of mathematics or any scientific theory may be axiomatized within set theory. The viewpoint which is expounded in detail in Chapter 12 is that the best way to axiomatize a theory is to define an appropriate predicate within set theory.

Since the beginning of this century philosophers have written a great deal about the structure of scientific theories but they have said lamentably little about the detailed structure of particular theories. The axiomatization of a theory within set theory is an important initial step in making its structure both exact and explicit. Once such an axiomatization is provided it is then possible to ask the kind of "structure" questions characteristic of modern mathematics. For instance, when are two models of a theory isomorphic, that is, when do they have exactly the same structure? Indeed, familiar philosophical problems like the reduction of one branch of empirical science to another may be made precise in terms of such set-theoretical notions as that of isomorphism. Application of these ideas to substantive examples from pure mathematics and the empirical sciences is given in Chapter 12.

The aim of both Parts I and II is to present logic as a part of mathematics and science and to show by numerous detailed examples how relevant logic is even to empirical sciences like psychology. For this reason it may be said that the emphasis in this book is on the systematic use and application of logic rather than on the development of logic as an autonomous discipline.

Finally, it should be remarked that no precise definition of *logic* is attempted in these pages. In the narrow sense, logic is the theory of valid arguments or the theory of deductive inference. A slightly broader sense includes the theory of definition. A still broader sense includes the general theory of sets. Moreover, the theory of definition together with the theory of sets provides an exact foundation for the axiomatic method, the study of which is informally considered part of logic by most mathematicians.

PART I

PRINCIPLES OF INFERENCE AND DEFINITION

CHAPTER 1

THE SENTENTIAL CONNECTIVES

To begin with, we want to develop a vocabulary which is precise and at the same time adequate for analysis of the problems and concepts of systematic knowledge. We must use vague language to create a precise language. This is not as silly as it seems. The rules of chess, for example, are a good deal more precise than those of English grammar, and yet we use English sentences governed by imprecise rules to state the precise rules of chess. In point of fact, our first step will be rather similar to drawing up the rules of a game. We want to lay down careful rules of usage for certain key words: 'not', 'and', 'or', 'if . . ., then . . .', 'if and only if', which are called *sentential connectives*. The rules of usage will not, however, represent the rules of an arbitrary game. They are designed to make explicit the predominant systematic usage of these words; this systematic usage has itself arisen from reflection on the ways in which these words are used in ordinary, everyday contexts. Yet we shall not hesitate to deviate from ordinary usage whenever there are persuasive reasons for so doing.

§ **1.1 Negation and Conjunction.** We deny the truth of a sentence by asserting its negation. For example, if we think that the sentence 'Sugar causes tooth decay' is false, we assert the sentence 'Sugar does not cause tooth decay'. The usual method of asserting the negation of a simple sentence is illustrated in this example: we attach the word 'not' to the main verb of the sentence. However, the assertion of the negation of a compound sentence is more complicated. For example, we deny the sentence 'Sugar causes tooth decay and whiskey causes ulcers' by asserting 'It is not the case that both sugar causes tooth decay and whiskey causes ulcers'. In spite of the apparent divergence between these two examples, it is convenient to adopt in logic a single sign for forming the negation of a sentence. We shall use the prefix '–', which is placed before the whole sentence. Thus the negation of the first example is written:

–(Sugar causes tooth decay).

The second example illustrates how we may always translate '–'; we may always use 'it is not the case that'.

The main reason for adopting the single sign '–' for negation, regardless of whether the sentence being negated is simple or compound, is that the meaning of the sign is the same in both cases. *The negation of a true sentence is false, and the negation of a false sentence is true.*

We use the word 'and' to conjoin two sentences to make a single sentence which we call the *conjunction* of the two sentences. For example, the sentence 'Mary loves John and John loves Mary' is the conjunction of the sentence 'Mary loves John' and the sentence 'John loves Mary'. We shall use the ampersand sign '&' for conjunction. Thus the conjunction of any two sentences P and Q is written

P & Q.

The rule governing the use of the sign '&' is in close accord with ordinary usage. *The conjunction of two sentences is true if and only if both sentences are true.* We remark that in logic we may combine any two sentences to form a conjunction. There is no requirement that the two sentences be related in content or subject matter. Any combinations, however absurd, are permitted. Of course, we are usually not interested in sentences like 'John loves Mary, and 4 is divisible by 2'. Although it might seem desirable to have an additional rule stating that we may only conjoin two sentences which have a common subject matter, the undesirability of such a rule becomes apparent once we reflect on the vagueness of the notion of common subject matter.

Various words are used as approximate synonyms for 'not' and 'and' in ordinary language. For example, the word 'never' in the sentence:

I will never surrender to your demands

has almost the same meaning as 'not' in:

I will not surrender to your demands.

Yet it is true that 'never' carries a sense of continuing refusal which 'not' does not.

The word 'but' has about the sense of 'and', and we symbolize it by '&', although in many cases of ordinary usage there are differences of meaning. For example, if a young woman told a young man:

I love you and I love your brother almost as well,

he would probably react differently than if she had said:

I love you but I love your brother almost as well.

In view of such differences in meaning, a natural suggestion is that different symbols be introduced for sentential connectives like 'never' and 'but'. There is, however, a profound argument against such a course of action. The rules of usage agreed upon for negation and conjunction make these two sentential connectives *truth-functional;* that is, the truth or falsity of the negation of a sentence P, or the truth or falsity of the conjunction of two sentences P and Q is a function just of the truth or falsity of P in the case of negation, and of P and Q in the case of conjunction. Clearly a truth-functional analysis of 'but' different from that given for 'and' is out of the question, but any venture into non-truth-functional analysis leads to considerations which are vague and obscure. Any doubt about this is quickly dispelled by the attempt to state a precise rule of usage for 'but' which differs from that already given for 'and'.

Of course, the rich, variegated character of English or any other natural language guarantees that in many contexts connectives are used in delicately shaded, non-truth-functional ways. Loss in subtlety of nuance seems a necessary concomitant to developing a precise, symbolic analysis of sentences. But this process of distorting abstraction is not peculiar to logic; it is characteristic of science in general. Few poets would be interested in a truth-functional analysis of language, and no naturalist would consider the physicist's concepts of position, velocity, acceleration, mass, and force adequate to describe the flight of an eagle. The concepts of logic developed in this book are useful in discovering and communicating systematic knowledge, but their relevance to other functions of language and thought is less direct.

§ 1.2 Disjunction. We use the word 'or' to obtain the *disjunction* of two sentences. In everyday language, the word 'or' is used in two distinct senses. In the so-called *non-exclusive* sense, the disjunction of two sentences is true if at least one of the sentences is true. In legal contracts this sense is often expressed by the barbarism 'and/or', illustrated in the following example:

> Before any such work is done or any such materials are furnished, the Lessee and any contractor or other person engaged to do such work and/or furnish such materials shall furnish such bond or bonds as the Lessor may reasonably require ....

We remark that in the above example there are no disjunctions of sentences, but disjunctions of clauses or terms which are not sentences. We shall find, however, that it is more convenient to treat such examples as disjunctions of sentences; this viewpoint reflects another divergence between logic and everyday language.

The Latin word 'vel' has approximately the sense of 'or' in the non-exclusive sense, and consequently we use the sign '$\vee$' for the disjunction of two sentences in this sense. Thus the disjunction of any two sentences

P and Q is written

$$P \vee Q.$$

We shall restrict our use of the word 'disjunction' to the non-exclusive sense, and our rule of usage is: *The disjunction of two sentences is true if and only if at least one of the sentences is true.*

When people use 'or' in the *exclusive* sense to combine two sentences, they are asserting that one of the sentences is true and the other is false. This usage is often made more explicit by adding the phrase 'but not both'. Thus a father tells his child, 'You may go to the movies or you may go to the circus this Saturday but not both'. We shall introduce no special sign for 'or' in the exclusive sense, for it turns out that in scientific discussions we can always get along with 'or' in the non-exclusive sense (which is also called the *inclusive* sense).

§ 1.3 Implication: Conditional Sentences. We use the words 'if ..., then ...' to obtain from two sentences a *conditional sentence.* A conditional sentence is also called an *implication.* As words are used in everyday language, it is difficult to characterize the circumstances under which most people will accept a conditional sentence as true. Consider an example similar to one we have already used:

(1) If Mary loves John, then John loves Mary.

If the sentence 'Mary loves John' is true and the sentence 'John loves Mary' is false, then everyone would agree that (1) is false. Furthermore, if the sentence 'Mary loves John' is true and the sentence 'John loves Mary' is also true, then nearly everyone would agree that (1) is true. The two possibilities of truth and falsity which we have just stated are the only ones that arise very often in the ordinary use of language. There are, however, two further possibilities, and if we ask the proverbial man in the street about them, there is no telling what his reply will be. These two further cases are the following. Suppose that the sentence 'Mary loves John' is false, then what do we say about the truth of (1): first, when the sentence 'John loves Mary' is also false; and second, when the sentence 'John loves Mary' is true? In mathematics and logic, this question is answered in the following way: sentence (1) is true if the sentence 'Mary loves John' is false, regardless of the truth or falsity of the sentence 'John loves Mary'.

To state our rule of usage for 'if ..., then ...', it is convenient to use the terminology that the sentence immediately following 'if' is the *antecedent* or *hypothesis* of the conditional sentence, and the sentence immediately following 'then' is the *consequent* or *conclusion.* Thus 'Mary loves John' is the antecedent of (1), and 'John loves Mary' is the consequent. The rule of usage is then: *A conditional sentence is false if the antecedent is true and the consequent is false; otherwise it is true.*

Intuitive objections to this rule could be made on two counts. First, it can be maintained that implication is not a truth-functional connective, but that there should be some sort of definite connection between the antecedent and the consequent of a conditional sentence. According to the rule of usage just stated, the sentence:

(2) If poetry is for the young, then $3 + 8 = 11$

is true, since the consequent is true. Yet many people would want to dismiss such a sentence as nonsensical; they would claim that the truth of the consequent in no way *depends* on the truth of the antecedent, and therefore (2) is not a meaningful implication. However, the logician's commitment to truth-functional connectives is not without its reasons. How is one to characterize such an obscure notion as that of dependence? This is the same problem we encountered in considering conjunctions. If you think an important, perhaps crucial problem is being dodged simply on the grounds that it is difficult, assurances will be forthcoming in the next chapter that truth-functional connectives are very adequate for both the theory and practice of logical inference.

Even if truth-functional commitments are accepted, a second objection to the rule of usage for implication is that the wrong stipulation has been made in calling any implication true when its antecedent is false. But particular examples argue strongly for our rule. For the case when the consequent is also false, consider:

(3) If there are approximately one hundred million husbands in the United States, then there are approximately one hundred million wives in the United States.

It is hard to imagine anyone denying the truth of (3). For the case when the consequent is true, consider the following modification of (3):

(4) If there are approximately one hundred million husbands in the United States, then the number of husbands in this country is greater than the number in France.

If (3) and (4) are admitted as true, then the *truth-functional* rule for conditional sentences with false antecedents is fixed.

It might be objected that by choosing slightly different examples a case could be made for considering any implication false when its antecedent is false. For instance, suppose that (3) were replaced by:

(5) If there are approximately one hundred million husbands in the United States, then there is exactly one wife in the United States.

Then within our truth-functional framework it might be maintained that an implication with false antecedent and false consequent is false, since it may be plausibly argued that in ordinary usage (5) is false. However, there are good grounds for choosing (3) rather than (5), for (3) has the property that its consequent follows from its antecedent on the basis of some familiar principles of arithmetic and marriage. With respect to (5) no such intuitive line of reasoning seems possible; in fact, it is the very absence of such a connection which makes us declare it false. Although we have already admitted that the notion of connection or dependence being appealed to here is too vague to be a formal concept of logic, in choosing examples which will force upon us, within our truth-functional framework, a truth value for implications with false antecedents it is reasonable to pick an example like (3) for which our intuitive feeling of dependence is strong rather than an example like (5) for which it is weak. The truth-functional demand that sentences like (5) be counted as true has no undesirable effects, since conditional sentences whose antecedents and consequents are unrelated and whose antecedents are false play no serious role in systematic arguments.

As a matter of notation, the conditional sentence formed from any two sentences P and Q is written

$$P \rightarrow Q.$$

The sign '→' is often called the sign of *implication*. Several other idioms in English have approximately the same systematic meaning as 'if ..., then ...'. We shall also write P → Q, for

P only if Q
Q if P
Q provided that P
P is a sufficient condition for Q
Q is a necessary condition for P

Of these five idioms, variant use of 'only if' is most pronounced. It is a common "mistake" to use 'only if' in the sense of 'if'. For example, the sentence:

(6) John dates Mary only if Elizabeth is mad at him

would not ordinarily be taken to mean:

If John dates Mary then Elizabeth is mad at him,

and it would be more accurate (but still not exactly idiomatically correct) to translate (6) as:

If Elizabeth is mad at him then John dates Mary.

The prevalence of sentences like (6) makes it difficult for many people first learning logic or mathematics to accept the stipulation that

(7) P only if Q

means the same as

(8) If P then Q.

Yet it is the case that in scientific discourse (7) and (8) are idiomatically equivalent and they will be treated as such throughout the rest of this book.

Concerning the last two idioms it is worth noting that they are widely used in mathematics. Thus the sentence 'If a triangle is equilateral then it is isosceles' may be rephrased:

In order for a triangle to be isosceles it is sufficient that it be equilateral

or:

It is necessary that an equilateral triangle be isosceles.

Notice that some grammatical changes in the component sentences P and Q are appropriate when we go from

If P then Q

to

(9) P is a sufficient condition for Q

so that (9) is not an exact formulation; but these changes are usually obvious and need not be pursued here.

§ 1.4 Equivalence: Biconditional Sentences. We use the words 'if and only if' to obtain from two sentences a *biconditional sentence*. A biconditional sentence is also called an *equivalence*, and the two sentences connected by 'if and only if' are called the *left* and *right members of the equivalence*. The biconditional

(1) P if and only if Q

has the same meaning as the sentence

(2) P if Q, and P only if Q

and (2) is equivalent to

(3) If P then Q, and if Q then P.

Our rules of usage for conjunction and implication tell us that (3) is true just when P and Q are both true or both false. Thus the rule: *A biconditional sentence is true if and only if its two members are either both true or both false.* As a matter of notation, we write

$$P \leftrightarrow Q$$

for the biconditional formed from sentences P and Q.

Corresponding to our remarks at the end of the last section it should be noted that (1) is equivalent to

Q is a necessary and sufficient condition for P.

§ 1.5 Grouping and Parentheses. In ordinary language the proper grouping of sentences which are combined into a compound sentence is indicated by a variety of linguistic devices. When symbolizing such sentences in logic, these devices may all be accurately translated by an appropriate use of parentheses.

For instance, the sentence:

If Showboat wins the race, then Shotless and Ursula will show

is symbolized by

$$S \rightarrow (H \& U), \tag{1}$$

where S is 'Showboat wins the race', H is 'Shotless will show', and U is 'Ursula will show'. We read (1)

If S then H and U.

On the other hand, we read

$$(S \rightarrow H) \& U \tag{2}$$

as

Both if S then H, and U.

It should thus be clear why (1) rather than (2) is the correct symbolization of the original sentence. The parentheses are used in a natural way, familiar from elementary algebra, to indicate which connective is dominant.

By adopting one natural convention concerning the relative dominance of the various connectives, a considerable reduction in the number of parentheses used in practice will be effected. The convention is '$\leftrightarrow$' and '$\rightarrow$' dominate '&' and '$\vee$'. Thus (1) may be written

$$S \rightarrow H \& U, \tag{3}$$

and

$$P \leftrightarrow Q \& R$$

means

$$P \leftrightarrow (Q \& R);$$

On the other hand, under this convention it is not clear what

$$P \,\&\, Q \vee R$$

is supposed to mean, and similarly for

$$P \leftrightarrow Q \rightarrow R.$$

§ **1.6 Truth Tables and Tautologies.** Our truth-functional rules of usage for negation, conjunction, disjunction, implication and equivalence may be summarized in tabular form. These *basic truth tables* tell us at a glance under what circumstances the negation of a sentence is true if we know the truth or falsity of the sentence, similarly for the conjunction of two sentences, and the disjunction or implication of two sentences as well.

Negation

P	−P
T	F
F	T

Conjunction

P	Q	P & Q
T	T	T
T	F	F
F	T	F
F	F	F

Disjunction

P	Q	P ∨ Q
T	T	T
T	F	T
F	T	T
F	F	F

Implication

P	Q	P → Q
T	T	T
T	F	F
F	T	T
F	F	T

Equivalence

P	Q	P ↔ Q
T	T	T
T	F	F
F	T	F
F	F	T

We may think of using the basic truth tables in the following manner. If N is the true sentence 'Newton was born in 1642' and G is the false sentence 'Galileo died in 1640', then we may compute the truth or falsity of a complicated compound sentence such as

(1) $$((N \vee G) \,\&\, -N) \rightarrow (G \rightarrow N).$$

Since N is true, we see from the disjunction table that $N \vee G$ is true, from the negation table that $-N$ is false, and hence from the conjunction table that the antecedent of (1) is false. Finally, from the implication table we conclude that the whole sentence is true. A more explicit application of the truth tables in a manner analogous to the use of a multiplication table is illustrated by the following diagrammatic analysis, which is self-explanatory.

$$((N \vee G) \ \& \ -N) \rightarrow (G \rightarrow N)$$

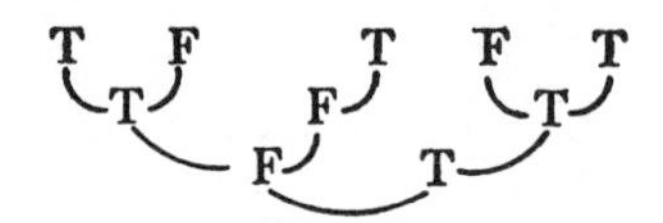

(Note that in this diagram the analysis proceeds from the inside out. The final loop connects the two members of the major connective.)

Let us call a sentence *atomic* if it contains no sentential connectives. Thus the sentence:

Mr. Knightley loved Emma

is atomic, while the sentences:

Emma did not love Frank Churchill

and

Mrs. Elton was a snob and Miss Bates a bore

are not atomic, for the first contains a negation and the second a conjunction.

We now use the concept of a sentence being atomic to define what is probably the most important notion of this chapter. The intuitive idea is that a compound sentence is a *tautology* if it is true independently of the truth values of its component atomic sentences. For instance for any atomic sentence P

$$P \vee -P$$

is a tautology. If P is true, we have:

$$P \vee -P$$

T T
F
T

If P is false, we have:

$$P \vee -P$$

F F
T
T

Thus whether P is true or false, P ∨ −P is true and hence a tautology. *Derived truth tables* are more convenient and compact than the diagrammatic analysis shown above when we want to know if a sentence is a tautology.

P	−P	P ∨ −P
T	F	T
F	T	T

The second column is obtained from the first by using the negation table, and the third column from the first two by using the disjunction table. Since both lines in the final column have the entry 'T', the whole sentence is a tautology. The idea of the derived truth table is that a sentence is a tautology if it is true for all combinations of possible truth values of its component atomic sentences. The number of such combinations depends on the number of component atomic sentences. Thus, if there are three distinct atomic sentences, there are eight distinct combinations of possible truth values, since each atomic sentence has exactly two possible truth values: truth and falsity. In general, if there are n component atomic sentences, there are 2^n combinations of possible truth values, which means that the derived truth table for a compound sentence having n distinct atomic sentences has 2^n lines. For instance, to show that P ∨ Q → P is not a tautology when P and Q are distinct atomic sentences, we need $2^2 = 4$ lines, as in the following truth table.

P	Q	P ∨ Q	P ∨ Q → P
T	T	T	T
T	F	T	T
F	T	T	F
F	F	F	T

In this table the third column is obtained from the first two by using the disjunction table, and the final column from the third and first by using the implication table. Since the third row of the fourth column has the entry 'F' for false, we conclude that P ∨ Q → P is not a tautology, for this third row shows that if P is false and Q true, then P ∨ Q → P is false. It should be emphasized that a sentence is a tautology if and only if every entry in the final column is 'T' (for true). The letter 'F' in a single row of the final column is sufficient to guarantee that the sentence being analyzed is not a tautology.

When we take as our formal definition

> *A sentence is a tautology if and only if the result of replacing any of its component atomic sentences (in all occurrences) by other atomic sentences is always a true sentence*

the relation of this definition to the truth table test for a tautology should be clear.* A given row of the table represents trying a particular combination of atomic sentences. Since only the truth or falsity of the atomic sentences effects the truth or falsity of the whole sentence in a truth-functional analysis, once all possible combinations of truth and falsity have been tested, the effects of all possible substitutions of atomic sentences have been tested. As we have seen, if a sentence contains just one atomic sentence, there are only two possibilities:

T
F

If it has two distinct component atomic sentences, there are four:

T T
T F
F T
F F

If it has three distinct component atomic sentences, there are eight:

T T T
T T F
T F T
T F F
F T T
F T F
F F T
F F F

And in general, as we have already remarked, if there are n distinct component atomic sentences there are 2^n possible combinations of truth values and thus 2^n rows to the truth table.

The phrase 'in all occurrences' is added parenthetically in the formal definition to make explicit that all occurrences of a given atomic sentence in a compound sentence are to be treated alike. Thus in the truth table for $P \vee Q \rightarrow P$ given above, one occurrence of P is given the value T when and only when the other is. This point is made clearer by using the following format (which is not as convenient for computational purposes):

* It is understood that we are considering only truth-functional sentences.

P	∨	Q	→	P
T	T	T	T	T
T	T	F	T	T
F	T	T	F	F
F	F	F	T	F

Note the identical columns for both occurrences of P.*

It is also important to notice that if the requirement that P and Q be distinct atomic sentences is lifted, instances of P and Q can be found for which $P \vee Q \rightarrow P$ is a tautology: e.g., let P be 'it is raining or it is not raining' and Q be 'it is hot'. Then it is easily shown by means of the appropriate derived truth table that the sentence 'if either it is raining or it is not raining or it is hot then either it is raining or it is not raining' is a tautology. Furthermore, if P and Q are the same sentence then $P \vee Q \rightarrow P$ is a tautology. As a second example, in general $P \rightarrow Q$ is not a tautology, that is, it is easy to find sentences P and Q such that $P \rightarrow Q$ is not a tautology, but if P and Q are the same sentence $P \rightarrow Q$ is a tautology.

Finally it should also be noticed that if a sentence is a tautology, we may substitute any compound sentence for a component atomic sentence (in all its occurrences) and the result will be a tautology. For example, the appropriate truth table quickly shows that if P is an atomic sentence $P \rightarrow P$ is a tautology, but once this is shown it easily follows that $P \rightarrow P$ is a tautology when P is any sentence whatsoever.

The last two paragraphs may be summarized in two useful rules:

(I) *A statement which is not a tautology may become one upon the substitution of compound sentences for atomic sentences or substitution of the same atomic sentence for distinct atomic sentences in the original statement.*†

(II) *A statement which is a tautology remains so when any sentences are substituted for its component atomic sentences in all occurrences.*

§ **1.7 Tautological Implication and Equivalence.** A sentence P is said to *tautologically imply* a sentence Q if and only if the conditional $P \rightarrow Q$ is a tautology. Thus the sentence 'Locke was a bachelor and Newton never married' tautologically implies 'Newton never married', since for any two sentences P_1 and P_2 we know that the sentence P_1 &

* Historical Note. The tabular test for tautology is essentially due to the American logician and philosopher Charles S. Peirce (1839–1914). The use in logic of 'tautology' is rather recent, being first introduced by Ludwig Wittgenstein in his book *Tractatus Logico-Philosophicus*, London, 1922. Some philosophers call any logical or mathematical truth a tautology, but this wider, somewhat obscure usage will not be needed in this book.

† It should be obvious that not every sentence can become a tautology by such substitution. In fact, any sentence which is the negation of a tautology cannot.

$P_2 \rightarrow P_2$ is a tautology. This example may falsely encourage the idea that the notion of tautological implication is trivial. As we shall see in the next chapter, it is in fact basic to the theory of logical inference. If P tautologically implies Q, then when P is true Q must be true. It can never happen that P is true and Q false, since it is required that $P \rightarrow Q$ be a tautology. Thus from the premise P we may safely infer Q without recourse to any other premises.

By use of truth tables it is easy to decide if one sentence tautologically implies another. The test is simply: any row which has the entry 'T' for the first sentence must have the same entry for the second. We may consider the above example; namely, the sentence P_1 & P_2 tautologically implies P_2.

P_1	P_2	P_1 & P_2
T	T	T
T	F	F
F	T	F
F	F	F

We notice that the sentence P_1 & P_2 has the entry 'T' only in the first row, and P_2 has the entry 'T' in this row also. On the other hand, P_2 does not tautologically imply P_1 & P_2, for P_2 has the entry 'T' in the third row, whereas P_1 & P_2 has the entry 'F'.

When two sentences tautologically imply each other, they are said to be *tautologically equivalent*. The notion of tautological equivalence is stronger than the notion of tautological implication; its role in inference is not as central as that of tautological implication, but it is important. The reason for this importance is not hard to find. If two sentences are tautologically equivalent, they express essentially the same facts, and consequently their roles in inference are nearly identical.

By way of example, let A be 'Aristotle was left-handed' and L be 'Leibniz was left-handed'. Then the sentence A & –L is tautologically equivalent to the sentence –(–A ∨ L). To see this, we may use truth tables.

A	L	–A	–L	–A ∨ L	–(–A ∨ L)	A & –L
T	T	F	F	T	F	F
T	F	F	T	F	T	T
F	T	T	F	T	F	F
F	F	T	T	T	F	F

The test is clear: the columns corresponding to the two sentences must agree row for row in their entries in order for the two sentences to be tautologically equivalent. This test is satisfied by Columns 6 and 7 of the above table, and we conclude $-(-A \vee L)$ is tautologically equivalent to $A \,\&\, -L$.

It is perhaps worth remarking that P and Q are tautologically equivalent when and only when the biconditional $P \leftrightarrow Q$ is a tautology.

EXERCISES

1. A classical example of a non-truth-functional connective is that of possibility. For example, the sentence:

(1) It is possible that there is life on Mars

is true under any liberal interpretation of the notion of possibility; but then so is the sentence:

(2) It is possible that there is not any life on Mars.

On the other hand, the sentence:

(3) It is possible that $2 + 2 = 5$

is ordinarily regarded as false. Using a diamond symbol '$\Diamond$' for 'it is possible that', M for 'there is life on Mars' and W for '$2 + 2 = 5$', we get the following tabular analysis of (1)–(3):

M	$-M$	W	$\Diamond M$	$\Diamond -M$	$\Diamond W$
T	F	F	T	T	F
F	T		T	T	

The analysis of $\Diamond M$ and $\Diamond -M$ entails that the only truth-functional analysis of the possibility connective is that for any sentence P, $\Diamond P$ is true, but the truth value of $\Diamond W$ controverts this; and we see that there is no appropriate truth-functional analysis.

Give examples and an analysis to show that the following are not truth-functional connectives:

(a) 'Mr. Smith believes that ...'
(b) 'It is necessary that ...'

2. Which of the truth-functional connectives introduced in this chapter is an approximate synonym of the connective 'unless'? (HINT: To say 'There will be peace unless there is a major war in the next five years' is equivalent to saying 'If there is not a major war in the next five years, then there will be peace'.)

3. Translate the following compound sentences into symbolic notation, using letters to stand for atomic sentences.

(a) Either the fire was produced by arson or it was produced by spontaneous combustion.
(b) If the water is clear, then either Henry can see the bottom of the pool or he is a nincompoop.
(c) Either John is not here or Mary is, and Helen certainly is.
(d) If there are more cats than dogs, then there are more horses than dogs and there are fewer snakes than cats.

(e) The man in the moon is a fake, and if the same is true of Santa Claus, many children are deceived.
(f) If either red-heads are lovely or blondes do not have freckles, then logic is confusing.
(g) If either housing is scarce or people like to live with their in-laws, and if people do not like to live with their in-laws, then housing is scarce.
(h) If John testifies and tells the truth, he will be found guilty; and if he does not testify, he will be found guilty.
(i) Either John must testify and tell the truth, or he does not have to testify.

4. In the following examples determine the truth value of the compound sentences from the given truth values of the component sentences (i)–(iv).

(i) 'Galileo was born before Descartes' is true.
(ii) 'Descartes was born in the sixteenth century' is true.
(iii) 'Newton was born before Shakespeare' is false.
(iv) 'Racine was a compatriot of Galileo' is false.

(a) If Galileo was born before Descartes, then Newton was not born before Shakespeare.
(b) If either Racine was a compatriot of Galileo or Newton was born before Shakespeare, then Descartes was born in the sixteenth century.
(c) If Racine was not a compatriot of Galileo, then either Descartes was not born in the sixteenth century or Newton was born before Shakespeare.

5. Let

N = 'New York is larger than Chicago';
W = 'New York is north of Washington';
C = 'Chicago is larger than New York'.
(Thus N and W are true and C is false.)

Which of the following sentences are true?

(a) $N \vee C$
(b) $N \& C$
(c) $-N \& -C$
(d) $N \leftrightarrow -W \vee C$
(e) $W \vee -C \rightarrow N$
(f) $(W \vee N) \rightarrow (W \rightarrow -C)$
(g) $(W \leftrightarrow -N) \leftrightarrow (N \leftrightarrow C)$
(h) $(W \rightarrow N) \rightarrow [(N \rightarrow -C) \rightarrow (-C \rightarrow W)]$

6. Let

P = 'Jane Austen was a contemporary of Beethoven';
Q = 'Beethoven was a contemporary of Gauss';
R = 'Gauss was a contemporary of Napoleon';
S = 'Napoleon was a contemporary of Julius Caesar'.
(Thus P, Q, and R are true, and S is false.)

Find the truth values of the following sentences:

(a) $(P \& Q) \& R$
(b) $P \& (Q \& R)$
(c) $S \rightarrow P$
(d) $P \rightarrow S$
(e) $(P \& Q) \& (R \& S)$
(f) $P \& Q \leftrightarrow R \& -S$
(g) $(P \leftrightarrow Q) \rightarrow (S \leftrightarrow R)$
(h) $(-P \rightarrow Q) \rightarrow (S \rightarrow R)$
(i) $(P \rightarrow -Q) \rightarrow (S \leftrightarrow R)$
(j) $(P \rightarrow Q) \rightarrow [(Q \rightarrow R) \rightarrow (R \rightarrow S)]$
(k) $P \rightarrow [Q \leftrightarrow (R \rightarrow S)]$

7. Let P be a sentence such that for any sentence Q the sentence $P \vee Q$ is true. What can be said about the truth value of P?

8. Let P be a sentence such that for any sentence Q the sentence $P \,\&\, Q$ is false. What can be said about the truth value of P?

9. If $P \leftrightarrow Q$ is true, what can be said about the truth value of $P \vee -Q$?

10. Let P, Q, and R be any three distinct atomic sentences. Decide by truth tables which of the following sentences are tautologies:

(a) $P \vee Q$
(b) $P \vee -P$
(c) $P \vee Q \rightarrow Q \vee P$
(d) $P \rightarrow (P \vee Q) \vee R$
(e) $P \rightarrow (-P \rightarrow Q)$
(f) $(P \rightarrow Q) \rightarrow (Q \rightarrow P)$
(g) $[(P \rightarrow Q) \leftrightarrow Q] \rightarrow P$
(h) $P \rightarrow [Q \rightarrow (Q \rightarrow P)]$
(i) $P \,\&\, Q \rightarrow P \vee R$
(j) $[P \vee (-P \,\&\, Q)] \vee (-P \,\&\, -Q)$
(k) $P \,\&\, Q \rightarrow (P \leftrightarrow Q \vee R)$
(l) $[P \,\&\, Q \rightarrow (P \,\&\, -P \rightarrow Q \vee -Q)] \,\&\, (Q \rightarrow Q)$

11. If P and Q are distinct atomic sentences, which of the following are tautologies?

(a) $P \leftrightarrow Q$
(b) $P \leftrightarrow P \vee P$
(c) $P \vee Q \leftrightarrow Q \vee P$
(d) $(P \rightarrow Q) \leftrightarrow (Q \rightarrow P)$
(e) $(P \leftrightarrow P) \leftrightarrow P$

12. On the basis of ordinary usage construct truth tables for the sentential connectives used in the following examples:

(a) Not both P and Q.
(b) Neither P nor Q.

13. Give examples of sentences P and Q (not necessarily atomic) such that the following compound sentences are tautologies.

(a) $P \,\&\, Q$
(b) $P \vee (P \,\&\, -Q)$
(c) $P \rightarrow P \,\&\, -Q$
(d) $P \rightarrow -P$

14. Is there any sentence P such that $P \,\&\, -P$ is a tautology?

15. If P and Q are distinct atomic sentences, the sentence $P \,\&\, Q$ tautologically implies which of the following?

(a) P
(b) Q
(c) $P \vee Q$
(d) $P \,\&\, -Q$
(e) $-P \vee Q$
(f) $-Q \rightarrow P$
(g) $P \leftrightarrow Q$

16. If P and Q are distinct atomic sentences, the sentence $-P \vee Q$ tautologically implies which of the following?

(a) P
(b) $Q \rightarrow P$
(c) $P \rightarrow Q$
(d) $-Q \rightarrow -P$
(e) $-P \,\&\, Q$

17. If P and Q are distinct atomic sentences, the sentence P is tautologically equivalent to which of the following?

(a) $P \vee Q$
(b) $P \vee -P$
(c) $P \,\&\, P$
(d) $P \rightarrow P$
(e) $-P \rightarrow P$
(f) $P \rightarrow -P$
(g) $Q \vee -Q \rightarrow P$

CHAPTER 2

SENTENTIAL THEORY OF INFERENCE

§ 2.1 Two Major Criteria of Inference and Sentential Interpretations. In this chapter we turn to the theory of logical inference. The rules of inference governing sentential connectives turn out to be quite simple. You may find it helpful to think of the rules introduced in this chapter and Chapter 4 as the elaborate statement of how to play a not-too-complicated game. The game shapes up as follows: we begin with a set of formulas which we call *premises.* The object of the game is to apply the rules so as to obtain some other *given* formula (the desired conclusion). The set of premises corresponds to the initial position of a player in a game. By a succession of moves, each move being sanctioned by a rule, we reach a winning position: the sought for conclusion. As in any game, the rules permit all kinds of silly moves; the problem is to learn how to make the right moves. (In learning the theory of inference it will be useful to remember that we ordinarily learn a game by example and illustration; we refer to the formal statement of the rules only to settle arguments or dispel confusions.)

Now for a game such as bridge or chess, rules are chosen which presumably yield something interesting or entertaining. The theory of logical inference, on the other hand, is more than entertaining. There are many considerations which guide the construction of a set of rules of inference, and certain aspects of the problem of giving such a set of rules are too technical to discuss here. However, there are two major criteria of construction which dominate all others.

> Criterion I. *Given a set of premises, the rules of logical derivation must permit us to infer* ONLY *those conclusions which logically follow from the premises.*
>
> Criterion II. *Given a set of premises, the rules of logical derivation must permit us to infer* ALL *conclusions which logically follow from the premises.*

But, it is proper to ask, how can criteria for adequate rules of inference be stated in terms of a conclusion *logically following* from premises? Is it not the very point of the rules of inference to characterize explicitly the notion of logical consequence? To avoid the charge of circularity an independent, intuitively plausible definition of logical consequence or logical validity needs to be given. A completely precise definition is somewhat technical; on the other hand, the basic idea is not too complicated and it introduces the notion of *interpretation* which will prove repeatedly useful. The idea is that Q logically follows from P when Q is true in every interpretation or model for which P is true. For our present purposes we may define the restricted notion of a *sentential interpretation:*

> *A sentence* P *is a sentential interpretation of a sentence* Q *if and only if* P *can be obtained from* Q *by replacing the component atomic sentences of* Q *by other (not necessarily distinct) sentences.*

If this definition sounds slightly bizarre an example will show how simple it is. Let Q_1 be

> 'If the sun is shining Marianne is happy'

and let P_1 be

> 'If either the battalion advances too fast or the general is wrong, then the battle is lost'.

Then P_1 is a sentential interpretation of Q_1, for 'the sun is shining' is replaced by 'either the battalion advances too fast or the general is wrong', and 'Marianne is happy' is replaced by 'the battle is lost'. To be a sentential interpretation of the particular sentence Q_1, P_1 need have just one characteristic: its major sentential connective must be an implication. In other words a sentential interpretation of a sentence must preserve its *sentential form.* For this reason Q_1 is not a sentential interpretation of P_1, for the disjunction in the antecedent of P_1 is missing in Q_1. Of course, the sentential interpretation may have more structure (thus P_1 has more structure than Q_1), but it must not have less.

It should be understood that if a component atomic sentence occurs more than once in a sentence, any sentential interpretation of that sentence must replace that component atomic sentence by the same thing in both of its occurrences. Thus if Q_2 is

> 'If the sun is shining then either it is raining or the sun is shining'

and P_2 is

> 'If $1 + 2 = 4$ then either snow is black or I am a fool',

then P_2 is not a sentential interpretation of Q_2, for the first occurrence of 'the sun is shining' in Q_2 is replaced by '$1 + 2 = 4$', and the second occurrence by a different sentence.

A sufficient (but not necessary) condition for one sentence to be a logical consequence of another may now be stated.

(I) Q *logically follows from* P *if every sentential interpretation of the implication* $P \rightarrow Q$ *is true.**

(As we shall see in Chapter 4, we obtain a complete characterization of logical consequence by omitting the restriction to *sentential* interpretations.) We may relate (I) to P tautologically implying Q by the following argument. First, we call a sentential interpretation *atomic* if the interpretation consists of replacing atomic sentences by atomic sentences. The definition of tautologies given in Chapter 1 may then be phrased:

(II) *A tautology is a sentence whose atomic sentential interpretations are all true.*

And as we implicitly observed in Chapter 1, it is not difficult to see that

(III) *If every atomic sentential interpretation of a sentence is true then every sentential interpretation of the sentence is true.*

From (I)–(III) we arrive at the workable criterion that Q logically follows from P if $P \rightarrow Q$ is a tautology, or in other words:

(IV) Q *logically follows from* P *if* Q *is tautologically implied by* P.

The whole sentential theory of inference is summarized by (IV). To determine if by virtue of the logical properties of the sentential connectives a given conclusion logically follows from a set of premises, we need only construct the appropriate truth table and see if the premises tautologically imply the conclusion. For example, suppose we want to know if the conclusion 'Peter is going to cry' logically follows from the two premises:

(1) Either Mary gives Peter his toy or Peter is going to cry.
(2) Mary does not give Peter his toy.

Let P be 'Peter is going to cry' and let M be 'Mary gives Peter his toy'. The argument may then be symbolized:

Premise 1: $M \vee P$
Premise 2: $-M$

Conclusion: P

* The use of 'if' without 'only if' indicates that a sufficient condition is given, but that no commitment regarding necessity is made. Compare § 1.3 for a discussion of the idiom of necessity and sufficiency.

To decide if the argument is valid we need to decide if the conjunction (M ∨ P) & −M tautologically implies P. For this task we construct the following four-line truth table:

M	P	M ∨ P	−M	(M ∨ P) & −M
T	T	T	F	F
T	F	T	F	F
F	T	T	T	T
F	F	F	T	F

And we observe that in the only case in which the premises are jointly true (third row, last column) the conclusion P is also true (third row, second column). We conclude that the argument is valid, that is, the conclusion is a logical consequence of the premises.

There are, however, two good reasons for developing a theory of sentential inference which does not simply consist of constructing a massive truth table to check the validity of an argument. In the first place, if a set of premises and the desired conclusion contain five or more distinct atomic sentences, the appropriate truth table must have at least 32 lines ($2^5 = 32$). It is not only tedious to construct a truth table with 32 lines but also difficult to avoid making mistakes when such a large number of elementary computations is involved. Secondly, the direct truth table approach is adequate to but a pitifully small fragment of logically valid arguments, as we shall see in Chapter 4. It is therefore desirable to develop a sentential theory of inference which has ready application in more general contexts. Before turning to this task in the next section, some further general remarks on the two criteria introduced at the beginning of this section are pertinent.

Criterion I says that the rules of inference must be *sound*, that is, they must not permit a fallacious inference. Combining the criterion with (I) we may obtain a working criterion for testing the validity of a proposed rule of inference. First we see that Q should be derivable from P by use of the rules only if every interpretation of P → Q is true. The test criterion is then:

> *If a new proposed rule of inference permits the derivation of a false conclusion from true premises, reject it.*

That is, if Q is derivable from P by use of the rule and some interpretation of P → Q is false, reject the rule.

For example, suppose someone proposes the rule:

(A) From a sentence Q and a sentence P → Q, we may infer P.

Then we may immediately construct the kind of counterexample required by the test criterion. Let

P = 'Lincoln was born in Illinois'
Q = 'Lincoln was born in the United States'.

Clearly both Q and $P \rightarrow Q$ are true, but P is false since Lincoln was born in Kentucky. By use of (A) we have derived a false conclusion from true premises, and our only alternative is to reject it.

Criterion II says that the rules of inference must be *complete*, that is, they must permit the derivation of every valid conclusion.*

The uses of Criterion II are more sophisticated, but a simple example can show why Criterion I alone is not sufficient.

Suppose some Simple Simon proposes as the only rule of inference:

(B) From any sentence P we may infer P.

Clearly (B) satisfies Criterion I, for if we begin with the true premise P we can only derive P itself. But to maintain that (B) is sufficient for all logical inference violates Criterion II. By use of (B) we cannot, for instance, infer the valid conclusion P from the premises $M \vee P$ and $-M$ (the example discussed above).

We shall refer to Criteria I and II a number of times in discussing the theory and practice of valid inference. In Chapter 4 particularly we shall use Criterion I to justify various restrictions on the rules of inference.

EXERCISES

1. Construct specific counterexamples for the following two fallacious rules of inference (for (*a*) give a different example than the one in the text). The traditional names of these two fallacies are indicated.

(a) Fallacy of Affirming the Consequent: From Q and $P \rightarrow Q$, we may derive P.
(b) Fallacy of Denying the Antecedent: From $-P$ and $P \rightarrow Q$, we may derive $-Q$.

2. Using Criterion I, decide which of the following are valid rules of inference. For those which you think are invalid, construct a specific counterexample to show it violates Criterion I; that is, give an example in which a false conclusion is derived from true premises by use of the invalid rule.

(a) From P and Q, we may derive P & Q.
(b) From P and $P \vee Q$, we may derive Q.
(c) From $-Q$ and $P \vee Q$, we may derive P.
(d) From $-P$ and $P \vee Q$, we may derive $-Q$.

* Although the set of rules introduced in this and the fourth chapter is sound and complete, it is beyond the scope of Part I to establish these facts. The notion of completeness is discussed in more detail in § 4.2.

3. Suppose someone proposed as the *only* rule of inference: From $\mathbf{P}$ we may derive $\mathbf{Q} \vee -\mathbf{Q}$. Does this rule violate Criterion I? Does it violate Criterion II? If so, explain why.

4. Construct a (non-valid) rule of inference which by itself will satisfy Criterion II but violates Criterion I.

5. Utilizing the discussion of sentential interpretations, explain exactly why the following rule of inference violates Criterion I.

> From 'Thomas Jefferson was President' we may infer 'Jane Austen wrote EMMA'.

(Note that this is not a case of a true premise and a false conclusion, since both statements are true.)

6. Which of the following sentences are sentential interpretations of the sentence 'It is raining or it is snowing'?

(a) If it is raining, then it is snowing.
(b) It is snowing.
(c) It is snowing or it is snowing.
(d) It is snowing or it is raining.
(e) Either it is snowing and it is raining or it is not raining.

7. Which of the following assertions are true? If false, give a counterexample.

(a) A sentential interpretation of a sentence $\mathbf{P}$ must have the same number of distinct atomic sentences as $\mathbf{P}$.
(b) A sentential interpretation of a non-atomic sentence $\mathbf{P}$ must have the same major sentential connective as $\mathbf{P}$.
(c) A sentential interpretation of a sentence $\mathbf{P}$ must have the same number of occurrences of sentential connectives as $\mathbf{P}$.
(d) A sentential interpretation of a sentence $\mathbf{P}$ must have at least as many occurrences of (not necessarily distinct) atomic sentences as $\mathbf{P}$.

§ **2.2 The Three Sentential Rules of Derivation.** For the reasons given in the previous section we replace the construction of a single truth table to test the validity of an argument by three rules of derivation. For the moment we consider only two of these rules. One permits us to introduce premises when needed, and the other permits piecemeal use of tautological implications.

Before giving a precise statement of these first two rules of inference, we may consider an example to show how they are used.

EXAMPLE 1. *If there are no government subsidies of agriculture, then there are government controls of agriculture. If there are government controls of agriculture, there is not an agricultural depression. There is either an agricultural depression or overproduction. As a matter of fact, there is no overproduction. Therefore, there are government subsidies of agriculture.*

We want to derive the conclusion 'There are government subsidies of agriculture' from the four premises given. For clarity, here and subsequently, we symbolize the argument; the meaning of the various numerals on the left is explained below.

1	(1) $-S \to C$	Premise
2	(2) $C \to -D$	Premise
3	(3) $D \vee O$	Premise
4	(4) $-O$	Premise
3, 4	(5) D	(3) & (4) tautologically imply (5)
2, 3, 4	(6) $-C$	(2) & (5) tautologically imply (6)
1, 2, 3, 4	(7) S	(1) & (6) tautologically imply (7)

The use of letters should be obvious: S is the sentence 'There are government subsidies of agriculture'; C is the sentence 'There are government controls of agriculture'; D is the sentence 'There is an agricultural depression'; O is the sentence 'There is overproduction'. There are seven lines to the derivation. The introduction of each line may be justified by one of the two rules. The first four lines are just the premises of the argument. And the last three lines are obtained by showing they are tautological implications of preceding lines. In the case of line (6), for example, it is easy to see that the conjunction of lines (2) and (5), that is, the conjunction

$$(C \to -D) \,\&\, D$$

tautologically implies

$$-C$$

In the case of line (7), the conjunction of lines (1) and (6), that is, the conjunction

$$(-S \to C) \,\&\, -C$$

tautologically implies

$$S$$

by use of a tautology similar to the one permitting us to infer line (6) from (2) and (5).

Notice how simple the three tautological implications used are, in comparison with the over-all implication needed to establish the conclusion at one stroke:

$$[(-S \to C) \,\&\, (C \to -D) \,\&\, (D \vee O) \,\&\, -O] \to S.$$

Moreover, a tedious sixteen-line truth table would be required to test this implication. The three tautological implications used are not only simple, but constantly recurring. About ten or twelve simple types of tautological implications are adequate for breaking down into parts most tautological inferences encountered in practice. (The next section will be concerned with these useful tautologies.)

We have not yet explained the listing of numbers on the left in the above derivation. For each line of the derivation, the list of numbers at the left corresponds to the *premises* on which that line depends. Thus line (1) depends only on itself, the first premise; similarly, line (2) depends only on itself, the second premise. But line (5) depends on the third and fourth premise; line (6) is derived from the second and fifth lines, hence it depends on the premises lines (2) and (5) depend on, that is, the second, third, and fourth premises; and line (7) is derived from lines (1) and (6), whence it depends on all four premises. Thus, two premises were used to derive line (5); three premises to derive line (6); and four premises to derive line (7). The intuitive significance of the numerals at the left should be emphasized: each line is a *logical consequence* of the set of premises corresponding to the numerals at the left. Thus, line (1) is trivially a logical consequence of itself, and line (6), for example, is a logical consequence of the second, third, and fourth premises. Note that the numerals at the left indicate the premises from which the line has been inferred, perhaps by a very complicated chain of inferences. On the other hand, the numerals at the right simply indicate what particular lines the line was immediately inferred from by the application of a single tautological implication.

A word of warning about how we enter the listing of numbers on the left. When we enter a premise as a line, we list at the left the number of *that line*, since it only depends on that line. We do not enter the number corresponding to the total number of premises now introduced. This is an important point to remember. We may make it clear by rewriting the previous derivation. Some useful abbreviations are also incorporated in this version.

{1}	(1) D ∨ O	P
{2}	(2) −O	P
{1, 2}	(3) D	1, 2 T
{4}	(4) C → −D	P
{1, 2, 4}	(5) −C	3, 4 T
{6}	(6) −S → C	P
{1, 2, 4, 6}	(7) S	5, 6 T

Thus, at line (4) we introduce the third premise, but the number we list at the left is 4, to indicate that this line depends only on itself; to list the number 3 instead would indicate to us that line (4) depended on line (3) which it does not. Similar remarks apply to line (6). We have also added braces around the numerals at the left. This additional notation makes it clearer that a given line is a logical consequence of the *set* of premises corresponding to the *set* of numbers attached to the line. (It is convenient and customary to describe a set by writing down the names of its members, separated by commas, and enclosing the whole in braces. Thus the set whose members are the numbers 1, 2, and 4 is described by writing: $\{1, 2, 4\}$.)

In the above rewrite of Example 1 we have abbreviated 'Premise' to 'P' and phrases like '(3) and (4) tautologically imply (5)' to '3, 4 T'. The numerals '3' and '4' indicate the previous lines used, and the letter 'T' refers to an application of a tautological implication. Rules P and T may be summarized:

> RULE P. *We may introduce a premise at any point in a derivation.*
>
> RULE T. *We may introduce a sentence* S *in a derivation if there are preceding sentences in the derivation such that their conjunction tautologically implies* S.*

Rule P permits us to introduce a new assumption or premise whenever we desire. This may sound absurd, since it would appear that with sufficient premises at hand, one could prove anything at all. The point is: exactly what premises we have used in an argument are explicitly indicated by the set of numbers at the left, and any logically correct argument is no better or worse than the premises on which it rests. A serious logical error is committed, of course, if a premise is used in an argument *without* explicit recognition. Moreover, it has not been shown that a conclusion is a logical consequence of a given set of premises if the set of numbers at the left contains a number referring to a premise not in the given set. Thus suppose someone said that the following derivation shows that from the premise $P \vee Q$ we may derive Q:

$\{1\}$	(1) $P \vee Q$	P
$\{2\}$	(2) $-P$	P (additional premise)
$\{1, 2\}$	(3) Q	1, 2 T

The derivation itself is correct, but it does not show that Q logically follows from $P \vee Q$, for line (3) has two numbers, not just the number 1, listed at the left, and consequently Q is a logical consequence of two premises, $P \vee Q$ and $-P$.

The third and last rule for sentential derivations is the rule of conditional proof, which we call *Rule C.P.* The general idea of this rule is that we may introduce a premise R conditionally, so to speak, use it in conjunction with the original premises to derive a conclusion S, and then assert that the implication $R \rightarrow S$ follows from the original premises alone. It should be intuitively clear that this rule does not violate Criterion I and is thus

* We permit the conjunction of any finite number of sentences, not just two. Thus we might have $((P \rightarrow Q) \& (Q \rightarrow R)) \& (R \rightarrow S)$, which tautologically implies $P \rightarrow S$.

acceptable. If we may validly infer S from premises $P_1, P_2, \ldots, P_n$ and R, then we may infer $R \rightarrow S$ from $P_1, P_2, \ldots, P_n$. For suppose the first inference were valid and the second not valid in some particular case. Then $R \rightarrow S$ would have to be false; but this could happen only if R were true and S false. However, if this were the case, S would be false, the premises $P_1, P_2, \ldots, P_n$, R true, and the first inference invalid also. Thus we may accept: *

RULE C.P. *If we can derive* S *from* R *and a set of premises, then we may derive* $R \rightarrow S$ *from the set of premises alone.*

Let us see how Rule C.P. works in an example. (We remark that this example, along with the previous one and any subsequent one, is not intended as a serious factual argument: we are not committed in this book to maintaining that the premises are factually true.)

EXAMPLE 2 (THE NATIONAL LEAGUE RACE). *If the Cards are third, then if the Dodgers are second the Braves will be fourth. Either the Giants will not be first or the Cards will be third. In fact, the Dodgers will be second. Therefore, if the Giants are first, then the Braves will be fourth.*

We use letters 'C', 'D', etc., in the obvious way; thus, C is the sentence 'The Cards are third'.

{1}	(1) $C \rightarrow (D \rightarrow B)$	P
{2}	(2) $-G \vee C$	P
{3}	(3) D	P
{4}	(4) G	P
{2, 4}	(5) C	2, 4 T
{1, 2, 4}	(6) $D \rightarrow B$	1, 5 T
{1, 2, 3, 4}	(7) B	3, 6 T
{1, 2, 3}	(8) $G \rightarrow B$	4, 7 C.P.

In this example, we wanted to obtain as a conclusion the conditional sentence $G \rightarrow B$. The derivation was greatly facilitated by introducing the sentence G as an additional premise and using this additional premise to help derive the sentence B. We then use the rule of conditional proof to assert that the sentence $G \rightarrow B$ follows from the three given premises, which is what is asserted in line (8). Notice that to the right of line (8) we list 4 and 7, which indicates that line (4) was our conditional premise and line (7) our conditional conclusion. By conditionalizing on line (4), when we pass from (7) to (8) we remove the number 4 from the set {1, 2, 3, 4} of numbers which correspond to the premises (7) depends on. The set of premises which (8) depends on then corresponds to the set

*The rule of conditional proof was first explicitly shown to be a valid rule of inference by Alfred Tarski in 1929, but was first proved for a specific system by Jacques Herbrand in 1928. When the rule is derived from other rules of inference rather than taken as primitive, it is usually called the *deduction theorem*.

$\{1, 2, 3\}$. It is important to realize:

(*a*) Rule C.P. is the only rule we shall introduce which permits us to *reduce* the set of numbers listed at the left.

(*b*) When Rule C.P. is applied to reduce the set of numbers listed at the left, only one number can be eliminated in a given application of the rule.

(*c*) Only a line which is a premise can be conditionalized on; in Example 2 we can conditionalize on line (4); it would be incorrect to conditionalize on line (5) and write as line (8)

(8′) $$C \to B,$$

for we would then not know what number to remove from the set $\{1, 2, 3, 4\}$.

The above remarks have been directed at showing you how the rule of conditional proof works, but nothing has been said about the *strategy* of applying it. The best general hint about strategy is that when you want to derive a conclusion which is an implication, always consider using a conditional proof. In Example 2, for instance, the desired conclusion is of the form $P \to Q$, so we assumed temporarily the premise corresponding to P.

A slightly more intricate example than the first two is desirable for reference in discussing useful tautologies in the next section. Notice that the conclusion of this example is an implication, and hence it is efficient to use a conditional proof.

EXAMPLE 3 (A HORSE RACE). *If* A *wins, then either* B *or* C *will place. If* B *places, then* A *will not win. If* D *places, then* C *will not. Therefore, if* A *wins,* D *will not place.*

$\{1\}$	(1) $A \to B \vee C$	P
$\{2\}$	(2) $B \to -A$	P
$\{3\}$	(3) $D \to -C$	P
$\{4\}$	(4) A	P
$\{1, 4\}$	(5) $B \vee C$	1, 4 T (Law of Detachment)
$\{2, 4\}$	(6) $-B$	2, 4 T (Laws of Double Negation and *modus tollendo tollens*)
$\{1, 2, 4\}$	(7) C	5, 6 T (*modus tollendo ponens*)
$\{1, 2, 3, 4\}$	(8) $-D$	3, 7 T (Laws of Double Negation and *modus tollendo tollens*)
$\{1, 2, 3\}$	(9) $A \to -D$	4, 8 C.P.

Since the application of particular tautologies is the core of sentential derivations, we shall not list any exercises until after the next section. In Example 3 the names of the particular tautologies used have been given.

Before turning to consideration of useful tautological implications, there are two general remarks about sentential derivations to be made. First, it should be clear by now what sort of thing a sentential derivation is. Roughly speaking, it is a sequence of sentences such that each sentence is either a premise, is tautologically implied by preceding sentences in the sequence, or is obtained from two preceding sentences in the sequence by conditionalization. The sentences or lines of a derivation correspond to the successive moves of a game, and each move must be sanctioned by one of the three rules given.

Second, although the virtues of sentential derivations in establishing the validity of a sentential argument are considerable, the alternative tedious one-shot truth table approach has one theoretical advantage which should be mentioned: it provides a mechanical test for the validity of a sentential argument. By a systematic check of the appropriate truth table we can always decide in a finite number of steps if the premises do tautologically imply the conclusion and thus if the conclusion is sententially valid.

On the other hand, a derivation which consists of a sequence of steps applying the three sentential rules only shows us if the conclusion is valid. If the conclusion of the argument is not valid, we could theoretically go on endlessly trying to find a valid derivation. At no point would we necessarily be able to decide that the conclusion was invalid by a piecemeal use of new premises, tautological implications and conditionalizations. However, when it appears intuitively that a conclusion is invalid, we may be able to find a sentential interpretation of the implication whose antecedent is the conjunction of the premises and whose consequent is the conclusion such that in this interpretation the premises are true and the conclusion false. This method is not mechanical like the truth table test, but it has the advantage of generalizing to the framework of inference considered in Chapter 4.

The application of the method of interpretation to show that an argument is invalid is straightforward. Suppose, for instance, someone claims that from the premises:

(1) Either Jane Austen wrote NORTHANGER ABBEY or Immanuel Kant wrote THE METAPHYSICAL FOUNDATIONS OF NATURAL SCIENCE

(2) Jane Austen wrote NORTHANGER ABBEY,

he may validly infer:

(3) Immanuel Kant wrote THE METAPHYSICAL FOUNDATIONS OF NATURAL SCIENCE.

We should not be misled by the fact that all the statements (1)–(3) are true into thinking that the argument is valid. We may symbolize the premises and conclusion in the obvious way:

$$\begin{array}{ll} (1') & \mathsf{A} \vee \mathsf{K} \\ (2') & \mathsf{A} \\ \hline (3') & \mathsf{K} \end{array}$$

To show that the argument is invalid we search for an interpretation of A and K such that (1′) and (2′) are true, but (3′) is false. In this case it is easy to find such an interpretation. Let A be 'George Washington was the first President of the United States' and let K be 'Andrew Jackson was the second President of the United States'. Under this interpretation, both $\mathsf{A} \vee \mathsf{K}$ and A are true since A is true, but K is false. Hence we conclude (3) does not logically follow from (1) and (2), since there is a sentential interpretation for which (1) and (2) are true and (3) is false.

The exercises at the end of the next section call for applications of the method of interpretation to show that some of the arguments given are invalid.

§ 2.3 Some Useful Tautological Implications. The tautological implication most frequently used in derivations is probably the *Law of Detachment*

$$\mathsf{P} \;\&\; (\mathsf{P} \rightarrow \mathsf{Q}) \rightarrow \mathsf{Q}.$$

This tautology corresponds to the rule that from P and $\mathsf{P} \rightarrow \mathsf{Q}$ we may infer Q. To each of the other tautological implications we consider in this section there is a corresponding rule of inference, which is simply a special case of Rule T. It is, of course, the rule corresponding to the tautology which we actually use in derivations. In Example 3 of the previous section the Law of Detachment is used to derive line (5) from (1) and (4):

$$\begin{array}{ll} (1) & \mathsf{A} \rightarrow \mathsf{B} \vee \mathsf{C} \\ (4) & \mathsf{A} \\ \hline (5) & \mathsf{B} \vee \mathsf{C} \end{array}$$

The Latin name for the Law of Detachment is *modus ponendo ponens.*

The tautology whose Latin name is *modus tollendo tollens* is similar to the Law of Detachment. In this case, instead of affirming (*ponendo*) the antecedent, we may by denying (*tollendo*) the consequent of an implication deny (*tollens*) the antecedent. Thus

$$-\mathsf{Q} \;\&\; (\mathsf{P} \rightarrow \mathsf{Q}) \rightarrow -\mathsf{P}.$$

Modus tollendo tollens is conveniently combined with the *Law of Double Negation:*

$$P \leftrightarrow --P.$$

For instance, in Example 3 we used these two tautologies to derive line (6) from lines (2) and (4):

(2) $B \rightarrow -A$
(4) A

---

(6) $-B$

We may expand this use of the two tautologies together to understand clearly the role of each:

(2)	$B \rightarrow -A$	
(4)	A	
(4′)	$--A$	4 T (Law of Double Negation)
(6)	$-B$	2, 4′ T (*modus tollendo tollens*)

These two tautologies are used in combination a great deal, and ordinarily we will not insert the step (4′) explicitly. There is, of course, no formal necessity of (4′), since $Q \,\&\, (P \rightarrow -Q) \rightarrow -P$ is a tautology. The point here is simply to name for useful reference the more familiar, recurrent tautologies, but it is important to realize there is nothing sacred nor special about these *named* tautologies, and you are free to use *any* tautologies that you find convenient. The listing of certain key ones is analogous to telling a beginner in chess about certain standard openings, or explaining when to bid a slam in bridge. Such hints are intended to help you develop a workable strategy.

In Example 3 we used the tautological implication *modus tollendo ponens* to derive line (7) from lines (5) and (6):

(5) $B \vee C$
(6) $-B$

---

(7) C

The principle of this tautology, as its name indicates, is that if we deny (*tollendo*) one member of a disjunction, we may then assert (*ponens*) the other. Stated as a rule of derivation it assumes the form:

From $-P$ *and* $P \vee Q$ *we may derive* Q.

We have named all the tautologies used in the three examples of derivations in the previous section. We shall now list some further tautologies and illustrate the use of some of them by simple examples. For completeness of reference we include the four tautologies already discussed.

TABLE OF USEFUL TAUTOLOGIES

TAUTOLOGICAL IMPLICATIONS

Law of Detachment	$P \,\&\, (P \rightarrow Q) \rightarrow Q$
Modus tollendo tollens	$-Q \,\&\, (P \rightarrow Q) \rightarrow -P$
Modus tollendo ponens	$-P \,\&\, (P \vee Q) \rightarrow Q$
Law of Simplification	$P \,\&\, Q \rightarrow P$
Law of Adjunction	$P \,\&\, Q \rightarrow P \,\&\, Q$
Law of Hypothetical Syllogism	$(P \rightarrow Q) \,\&\, (Q \rightarrow R) \rightarrow (P \rightarrow R)$
Law of Exportation	$[P \,\&\, Q \rightarrow R] \rightarrow [P \rightarrow (Q \rightarrow R)]$
Law of Importation	$[P \rightarrow (Q \rightarrow R)] \rightarrow [P \,\&\, Q \rightarrow R]$
Law of Absurdity	$[P \rightarrow Q \,\&\, -Q] \rightarrow -P$
Law of Addition	$P \rightarrow P \vee Q$

TAUTOLOGICAL EQUIVALENCES

Law of Double Negation	$P \leftrightarrow --P$
Law of Contraposition	$(P \rightarrow Q) \leftrightarrow (-Q \rightarrow -P)$
De Morgan's Laws	$-(P \,\&\, Q) \leftrightarrow -P \vee -Q$
	$-(P \vee Q) \leftrightarrow -P \,\&\, -Q$
Commutative Laws	$P \,\&\, Q \leftrightarrow Q \,\&\, P$
	$P \vee Q \leftrightarrow Q \vee P$
Law of Equivalence for Implication and Disjunction	$(P \rightarrow Q) \leftrightarrow -P \vee Q$
Law of Negation for Implication	$-(P \rightarrow Q) \leftrightarrow P \,\&\, -Q$
A Law for Biconditional Sentences	$(P \leftrightarrow Q) \leftrightarrow (P \rightarrow Q) \,\&\, (Q \rightarrow P)$
Another Law for Biconditional Sentences	$(P \leftrightarrow Q) \leftrightarrow (P \,\&\, Q) \vee (-P \,\&\, -Q)$

TWO FURTHER TAUTOLOGIES

Law of Excluded Middle	$P \vee -P$
Law of Contradiction	$-(P \,\&\, -P)$

Given the premise:

(1) If Fillmore was born in New York then he was born in the United States,

we derive by the Law of Contraposition:

(2) If Fillmore was not born in the United States then he was not born in New York,

and by the Law of Equivalence for Implication and Disjunction:

(3) Either Fillmore was not born in New York or he was born in the United States;

and from (3) by De Morgan's Laws, we derive:

(4) It is not the case that Fillmore was born in New York and not born in the United States.

Given the premise:

(5) If the store is open and the sale is still on, then I will buy a rowboat,

we derive by the Law of Exportation:

(6) If the store is open, then if the sale is still on I will buy a rowboat,

and by the Law of Contraposition we obtain from (6):

(7) If it is not the case that if the sale is still on then I will buy a rowboat, then the store is not open.

The above illustrations are limited in value because they are not embedded in logical derivations of any complexity. As we come to use tautologies from our table in the next section and also in Chapter 4, we shall explicitly mention them.

EXERCISES

1. State the rule of derivation corresponding to each of the tautologies in the above table, except for the Laws of Excluded Middle and Contradiction. When the tautology is an equivalence rather than just an implication, the rule is stated as in the following for the Law of Double Negation:

From P *we may derive* $--$P, *and conversely.*

We add 'and conversely' to indicate that the inference goes both ways.

For the following arguments (Exercises 2–14) try to construct sentential derivations like those for Examples 1–3 in the previous section. In some cases the argument is logically invalid; in such cases, write 'C.D.N.F.' (conclusion does not follow) and then give a sentential interpretation which will show its invalidity. Use the letters indicated, and identify by name the tautologies you use.

2. If the market is perfectly free, then a single supplier cannot affect prices. If a single supplier cannot affect prices, then there are a large number of suppliers. Moreover, there are a large number of suppliers. Therefore, the market is perfectly free. (F, S, N)

3. If prices are high, then wages are high. Prices are high or there are price controls. Also, if there are price controls, then there is not an inflation. However, there is an inflation. Therefore, wages are high. (P, W, C, I)

4. If either wages or prices are raised, there will be inflation. If there is inflation, then either Congress must regulate it or the people will suffer. If the people suffer, Congressmen will be unpopular. Congress will not regulate inflation, and Congressmen will not be unpopular. Therefore, wages will not rise. (W, P, I, C, S, U)

5. Either logic is difficult, or not many students like it. If mathematics is easy, then logic is not difficult. Therefore, if many students like logic, mathematics is not easy. (D, L, M)

6. If Algernon is in jail, then he is not a nuisance to his family. If he is not in jail, then he is not a disgrace. If he is not a disgrace, then he is in the army. If he is drunk, he is a nuisance to his family. Therefore, he is either not drunk or in the army. (J, N, D, A, R)

7. If Algernon is a nuisance, then he is not in jail. If he is in jail, then he is a disgrace. If he is a disgrace, then he is not in the army. Therefore, he is either not in the army or not a nuisance. (N, J, D, A)

8. Either John and Henry are the same age, or John is older than Henry. If John and Henry are the same age, then Elizabeth and John are not the same age. If John is older than Henry, then John is older than Mary. Therefore, either Elizabeth and John are not the same age or John is older than Mary. (S, O, E, M)

9. Marianne believed that Colonel Brandon was too old to marry. If Marianne's conduct was always consistent with her beliefs, and if she believed that Colonel Brandon was too old to marry, then she did not marry Colonel Brandon. But Marianne married Colonel Brandon. Therefore, Marianne's conduct was not always consistent with her beliefs. (B, C, M)

10. If this is December, then last month was November. If last month was November, then six months ago it was June. If six months ago it was June, then eleven months ago it was January. If next month will be January, then this is December. Last month was November. Therefore, this is December. (D, N, J, A, X)

11. If Mary is a true friend, then John is telling the truth. If John is telling the truth, then Helen is not a true friend. If Helen is not a true friend, then Helen is not telling the truth. If Helen is not telling the truth, then Mary is a true friend. But if Mary is a true friend, then Helen is not a true friend. Therefore, Helen is not telling the truth. (M, J, H, T)

12. If, and only if, Roger has entered into the contract, *and* the contract is legal, *and* Roger has not performed the contract, Jones will win the lawsuit. If Roger has not accepted Jones' offer, Roger has not entered into the contract. The fact is that Roger has not accepted Jones' offer. Therefore, Jones will not win the lawsuit. (R, L, P, J, A)

13. If Brown entered into the contract, *or* if Brown received substantial benefits from acts performed by Smith, Brown will *not* win the lawsuit. If Brown revoked his offer before Smith accepted it, Brown did not enter into the contract. The fact is that Brown did not revoke his offer before Smith accepted it. Therefore, Brown will not win the lawsuit. (C, B, W, R)

14. If Brown did not enter into the contract, *or* if Brown performed the contract, Smith will not win the lawsuit. If Brown failed to deliver the goods on the due date, Brown did not perform the contract. The fact is that Brown did enter into the contract and failed to deliver the goods on the due date. Therefore, Smith will win the lawsuit. (C, P, W, D)

§ 2.4 Consistency of Premises and Indirect Proofs. Sometimes we are not interested in deriving a particular conclusion from a set of premises, but in deciding if the premises are consistent. This is often the prime objective of a lawyer cross-examining a witness for the other side. If he can show that the testimony is inconsistent, he has gone far toward discrediting the evidence presented by that witness. The intuitive notion of inconsistency is that a set of premises is inconsistent if the premises cannot be true together. For example, if one witness testifies that

(1) Blenker was in Washington, D. C., on the night of the murder,

and Blenker asserts that

(2) He, Blenker, was not in Washington, D. C., on the night of the murder,

then we know at once that someone is lying, since (1) and (2) cannot be true together, that is, the statements of Blenker and the witness are inconsistent.

In many cases, it is not easy to decide if a set of premises is consistent simply by "looking" at them, and consequently it is desirable to develop an analytical technique for investigating consistency. To begin with, two sentences are said to be *contradictory* if one is the negation of the other; a *contradiction* is a conjunction of two contradictory sentences, that is, it is a conjunction of the form $S \,\&\, -S$. Now it is easy to see that a set of premises is inconsistent if a contradiction can be logically derived, for if the premises could all be true together, we could construct an example violating Criterion I—that is, true premises and the necessarily false conclusion $S \,\&\, -S$. Our technique for investigating the consistency of a set of premises is thus to attempt to derive a contradiction. We approach the problem of attempting to derive a contradiction in the same general way we approach the problem of deriving a given conclusion. The essential difference is that in deriving a given conclusion the terminal point of the derivation is fixed in advance, whereas in deriving a contradiction, the terminal point is *any* contradiction—it does not matter what particular one.

The purpose of the following example is to illustrate how the rules of logical inference may be used to show that a set of premises is inconsistent.

EXAMPLE 4. *If the contract is valid, then Horatio is liable. If Horatio is liable he will go bankrupt. If the bank will loan him money, he will not go bankrupt. As a matter of fact, the contract is valid and the bank will loan him money.*

{1}	(1) $V \rightarrow L$	P
{2}	(2) $L \rightarrow B$	P
{3}	(3) $M \rightarrow -B$	P
{4}	(4) $V \,\&\, M$	P
{4}	(5) V	4 T (Law of Simplification)
{1, 2}	(6) $V \rightarrow B$	1, 2 T (Law of Hypothetical Syllogism)
{1, 2, 4}	(7) B	5, 6 T
{4}	(8) M	4 T
{3, 4}	(9) $-B$	3, 8 T
{1, 2, 3, 4}	(10) $B \,\&\, -B$	7, 9 T (Law of Adjunction)

Notice that in the proof that the four premises of Example 4 imply a contradiction, we have used from the list in the previous section three tautological implications not previously used.

Once a logical technique for showing that a set of premises is inconsistent has been described, it is natural to ask about the existence of a similar technique for proving a set of premises consistent. The method of interpretation is the appropriate device. If a true sentential interpretation of the conjunction of the premises can be found, then the premises are consistent; for if a contradiction could also be derived from them, then in the interpretation we would have true premises and a valid false conclusion (the contradiction) in violation of Criterion I. As an application of this method, consider the premises:

> *If war is near then the army has mobilized. If the army has mobilized then labor costs are high. However, war is not near and yet labor costs are high.*

We may symbolize these premises:

(1) $W \rightarrow A$

(2) $A \rightarrow L$

(3) $-W \,\&\, L$

To show that the three premises are consistent, we want to find an interpretation of W, A and L such that under this interpretation all three premises will be true. This is an easy task. Let

$$W = \text{`}2 + 2 = 5\text{'}$$
$$A = \text{`}1 + 1 = 2\text{'}$$
$$L = \text{`}2 + 2 = 4\text{'}$$

Then W is false, A and L are true, and a simple truth-functional analysis shows that (1)–(3) are true. Hence the given premises are consistent.

It should be noticed, of course, that the consistency of a set of premises whose logical structure may be expressed by sentential connectives alone may be determined directly by a mechanical truth table test. The truth table for the conjunction of the premises is constructed. If *every* entry in the final column is 'F' then the premises are inconsistent, for there is no interpretation of their component atomic sentences which will render them true. If *at least* one entry is 'T' the premises are consistent, for this particular row of the truth table provides the basis for a true interpretation. The difficulty with this test, like the difficulty with the mechanical truth table test for logical validity, is that it cannot be used once the additional logical apparatus of Chapter 4 is introduced.

We may use the rule of conditional proof and the notion of an inconsistent set of premises to introduce the important method of *indirect proof* (also called: proof by contradiction, *reductio ad absurdum* proof). The use of indirect arguments is perhaps familiar from elementary geometry. The technique of such proofs runs as follows:

(1) Introduce the negation of the desired conclusion as a new premise.

(2) From this new premise, together with the given premises, derive a contradiction.

(3) Assert the desired conclusion as a logical inference from the premises.

We may show schematically how these three steps fall into the pattern of a formal derivation. Let $\mathscr{P}$ be the conjunction of the premises, and C the desired conclusion.

{1}	(1) $\mathscr{P}$	P
{2}	(2) $-C$	P
⋮	⋮	⋮
{1, 2}	(n) $S \& -S$	By rules of derivation
{1}	($n + 1$) $-C \rightarrow (S \& -S)$	2, n C.P.
{1}	($n + 2$) C	$n + 1$ T (Law of Absurdity)

To illustrate this schema and to give a particular example of an indirect proof, we may re-prove the validity of the conclusion of Example 3 of § 2.2. Notice that line (4) is tautologically equivalent to the negation of the desired conclusion $A \rightarrow -D$. (The fact that $A \& D \leftrightarrow -(A \rightarrow -D)$ is a tautology follows from the Law of Negation for Implication.)

{1}	(1) $A \rightarrow B \vee C$	P
{2}	(2) $B \rightarrow -A$	P
{3}	(3) $D \rightarrow -C$	P
{4}	(4) $A \& D$	P
{1, 4}	(5) $B \vee C$	1, 4 T
{3, 4}	(6) $-C$	3, 4 T
{1, 3, 4}	(7) B	5, 6 T
{1, 2, 3, 4}	(8) $-A$	2, 7 T
{1, 2, 3, 4}	(9) $A \& -A$	4, 8 T
{1, 2, 3}	(10) $A \& D \rightarrow A \& -A$	4, 9 C.P.
{1, 2, 3}	(11) $A \rightarrow -D$	10 T

In this example, line (4) corresponds to (2) of the schema, line (9) to line (n), line (10) to line ($n + 1$), and line (11) to line ($n + 2$). Note that (11) follows from (10) by the Law of Absurdity and the Law of Negation for Implication.

It is clear that the line corresponding to line ($n + 1$) of the schema (line (10) of the above example) is going to have the same form in every indirect proof. The natural suggestion is to introduce a new rule permitting the elimination of this intuitively redundant line and sanctioning direct passage from line (n) to ($n + 2$) of the schema (from line (9) to (11) of the example). Since this new rule will shorten derivations by only one line, it is not of great practical importance. However, in Chapter 5 a number of such "short cut" rules are introduced, and some of them lead to considerable simplifications of derivations. Thus, for the moment the important thing is to be very clear about the general status of this new rule for indirect proofs. The most significant fact about it is that it is not on the same footing with the three rules already introduced. Why? Because we use the three rules already introduced to establish its logical validity. In other words, the rule for indirect proofs is a *derived* rule rather than an original rule.*

DERIVED RULE FOR INDIRECT PROOFS: R.A.A. *If a contradiction is derivable from a set of premises and the negation of a formula* S, *then* S *is derivable from the set of premises alone.*

We use the expression 'R.A.A.' to stand for '*reductio ad absurdum*'.

PROOF OF DERIVED RULE: R.A.A. Let $\mathscr{P}$ be the set of premises. By hypothesis we derive from $\mathscr{P}$ and $-T$ a contradiction, i.e., a sentence $S \& -S$. By the rule for conditional proof we then obtain that the sentence $-T \rightarrow S \& -S$ is derivable from $\mathscr{P}$. Using next the tautological implication which we called the Law of Absurdity in § 2.4, we derive T.

We have written this proof in the above form to emphasize that the proof of a derived (i.e., short cut) rule is *not* a derivation. The proof cannot be a derivation, for a derivation is concerned with definite formulas, whereas the proof of a derived rule must be given in a manner which will apply to any formulas whatsoever.

The use of Rule R.A.A. is illustrated in the following example.

EXAMPLE 5. *If twenty-five divisions are enough, then the general will win the battle. Either three wings of tactical air support will be provided, or the general will not win the battle. Also, it is not the case that twenty-five divisions are enough and that three wings of tactical air support will be provided. Therefore, twenty-five divisions are not enough.*

{1}	(1) $D \rightarrow W$	P
{2}	(2) $A \vee -W$	P
{3}	(3) $-(D \& A)$	P
{4}	(4) D	P
{1, 4}	(5) W	1, 4 T
{1, 2, 4}	(6) A	2, 5 T
{3}	(7) $-D \vee -A$	3 T (De Morgan's Laws)
{1, 2, 3, 4}	(8) $-D$	6, 7 T
{1, 2, 3, 4}	(9) $D \& -D$	4, 8 T
{1, 2, 3}	(10) $-D$	4, 9 R.A.A.

* In this context, the reader might wonder if any one of the three original rules could have been derived from the other two. As they are formulated in the text, the answer is negative; each is independent of the other two. This point is returned to in Chapter 5.

In this example, Rule R.A.A. is used to derive line (10) directly from lines (4) and (9). Since a *reductio ad absurdum* argument is a special case of the rule of conditional proof, it permits a reduction in the set of numbers attached to a given line. Thus in the above example, $\{1, 2, 3, 4\}$ is attached to (9), and the reduced set $\{1, 2, 3\}$ is attached to (10). Intuitively this situation reflects the fact that the contradiction of the conclusion, line (4), is a premise introduced for working purposes only.

The decision to attempt an indirect proof rather than a direct proof is always determined by a number of factors. An indirect proof is immediately suggested by a situation where there do not seem to be enough premises, for by venturing to find an indirect proof, we are permitted to introduce the negation of the conclusion as another premise. Although Example 5 involves a very simple inference, it illustrates this point. The conclusion is represented by a simple negation, so a direct conditional proof is not convenient. Moreover, none of the three premises is a simple atomic sentence. Faced with an implication and two disjunctions, it is hard to see how to begin. Negate the conclusion, obtain an atomic sentence as an extra premise, and the deductive machinery is ready to grind out an answer.

EXERCISES

1. Which of the following sets of premises are inconsistent? If a set is inconsistent, derive a contradiction. ~~If consistent, give a true sentential interpretation to prove it.~~

(a) If the contract is valid, then Horatio is liable. If Horatio is liable he will go bankrupt. Either Horatio will go bankrupt or the bank will lend him money. However, the bank will definitely not lend him money. (V, L, B, M)

(b) If Jones committed the murder, then he was in the victim's apartment and he did not leave before eleven. In fact, he was in the victim's apartment. If he left before eleven, then the doorman saw him. But it is not the case either that the doorman saw him or that he committed the murder. (M, A, L, D)

(c) The contract is satisfied if and only if the building is completed by November 30. The building is completed by November 30 if and only if the electrical subcontractor completes his work by November 10. The bank loses money if and only if the contract is not satisfied. Yet the electrical subcontractor completes his work by November 10 if and only if the bank loses money. (C, B, E, L)

(d) Smiling Pete will win the fourth race if and only if Crazy Cat comes in to show. If Crazy Cat comes in to show, then Dapper Dan will come in last. Either Dapper Dan will not come in last or Wildwood will tie for first. If Smiling Pete does not win the fourth race then Wildwood will tie for first. Furthermore, Wildwood will not tie for first. (S, C, D, W)

(e) $P \rightarrow Q$
$Q \leftrightarrow R$
$R \vee S \leftrightarrow -Q$

(f) $-(-Q \vee P)$
$P \vee -R$
$Q \rightarrow R$

2. Show that any formula may be derived from an inconsistent set of premises.

3. Prove if possible by *reductio ad absurdum:*

(a) If John plays first base, and Smith pitches against us, then Winsocki will win. Either Winsocki will not win, or the team will end up at the bottom of the league. The team will not end up at the bottom of the league. Furthermore, John will play first base. Therefore, Smith will not pitch against us. (J, S, W, T)

(b) If hedonism is not correct then eroticism is not virtuous. If eroticism is virtuous, then either duty is not the highest virtue or the supreme duty is the pursuit of pleasure. But the supreme duty is not the pursuit of pleasure. Therefore, either duty is not the highest virtue or hedonism is not correct. (H, E, D, S)

(c) If a declaration of war is a sound strategy then either fifty divisions are poised at the border or twenty wings of long-range bombers are ready to strike. However, fifty divisions are not poised at the border. Therefore, if twenty wings of long-range bombers are not ready to strike, then either a declaration of war is not a sound strategy or new secret weapons are available. (D, F, T, S)

4. Prove by *reductio ad absurdum:*

(a) Exercise 5 of § 2.3.

(b) Exercise 8 of § 2.3.

CHAPTER 3

SYMBOLIZING EVERYDAY LANGUAGE

§ **3.1 Grammar and Logic.** In the previous two chapters we have developed the logic of the sentential connectives. Our notation has consisted just of '$-$', '&', '$\vee$', '$\rightarrow$', '$\leftrightarrow$', parentheses, and single letters standing for sentences. A moment's reflection is enough to make anyone realize that an apparatus as meager as this is not sufficient to express logical distinctions of the most obvious and elementary kind. For example, simple truths of arithmetic cannot be formulated within this framework. In fact, our logical apparatus would not be adequate for the needs of a three-year-old child. There are no means of symbolizing common and proper nouns, pronouns, verbs, adjectives, or adverbs; most common grammatical distinctions cannot be indicated. Of course, we are not committed to developing a logic adequate to all the nuances of everyday language. What we do want is enough logical notions to express any set of systematic facts. Specifically we shall introduce three kinds of expressions corresponding to the many kinds distinguished in the grammar of ordinary language. We shall consider *terms, predicates,* and *quantifiers;* and we shall be concerned to see how well we can translate sentences of everyday language into a language consisting just of sentential connectives, terms, predicates, quantifiers, and parentheses. We shall also be interested in comparing the grammar of our logical symbolism with the grammar of everyday language.

It should be realized at the very beginning of this discussion that the usage of ordinary language is not sufficiently uniform and precise to permit the statement of unambiguous and categorical rules of translation. In translating a sentence of English into our symbolism, it is often necessary carefully to consider several possibilities.

§ **3.2 Terms.** A precise definition of 'term' will now be given. Perhaps the best place to begin is by considering the most important kind of terms in logic and mathematics, namely, *variables.* Beginning students of logic sometimes find the notion of variables rather confusing. The important thing to remember is that variables are simply letters, such as, 'x', 'y', 'z', or letters with subscripts, such as 'x_1', 'x_2', 'x_3', ... The grammatical function of variables is similar to that of *pronouns* and *common nouns* in everyday language. Thus, the sentence:

(1) Everything is either red or not red

is equivalent to the sentence:

(2) For every x, x is red or x is not red.

In this instance the use of the variable 'x' in (2) corresponds to the use of 'thing' in (1). The particular variable 'x' has no special significance, and (2) can be rewritten using 'y' in place of 'x':

For every y, y is red or y is not red.

From a literary standpoint (2) seems to be a rather barbarous version of (1). However, variables come into their own proper place when even the simplest arithmetical statements are considered. For example, we all know that for any numbers x, y, and z,

(3) $$x \cdot (y + z) = x \cdot y + x \cdot z.$$

However, a literary rendering of (3) is awkward and far from perspicuous. It would run something like:

(4) If a number is multiplied by the sum of a second and third number then the result is equal to the sum of the number multiplied by the second number, and the number multiplied by the third number.

Even more awkward than (4) is the statement of a piece of reasoning in arithmetic without the use of variables.

Proper names form another important class of terms. In order to have a compact notation, we use lower-case letters at the beginning of the alphabet to stand for proper names. Thus if we let

a = Isaac Newton

we may translate:

(5) Isaac Newton is the greatest mathematician of the last three centuries,

by:

(6) a is the greatest mathematician of the last three centuries.

Now (6) does not seem to be much more compact than (5), but a further substantial abbreviation is permitted when we notice that definite descriptions such as 'the greatest mathematician of the last three centuries' function just like proper names and thus form another class of terms. Since we also use lower-case letters standing at the beginning of the alphabet to stand for definite descriptions, if we let

b = the greatest mathematician of the last three centuries,

we may now translate (5) by:

(7) a is b.

The meaning of 'is' in (5) and (7) is that of identity and we may in fact write (7):

$$a = b.$$

We remark that names or descriptions of objects are sometimes designated 'constants'. For example, 'a', 'b' and 'Isaac Newton' are constants.

Thus far we have considered three kinds of terms: variables, proper names, and definite descriptions. The intuitive idea of a term is that it designates or names an entity. On the other hand, variables do not name anything. For example, in the expression:

$$x > 3$$

the variable 'x' does not name some unique number like 5 or 17. None the less, this intuitive idea of terms naming is a sound one, and we may use it to provide a general definition.

DEFINITION. *A* TERM *is an expression which either names or describes some object, or results in a name or description of an object when the variables in the expression are replaced by names or descriptions.*

Thus 'x' is a term according to the definition because when we replace 'x' by the Arabic numeral '3' we obtain an expression which designates the number three. The definition also classifies as terms such expressions as '$x + y$', for when we replace 'x' by '2' and 'y' by '3', we obtain '$2 + 3$', an expression which designates (i.e., names) the number 5. Some other examples of terms are:

$x + 3$
$x^2 + y^2 - 1$
the fattest man in township x.

§ 3.3 Predicates. In addition to our notation for sentential connectives and terms, we shall use capital letters to stand for predicates. Thus, we may use the letter 'R' in place of the predicate 'is red' and symbolize (2) of the previous section:

For every x, $Rx \vee -Rx$.

It should be clear that '$-Rx$' is read 'x is not red'; no new use of negation is involved, since '$-Rx$' is simply the negation of the formula 'Rx'.

Now in traditional grammar a predicate is the word or words in a sentence or clause which express what is said of the subject. In English, predicates may be formed in several ways: a verb standing alone, a verb together with an adverb or adverbial clause modifying it, a copula verb with a noun or an adjective. Thus, examples of predicates are 'swims', 'is swimming rapidly', 'is a man', 'is cantankerous'. In logic we do not distinguish predicates according to the grammatical parts out of which they are constructed; in fact, no place is even accorded to adverbs, adjectives, and common nouns. Single letters alone stand for predicates, and there are no symbols admitted for adverbs and the like. Furthermore, predicates are given a broader role in logic than in ordinary usage. For example, since a notation for common nouns is not introduced, common nouns which stand as the subjects of sentences are symbolized by means of variables and predicates. Consider, for instance, the sentence:

(1) Every man is an animal.

We translate this:

(2) For every x, if x is a man then x is an animal.

In ordinary grammar 'is an animal' is the predicate of (1). Its translation (2), has the additional predicate 'is a man' which replaces the common noun 'man' in (1). Using 'M' for the predicate 'is a man' and 'A' for the predicate 'is an animal', we may then symbolize (1):

(3) For every x, $Mx \rightarrow Ax$.

Sentence (3) in fact exemplifies the standard form for sentences of the type 'Every such and such is so and so'.

There is a useful logical distinction concerning predicates which may be illustrated by symbolizing the sentences:

(4) Emma was very gay

and:

(5) Mr. Knightley was considerably older than Emma.

Letting

$$e = \text{Emma}$$

and

$$k = \text{Mr. Knightley},$$

and using 'G' for the predicate 'was very gay' and 'O' for the predicate 'was considerably older than', we may symbolize (4) and (5):

$$Ge,$$
$$Oke.$$

For obvious reasons we refer to 'G' as a *one-place* predicate and to 'O' as a *two-place* predicate. A simple example of a three-place predicate is suggested by the notion of betweenness. An example of a four-place predicate is suggested by the relation of two objects being the same distance apart as two other objects. (Euclidean geometry, by the way, can be axiomatized in terms of just these two notions of betweenness and equidistance.)

In the sentence:

(6) Emma was very gay and beautiful

the predicate 'was very gay and beautiful' is called in grammar a *compound* predicate. Using 'B' for the predicate 'was beautiful', we may symbolize (6):

$$Ge \,\&\, Be. \tag{7}$$

On the other hand, using 'W' for 'was very gay and beautiful', we may also symbolize (6):

$$We. \tag{8}$$

In logic there is no commitment to simple predicates and hence the choice between (7) and (8) should turn on questions of convenience and context. If in a given argument gaiety and beauty are always associated, then (8) would probably be the best symbolization. For an argument which dissociates these two properties, (7) is to be preferred.

§ **3.4 Quantifiers.** Certain formulas containing variables are neither true nor false. Thus:

$$x \text{ loved } y \tag{1}$$

and:

$$x + y = z + 2 \tag{2}$$

are neither true nor false. However, if we replace 'x' by 'Emma' and 'y' by 'Mr. Knightley' we obtain from (1) a true sentence (true at least in the world of Jane Austen). And if in (2) we replace 'x' by '2', 'y' by '3' and 'z' by '4', we obtain a false sentence of arithmetic.

But one of the profoundest facts of logic is that we do not have to replace variables by names in order to get true or false sentences from (1) and (2). Another and equally important method is to prefix to expressions such as (1) and (2) phrases like 'for every x', 'there is an x such that', 'for all y'.

Thus from the formula:

$$x \text{ is a miser}$$

we obtain the true sentence:

There exists an x such that x is a miser,

which has the same meaning as the less strange-looking sentence:

There are misers;

and we can also obtain from this same formula the false sentence:

(3) For every x, x is a miser,

which has the same meaning as the more usual form:

Everyone is a miser.

The phrase 'for every x' is called a *universal quantifier*. The logical symbol we use for the universal quantifier is illustrated by rewriting (3):

$$(x)(x \text{ is a miser}).$$

In everyday language and in mathematics, words and phrases like 'some', 'there exists an x such that', 'there is at least one x such that' are called *existential quantifiers*. Thus the formula:

$$x > 0$$

becomes a true sentence when the existential quantifier 'there is an x such that' is prefixed to it. The logical symbol we use is illustrated by writing this sentence:

$$(\exists x)(x > 0).$$

It is important to realize that a formula which is neither true nor false is not necessarily made so by simply adding a single quantifier. The appropriate number of quantifiers must be added. For example, if we add an existential quantifier to (2), we obtain:

$$(\exists x)(x + y = z + 2), \tag{4}$$

and (4) is neither true nor false. However, if we also add '$(\exists y)$' and '$(\exists z)$' then we obtain the true sentence:

$$(\exists x)(\exists y)(\exists z)(x + y = z + 2). \tag{5}$$

Put in ordinary mathematical language, (5) says:

There are numbers x, y, and z such that $x + y = z + 2$.

With a logical notation for quantifiers available, we may now completely symbolize sentence (1) of the previous section:

(6) $(x)(Mx \rightarrow Ax).$

As already remarked, (6) exemplifies the appropriate symbolic formulation of sentences of the type 'Every such and such is so and so', or what is equivalent, sentences of the type 'All such and such are so and so'. Thus, as a second example, we would symbolize the sentence:

(7) All freshmen are intelligent

by:

(8) $(x)(Fx \rightarrow Ix),$

where we use 'F' for the predicate 'is a freshman' and 'I' for the predicate 'is intelligent'.

Consider now the sentence:

(9) No freshmen are intelligent.

We first obtain as a partial translation of (9):

(10) For all x, if x is a freshman then x is not intelligent.

And (10) is translated into symbols by:

(11) $(x)(Fx \rightarrow -Ix).$

The only difference between (8) and (11) is the presence in (11) of the negation sign before the predicate 'I', but this difference is, of course, crucial. Sentence (11) exemplifies the standard form for sentences of the type 'No such and such are so and so'.

We now turn to some sentences whose translations use existential quantifiers.*

(12) Some freshmen are intelligent.

We translate (12) by:

(13) For some x, x is a freshman and x is intelligent.

And (13) is easily translated into symbols by:

(14) $(\exists x)(Fx \;\&\; Ix).$

There is a very common mistake which beginners make in translating (12), and which should be carefully avoided. The translation of (7) as (8) falsely suggests the translation of (12) by:

(15) For some x, if x is a freshman then x is intelligent.

* It is also possible to use an existential quantifier to translate (11) by: $-(\exists x)(Fx \;\&\; Ix)$. By use of the rules of inference given in the next chapter we may establish the logical equivalence of these two translations.

Sentence (15) is rendered in symbols by:

(16) $(\exists x)(Fx \rightarrow Ix)$.

It is important to understand why (14) rather than (16) is the correct symbolization of (12). To see that (16) is incorrect, let us see what it really asserts. If in (16) we replace the formula '$Fx \rightarrow Ix$' by the tautologically equivalent formula '$-Fx \vee Ix$', we obtain:

(17) $(\exists x)(-Fx \vee Ix)$,

which we may translate in words by:

(18) For some x, either x is not a freshman or x is intelligent,

or, using another idiom for the existential quantifiers, by:

(19) There is something which is either not a freshman or is intelligent.

It should be obvious that (12) and (19) are not logically equivalent. If all freshmen were completely stupid, (12) would be false, but (19) would still be true as long as there is something in the universe which is not a freshman.

Statements of the form of (16) are nearly always true but utterly trivial. Consequently, such statements are very seldom the correct translation of any sentence of ordinary language in which we are interested.

Another example may help to underscore the triviality of (16). Let us translate the sentence:

(20) Some men are three-headed.

The correct symbolization is:

(21) $(\exists x)(Mx \,\&\, Tx)$,

and (20) and (21) are, to the best of our knowledge, false. If we had translated (15) by:

$$(\exists x)(Mx \rightarrow Tx),$$

and thus committed ourselves to the logically equivalent statement:

(22) $(\exists x)(-Mx \vee Tx)$,

we would be affirming the absurdity that (20) is true, as long as there are objects in the universe which are not men; for this trivial condition guarantees the truth of (22).

From the above discussion, the correct translation of such related sentences as:

Some freshmen are not intelligent

should be obvious. It is:

$$(\exists x)(Fx \ \& -Ix).$$

A useful extension of the kinds of sentences considered so far is to sentences using both variables and proper names. Thus we symbolize:

Adams is not married to anyone,

by:

$$(x)-(Max),$$

where 'Max' is read: Adams is married to x. As another example, we symbolize:

Some freshmen date Elizabeth

by:

$$(\exists x)(Fx \ \& \ Dxe),$$

where the one-place predicate 'F' and the two-place predicate 'D' have the obvious meaning, and e = Elizabeth.

EXERCISES

1. State without the use of variables the following two true sentences of arithmetic:

(a) $(\exists x)(\exists y)(x + y \neq x \cdot y)$.
(b) $(\exists x)(\exists y)(\exists z)(x - (y - z) \neq (x - y) - z)$.

2. Translate the following sentences from everyday language into a notation using terms, predicates, quantifiers, and sentential connectives. Use the letters indicated for predicates and proper names. Note that plural nouns not preceded by 'some' or 'all' sometimes are to be translated using universal quantifiers and sometimes using existential quantifiers.

(a) All seniors are dignified. (Sx, Dx)
(b) Some juniors are pretty. (Jx, Px)
(c) No freshmen are dignified. (Fx)
(d) Some freshmen are not pretty.
(e) Some seniors are both pretty and dignified.
(f) Betty is pretty but not dignified. (b)
(g) Not all juniors are pretty.
(h) Some seniors who are not serious like Greek. (Tx, Gx)
(i) No freshmen are not serious.
(j) Some juniors who like Greek are not pretty.
(k) All freshmen who like Greek are pretty.
(l) Anyone who likes Greek is either a senior or a junior.
(m) Some seniors like both Greek and mathematics. (Mx)
(n) No juniors like both Greek and mathematics.
(o) There are some seniors who like Greek but not mathematics.
(p) Every freshman dates some junior. (Dxy)
(q) Elizabeth does not date any freshmen. (e)

(r) Some juniors date only seniors.
(s) Some freshmen date seniors.
(t) Some seniors date no juniors.
(u) There are both seniors and juniors who date Betty.
(v) Some freshmen date only seniors who like Greek.
(w) Only freshmen date freshmen.
(x) Freshmen date only freshmen.
(y) Adams dates freshmen only if they are pretty. (*a*)
(z) Adams dates a junior.

§ 3.5 Bound and Free Variables. An occurrence of a variable in a formula can be one of two types: either that occurrence is controlled by a quantifier, or it is not. For the exact statement of the rules of inference governing quantifiers it is important to make precise this notion of an occurrence of a variable "being controlled" by a quantifier. To begin with, we need to make the notion of a formula more precise. An *atomic formula* is a predicate followed or flanked by the appropriate number of terms as arguments. Thus the expressions:

$$Ra, \qquad x + y = z, \qquad x \text{ is blue}$$

are atomic formulas. We may then give what is called a recursive definition of *formulas:*

(a) *Every atomic formula is a formula.*
(b) *If* S *is a formula, then* $-$(S) *is a formula.*
(c) *If* R *and* S *are formulas then* (R) & (S), (R) $\vee$ (S), (R) $\rightarrow$ (S), *and* (R) $\leftrightarrow$ (S) *are formulas.*
(d) *If* R *is a formula and* v *is any variable then* (v)(R) *and* ($\exists$v)(R) *are formulas.*
(e) *No expression is a formula unless its being so follows from the above rules.*

This rather elaborate definition makes explicit what should be intuitively obvious.* (In practice we omit the indicated parentheses when they are unnecessary.) Giving this definition corresponds to listing in the rules of chess, checkers, or bridge exactly what positions make sense in the game: for example, no one player may have six cards and his partner eight at any point in a bridge game. And the expression:

$$x = y \rightarrow \vee(x) \leftrightarrow$$

is not a formula, but a meaningless string of symbols. (Here the word

* The definition given here is not completely precise, for no explicit list of primitive terms and predicates is given, as would be required in constructing a completely formalized language. Moreover, in a formalized language admitting terms like '$x + y$', a recursive definition of terms would also be needed.

'meaningless' should be taken in a formal sense. A formula is not required to be meaningful in the sense of saying something significant; it simply has to have a certain formal or *syntactical* structure.)

With a relatively precise idea of formula before us, we may now define the important notion of the *scope* of a quantifier.

DEFINITION. *The* SCOPE *of a quantifier occurring in a formula is the quantifier together with the smallest formula immediately following the quantifier.*

In the following examples the scope of the quantifier '$(\exists x)$' is indicated by the line underneath the formula:

(1) $$\underbrace{(\exists x)Mx}\vee Rx$$

(2) $$(y)\underbrace{(\exists x)(x > y \mathbin{\&} (z)(z = 2))}_{\text{scope}}$$

(3) $$\underbrace{(\exists x)(y)((xy = 0) \mathbin{\&} (x)(z)(x + z = z + x))}_{\text{scope}}$$

(4) $$\underbrace{(\exists x)(y)(xy = 0)}_{\text{scope}} \mathbin{\&} (x)(z)(x + z = z + x)$$

The parentheses in a formula always make clear in a perfectly natural way what expression is the smallest formula following a quantifier. Thus in (1) 'Mx' is the smallest formula following '$(\exists x)$'. In (2) the mate of the left parenthesis immediately following '$(\exists x)$' occurs at the end of the formula, which determines that the smallest formula is in this case the whole remaining formula. But as (3) shows, it is not sufficient to look for the mate of the left parenthesis immediately following the quantifier, for in the case of (3) this would falsely determine '(y)' as the smallest formula following '$(\exists x)$', but the quantifier '(y)' is not a formula. However, if such quantifiers are ignored the hunt-for-the-mate rule always works.

We are now in a position to define *bound* and *free* occurrences of variables.

DEFINITION. *An occurrence of a variable in a formula is* BOUND *if and only if this occurrence is within the scope of a quantifier using this variable.*

The phrase 'quantifier using this variable' should be clear. Thus, the quantifier '$(\exists x)$' uses the variable 'x'; the quantifier '(y)' uses the variable 'y'. In formulas (2)–(4) above, every occurrence of variables is bound.

DEFINITION. *An occurrence of a variable is* FREE *if and only if this occurrence of the variable is not bound.*

In the formula:

$$(5) \qquad (\exists y)(x > y)$$

both occurrences of the variable 'y' are bound and the single occurrence of 'x' is free. In (1) the first two occurrences of 'x' are bound and the third occurrence is free.

DEFINITION. *A variable is a* FREE VARIABLE *in a formula if and only if at least one occurrence of the variable is free, and a variable is a* BOUND VARIABLE *in a formula if and only if at least one occurrence of the variable is bound.*

We notice that a variable may be both free and bound in the same formula, but any given occurrence of the variable is either bound or free but not both. Thus in the formula:

$$y > 0 \vee (\exists y)(y < 0)$$

the variable 'y' is both free and bound; its first occurrence is free, and the second and third occurrences are bound.

Without explicitly giving notice, we have reserved the word 'sentence' to apply to formulas which are true or false, but a formula is true or false if and only if it has no free variables. Consequently a precise syntactical definition is simply:

DEFINITION. *A* SENTENCE *is a formula which has no free variables.*

Thus formulas (2)–(4) of this section are sentences, but (5) is not, because of the free occurrence of 'x'.

With respect to the problem of correctly symbolizing sentences of everyday language, it should be emphasized that the end result should contain no free variables. In other words, the symbolized expression should also be a sentence. In Chapter 7, on the other hand, we shall discuss the universal tendency in mathematics to use formulas which have free variables.

EXERCISES

1. Classify the following expressions as: (i) terms, (ii) formulas, or (iii) neither terms nor formulas.

(a) Thomas Jefferson
(b) slowly
(c) the number six
(d) x is purple
(e) the integer y
(f) the number x such that $x = x + x$
(g) $x + y = z$

(h) $(x + y) + z^2$
(i) every man x
(j) that man who will be the next President
(k) x hates y, but loves z
(l) $(3 + 1) + 10$
(m) very pretty

2. How many free occurrences of variables and how many free variables are there in each of the following formulas? Which of the formulas are sentences?

(a) $(x)(y)(z)(x > y \ \& \ y > z) \rightarrow (\exists w)(w > w)$
(b) $(\exists x)(x$ is red$) \vee (y)(y$ is blue and x is purple$)$
(c) $x + x = x + x$
(d) $(\exists y)(x + x = x + x)$
(e) $(\exists x)(\exists y)(x$ is married to y and z is their child$)$

§ 3.6 A Final Example. The sentences considered so far in this chapter are relatively simple in structure. It is desirable to consider a relatively more complicated example:

(1) If one instant of time is after a second, then there is an instant of time after the second and before the first.

The best way to begin a translation is to introduce variables before we symbolize the sentence. It is clear that we need three distinct variables in translating (1):

(2) If instant of time x is after instant of time y, then there is an instant of time z such that z is after y and z is before x.

It is not difficult to pass from (2) to a complete symbolization of (1). The main change is to replace the descriptive adjectival phrase 'instant of time' by the corresponding predicate. The result of this substitution is a more awkward English sentence, but one whose logical structure is simpler than that of (2):

(3) If x is an instant of time, y is an instant of time, and x is after y, then there is a z such that z is an instant of time, z is after y, and z is before x.

It is a routine task to symbolize (3):

(4) $$Tx \ \& \ Ty \ \& \ Axy \rightarrow (\exists z)(Tz \ \& \ Azy \ \& \ Bzx),$$

where 'T' stands for 'is an instant of time', 'A' for the two-place predicate 'is after', and 'B' for the two-place predicate 'is before'.

Several remarks are in order concerning the formulas (1) to (4). Formula (4) has two free variables and consequently is not strictly a sentence which is either true or false. On the other hand, it seems natural and intuitively correct to regard (1) as a sentence. In (1) the words 'one' and 'second' function as universal quantifiers. You will perhaps have a better sense of their role if you insert the word 'arbitrary' after 'one' and 'second', and change (1) to read:

> If one arbitrary instant of time is after a second arbitrary instant of time, then ....

On the other hand, we easily obtain a sentence (in the strict sense of the word) from (4) by adding two universal quantifiers:

$$(5) \qquad (x)(y)[Tx \; \& \; Ty \; \& \; Axy \rightarrow (\exists z)(Tz \; \& \; Azy \; \& \; Bzx)]$$

which is the final step in completely symbolizing (1).

Comparison of (1) and (5) emphasizes the point that universal quantifiers are expressed in idiomatic English usage in a variety of ways. It might seem that it would be difficult to ferret out and recognize these different idioms, but in practice it turns out to be fairly simple: after giving a translation which seems intuitively appropriate, add however many universal quantifiers are necessary to obtain a sentence, as we did in obtaining (5) from (4). As already remarked, in the next chapter we shall see that formulas such as (4), which contain free variables and are not sentences, play an important role in the theory of inference.

It perhaps needs to be repeated that no hard and fast rules can be given for correctly symbolizing sentences of ordinary language. The correctness or incorrectness of a proposed translation must be decided by a variety of informal, intuitive considerations. In certain cases several non-equivalent translations seem equally correct, but it is usually the case that if two non-equivalent translations are put forth, one can be shown to be incorrect.

EXERCISES

1. Using the letters indicated for predicates and constants, translate the following. The variables following the predicates are not necessarily the ones you should use in symbolizing the sentence.

(a) Sophomores like Greek only if they like mathematics. (Sx, Gx, Mx)
(b) Seniors date only juniors. (Sx, Dxy, Jx)
(c) Some seniors like Greek, but no seniors like both French and mathematics. (Fx)
(d) Every instant of time is after some instant. (Ix, Axy)
(e) If two instants of time are not identical, then one is after the other. ($x = y$)
(f) There is no instant of time such that every instant is after it.
(g) There is no instant of time such that no instant is after it.
(h) If one instant is after a second instant, then the second is before the first. (Bxy)
(i) The only sophomores who date Betty are those who like Greek. (Dxy, b)
(j) Some seniors who like mathematics do not date Elizabeth. (e)

(k) Some juniors who do not like French date both Betty and Elizabeth.
(l) Some freshmen who like both Greek and mathematics date neither Betty nor Elizabeth.
(m) If all sophomores like Greek then some freshmen do.
(n) Either all juniors like mathematics or some sophomores like Greek.
(o) If seniors date only juniors then some seniors date no one.
(p) If no senior dates Betty then either some sophomore or some junior does.

2. Using the logical notation developed, and the standard symbols of arithmetic, such as '$<$' for 'less than', symbolize the following sentences.

(a) There is a number x less than 5 and greater than 3.
(b) Given any number x there is a smaller number y.
(c) There is no largest number.
(d) For any two numbers x and y the sum of x and y is the same as the sum of y and x.
(e) There are numbers x, y, and z such that the difference of x and y is less than the product of x and z.
(f) For every number x there is a number y such that for every number z if the difference of z and 5 is less than y then the difference of x and 7 is less than 3.
(g) If the sum of two numbers which are neither zero is zero then one of the numbers is greater than zero.

CHAPTER 4

GENERAL THEORY OF INFERENCE

§ 4.1 Inference Involving Only Universal Quantifiers. It is advisable to learn in piecemeal fashion the somewhat complex rules of inference which go beyond the truth-functional methods of Chapter 2 and depend upon the logical properties of quantifiers. Four new rules are needed: a pair for dropping and adding universal quantifiers, and a like pair for existential quantifiers. Roughly speaking, the *strategy* for handling inferences involving quantifiers falls into four parts:

I. Symbolize premises in logical notation.
II. Drop quantifiers according to rules introduced in this chapter.
III. Apply sentential methods of derivation to obtain conclusion without quantifiers.
IV. Add quantifiers to obtain conclusion in final form.

(This statement about strategy is to be regarded as a useful hint; sometimes it is not applicable.)

In order to begin in a simple way, in this section we consider formulas which have only universal quantifiers with scopes running over the entire formula. A classical syllogism may serve as the first example.

All animals are mortal.
All human beings are animals.
Therefore, all human beings are mortal.

Using the methods of translation of the previous chapter we symbolize this argument:

$$(x)(Ax \rightarrow Mx)$$
$$(x)(Hx \rightarrow Ax)$$
$$\overline{(x)(Hx \rightarrow Mx)}$$

To construct a derivation corresponding to this simple argument we introduce two new rules of inference. The first permits us to drop a universal quantifier whose scope runs over the whole formula. The intuitive idea of the rule is that whatever is true for every object is true for any given object. The second rule permits us to add a quantifier; in this case we infer a truth about everything from a truth about an arbitrarily selected object.

The derivation goes as follows:

EXAMPLE 1.

{1}	(1) $(x)(Ax \rightarrow Mx)$	P
{2}	(2) $(x)(Hx \rightarrow Ax)$	P
{1}	(3) $Ax \rightarrow Mx$	Drop universal quantifier of (1)
{2}	(4) $Hx \rightarrow Ax$	Drop universal quantifier of (2)
{1, 2}	(5) $Hx \rightarrow Mx$	3, 4 T (Law of Hypothetical Syllogism)
{1, 2}	(6) $(x)(Hx \rightarrow Mx)$	Add universal quantifier to (5)

The analysis of this example in terms of the four-part strategy suggested above is straightforward:

STEP I. *Symbolize premises:* lines (1) & (2).
STEP II. *Drop quantifiers:* lines (3) & (4).
STEP III. *Apply sentential methods:* line (5).
STEP IV. *Add quantifier:* line (6).

We now want to give an exact statement of the two rules governing universal quantifiers. We call the first one, the *rule of universal specification,* because a statement true of everything is true of any arbitrarily *specified* thing.* The exact formulation is:

RULE OF UNIVERSAL SPECIFICATION: *US. If a formula* **S** *results from a formula* **R** *by substituting a term* **t** *for every free occurrence of a variable* **v** *in* **R** *then* **S** *is derivable from* (**v**)**R**.†

In the above derivation, the term **t** is simply 'x' in both cases of universal specification, which is why we spoke of *dropping* quantifiers, but in general

* The phrase 'universal instantiation' is often used. I prefer 'specification' to 'instantiation' on the principle that peculiar words unused in everyday language should be introduced only as a last resort.

† Here and in the subsequent rules, to avoid any possible ambiguity in the scope of the quantifier we may treat '**R**' as if it were flanked by parentheses. Thus we read '**S** is derivable from (**v**)**R**' as if it were '**S** is derivable from (**v**)(**R**)'.

we want to be able to replace the quantified variable by any term.* Thus if 'a' is the name of John Quincy Adams, we may infer '$Aa \rightarrow Ma$', as well as '$Ay \rightarrow My$', from '$(x)(Ax \rightarrow Mx)$'. And in the case of arithmetic from the statement '$(x)(x + 0 = x)$' we may infer by universal specification '$(x + y) + 0 = x + y$'; in this case the term t is '$x + y$'.

The second rule, which permits the addition of quantifiers, is complicated by the necessity of a restriction on its range of applicability, even for the limited arguments of this section. The character of this restriction may be elucidated by the following simple argument. We begin with the premise that some arbitrary x is human, and we conclude fallaciously that everything is human.

{1}	(1) Hx	P
{1}	(2) $(x)Hx$	Add universal quantifier to (1)

The argument says that for every object x if x is human then everything is human. If no objects were human the antecedent would be always false, but since there are humans there are cases of x for which the antecedent is true and obviously the conclusion is false. But a restriction which blocks this fallacious inference is not hard to find: *do not universally generalize*, that is, do not universally quantify, a variable which is free in a premise. In the above fallacious argument we universally generalized on the variable 'x' free in the premise 'Hx'. For compactness of reference we shall say that a variable free in a premise is *flagged*. Moreover, a variable which is free in a premise is also *flagged* in any line in which it is free and which depends on the premise. We list the flagged variables in a formula at the right in a derivation (see Example 2 below).

The exact statement of the rule of inference for adding a universal quantifier is then:

RULE OF UNIVERSAL GENERALIZATION: *UG*. *From formula* S *we may derive* $(\mathsf{v})\mathsf{S}$, *provided the variable* v *is not flagged in* S.†

The following example applying the two new rules of inference illustrates three things: (i) the abbreviated notation: *US* for universal specification (lines (4)–(6)) and the abbreviated notation: *UG* for universal generalization (line (12)); (ii) how flagged variables are indicated (lines (7)–(10)); (iii) how universal specification is applied when the premises happen to be symbolized by use of different variables.

* In connection with the systematic consideration of terms in the theory of derivation, there is one informal restriction that needs to be made clear. We always assume that the object designated by a term exists and is unique. In other words, names and descriptions of fictitious objects are excluded. We thus prohibit such terms as 'the only son of Franklin Delano Roosevelt', 'the eleventh wife of Napoleon', 'the prime number between eight and ten'. We return to these matters in Chapter 8.

† The flagging restriction is also discussed in the next section. Its intuitive significance is further expounded in Chapter 7.

EXAMPLE 2 (LEWIS CARROLL). *No ducks are willing to waltz. No officers are unwilling to waltz. All my poultry are ducks. Therefore, none of my poultry are officers.*

{1}	(1) $(x)(Dx \rightarrow -Wx)$	P
{2}	(2) $(y)(Oy \rightarrow Wy)$	P
{3}	(3) $(z)(Pz \rightarrow Dz)$	P
{1}	(4) $Dx \rightarrow -Wx$	1 *US*
{2}	(5) $Ox \rightarrow Wx$	2 *US*
{3}	(6) $Px \rightarrow Dx$	3 *US*
{7}	(7) Px	x P
{3, 7}	(8) Dx	x 6, 7 T
{1, 3, 7}	(9) $-Wx$	x 4, 8 T
{1, 2, 3, 7}	(10) $-Ox$	x 5, 9 T
{1, 2, 3}	(11) $Px \rightarrow -Ox$	7, 10 C.P.
{1, 2, 3}	(12) $(x)(Px \rightarrow -Ox)$	11 *UG*

Notice that 'x' is first flagged in line (7) because it is free in 'Px' and 'Px' is a premise. It remains flagged in lines (8)–(10) because these three lines depend on (7), as indicated by the numbers listed at the left. The conditionalization resulting in line (11) ends the flagging of 'x' and permits universal generalization on (11). In working out derivations it is a mistake to worry very much about the flagging restriction. Ordinarily one is only inclined to universally generalize on a variable when it is both natural and correct to do so. Hence the practical recommendation is: construct a derivation without regard to flagging, at the end quickly inspect each line, flag the appropriate variables and check that no flagged variable has been universally generalized. In derivations involving manipulation of quantifiers, almost any application of the rule of conditional proof involves routine flagging of at least one variable for a few lines.

The symbolization of the three premises of the above argument by use of three different variables was done only for purposes of illustration, showing how the term 'x' is substituted in all three by universal specification. Ordinarily the natural thing to do is to use the same variable as much as possible—thus here, to use 'x' in symbolizing all three premises.

An analysis of the general strategy used in Example 2 shows the same four-step development as in Example 1.

STEP I. *Symbolize premises:* lines (1)–(3).
STEP II. *Drop quantifiers:* lines (4)–(6).
STEP III. *Apply sentential methods:* lines (7)–(11).
STEP IV. *Add quantifier:* line (12).

It should be obvious that under the restriction of this section (only universal quantifiers standing in front) the exercise of logical acumen is almost wholly concentrated in Step III, application of sentential methods. The general importance of a firm mastery of tautological implications can scarcely be emphasized too much.

A final example for this section will be useful for illustrating inferences involving proper names.

EXAMPLE 3. *No Episcopalian or Presbyterian is a Unitarian. John Quincy Adams was a Unitarian. Therefore, he was not an Episcopalian.*

{1}	(1) $(x)[(Ex \vee Px) \rightarrow -Ux]$	P
{2}	(2) Ua	P
{1}	(3) $(Ea \vee Pa) \rightarrow -Ua$	1 *US*
{1, 2}	(4) $-(Ea \vee Pa)$	2, 3 T
{1, 2}	(5) $-Ea \,\&\, -Pa$	4 T
{1, 2}	(6) $-Ea$	5 T

We derive line (3) from (1) by *US*, replacing 'x' by the proper name 'a', where a = John Quincy Adams. Notice the application of De Morgan's Laws in obtaining (5) from (4).

EXERCISES

Construct (if possible) a derivation corresponding to the following arguments. The variables used to indicate the number of places of the predicates are not necessarily the variables you should use in symbolizing the premises. Only universal quantifiers standing in front are required to symbolize all sentences.

1. All scientists are rationalists. No British philosophers are rationalists. Therefore, no British philosophers are scientists. (Sx, Rx, Bx)
2. No existentialist likes any positivist. All members of the Vienna Circle are positivists. Therefore, no existentialist likes any member of the Vienna Circle. (Ex, Lxy, Px, Mx)
3. If one man is the father of a second, then the second is not father of the first. Therefore, no man is his own father. (Fxy)
4. For every x and y either x is at least as heavy as y or y is at least as heavy as x. Therefore, x is at least as heavy as itself. (Hxy)
5. Adams is a boy who does not own a car. Mary dates only boys who own cars. Therefore, Mary does not date Adams. (Bx, Ox, Dxy, a, m)
6. Every member of the City Council lives within the city limits. Mr. Fairman does not live within the city limits. Therefore, Mr. Fairman is not a member of the City Council. (Mx, Lx, f)
7. Every member of the policy committee is either a Democrat or a Republican. Every Democratic member of the policy committee lives in California. Every Republican member of the policy committee is a member of the tax committee. Therefore, every member of the policy committee who is not also on the tax committee lives in California. (Px, Dx, Rx, Lx, Tx)
8. Anyone who works in the factory is either a union man or in a managerial position. Adams is not a union man, and he is not in a managerial position. Therefore, Adams does not work in the factory. (Wx, Ux, Mx, a)

9. Ptah is an Egyptian god, and he is the father of all Egyptian gods. Therefore, he is the father of himself. (p, Gx, Fxy)

10. Given: (i) for any numbers x, y, and z, if $x > y$ and $y > z$ then $x > z$; (ii) for any number x, it is not the case that $x > x$. Therefore, for any two numbers x and y, if $x > y$ then it is not the case that $y > x$. $(Nx, x > y)$

11. In the theory of rational behavior, which has applications in economics, ethics and psychology, the notion of an individual preferring one object or state of affairs to another is of importance. We may say that an individual *weakly* prefers x to y if he does not strictly prefer y to x. We use the notion of weak preference for formal convenience, for if we use strict preferences, we also need a notion of *indifference*. The point of the present exercise is to ask you to show that on the basis of two simple postulated properties of weak preference and the appropriate definitions of strict preference and indifference in terms of weak preference, we may logically infer all the expected properties of strict preference and indifference. Let us use 'Q' for weak preference, 'P' for strict preference and 'I' for indifference. Our two postulates or premises on the predicate 'Q' just say that it is transitive and that of any two objects in the domain of objects under consideration, one is weakly preferred to the other. In symbols:

(1) $(x)(y)(z)(xQy \;\&\; yQz \rightarrow xQz)$,
(2) $(x)(y)(xQy \vee yQx)$.

As additional premises we introduce the two obvious definitions:

(3) $(x)(y)(xIy \leftrightarrow xQy \;\&\; yQx)$,
(4) $(x)(y)(xPy \leftrightarrow -yQx)$.

Derive the following conclusions from these four premises:

(a) $(x)(xIx)$
(b) $(x)(y)(xIy \rightarrow yIx)$
(c) $(x)(y)(z)(xIy \;\&\; yIz \rightarrow xIz)$
(d) $(x)(y)(xPy \rightarrow -yPx)$
(e) $(x)(y)(z)(xPy \;\&\; yPz \rightarrow xPz)$
(f) $(x)(y)(xIy \rightarrow -(xPy \vee yPx))$
(g) $(x)(y)(z)(xIy \;\&\; yPz \rightarrow xPz)$
(h) $(x)(y)(z)(xIy \;\&\; zPx \rightarrow zPy)$

12. E. V. Huntington gave a list of axioms for the "informal" part of Whitehead and Russell's *Principia Mathematica*, which depends on three primitive symbols, a one-place relation symbol 'C', a binary operation symbol '$+$', and a unary operation symbol '$^{\prime}$'.* The main intended interpretation of the theory is that the variables take as values elementary propositions (or sentences if you wish); a proposition has the property C if it is true, $x + y$ is the disjunction of the two propositions x and y; $x^{\prime}$ is the negation of the proposition x.† (Thus this postulate set is closely connected with the logic of sentential connectives considered in Chapter 1.) The five axioms are:

A1. $(x)(y)[C(x + y) \rightarrow C(y + x)]$
A2. $(x)(y)[C(x) \rightarrow C(x + y)]$
A3. $(x)[C(x^{\prime}) \rightarrow -C(x)]$
A4. $(x)[-C(x^{\prime}) \rightarrow C(x)]$
A5. $(x)(y)[C(x + y) \;\&\; C(x^{\prime}) \rightarrow C(y)]$

* *Bulletin of the American Mathematical Society*, Vol. 40 (1934) pp. 127–136. Our formalization reduces the number of axioms from eight to five.

† Here the symbol '$+$' has no connection with ordinary arithmetic.

Construct derivations to show that the following theorems are logical consequences of the axioms. Since some of the theorems are useful in proving others, one long derivation for all parts is recommended. (Later we develop rules for inserting previously proved results in new derivations.) If one long derivation is given, indicate at the end which line of the derivation corresponds to each theorem. Make sure that no theorem depends on more than the five given axioms as premises.

THEOREM 1. $(x)[-C(x) \rightarrow C(x^{\prime})]$
THEOREM 2. $(x)[C(x) \rightarrow -C(x^{\prime})]$
THEOREM 3. $(x)(y)[-C(x+y) \rightarrow -C(y+x)]$
THEOREM 4. $(x)(y)[C(x+y) \,\&\, -C(x) \rightarrow C(y)]$
THEOREM 5. $(x)(y)[-C(x+y) \rightarrow -C(x) \,\&\, -C(y)]$
THEOREM 6. $(x)(y)[-C(x) \,\&\, -C(y) \rightarrow -C(x+y)]$
THEOREM 7. $(x)[C((x^{\prime})^{\prime}) \leftrightarrow C(x)]$
THEOREM 8. $(x)(y)[C(x) \,\&\, C(x^{\prime}+y) \rightarrow C(y)]$
THEOREM 9. $(x)C((x+x)^{\prime}+x)$
THEOREM 10. $(x)(y)C(y^{\prime}+(x+y))$
THEOREM 11. $(x)(y)C((x+y)^{\prime}+(y+x))$
THEOREM 12. $(x)(y)(z)C[(x^{\prime}+y)^{\prime}+[(z+x)^{\prime}+(z+y)]]$ *

§ 4.2 Interpretations and Validity. At the beginning of Chapter 2 the notion of *interpretation* of a sentence or set of sentences was introduced, and the more special notion of a *sentential* interpretation was explicitly defined. It was also pointed out in that chapter how sentential interpretations may be used to prove that a truth-functional argument is invalid or that a set of premises is consistent. Before proceeding to the rules of inference for existential quantifiers, we want to define the general notion of interpretation and use it to obtain general definitions of validity, invalidity, and consistency. This discussion is interjected here because it will prove useful to have the notion of an interpretation at hand in discussing the rules governing existential quantifiers.

In giving a particular interpretation of a formula or sentence we fix on a *domain of individuals*, such as the set of all men, or the set of positive integers. Thus, given the sentence '$(x)(Hx \rightarrow Mx)$', which occurs in Example 1 of this chapter, an interpretation of it in the domain of positive integers, which is often called an *arithmetical interpretation*, is:

$$(x)(x > 3 \rightarrow x > 2).$$

Here the predicate 'H' is replaced by the predicate '>3' (meaning 'is

* The last four theorems correspond to four of the five axioms used by Whitehead and Russell (*Principia Mathematica*, first edition) in presenting their axiomatized form of sentential logic. (The fifth axiom was later derived from these four by P. Bernays.) Translated into the logical notation of Chapter 1 and using the fact that $-P \vee Q$ is tautologically equivalent to $P \rightarrow Q$, these four theorems become

AXIOM 1. $P \vee P \rightarrow P$
AXIOM 2. $P \rightarrow P \vee Q$
AXIOM 3. $P \vee Q \rightarrow Q \vee P$
AXIOM 4. $(P \rightarrow Q) \rightarrow [R \vee P \rightarrow R \vee Q]$

greater than three') and the predicate 'M' by '>2' (meaning 'is greater than two'). For a second example, consider the two postulates * for weak preference given in the next to last exercise of the preceding section:

$$(x)(y)(z)(xQy \ \& \ yQz \rightarrow xQz)$$
$$(x)(y)(xQy \vee yQx).$$

A simple arithmetical interpretation of the conjunction of these two postulates is to interpret the predicate 'Q' as '$\geq$'. Notice that if we demand a *true* arithmetical interpretation we may not interpret 'Q' as '$>$', i.e., we may not interpret weak preference as greater than, for given any two numbers it is not always the case that

$$x > y \vee y > x. \tag{1}$$

Obviously (1) does not hold when x and y are the same number.

The two examples considered are not intended to foster the idea that interpretations should always be arithmetical in character. Consider this sentence about strict preference, where as in the exercise just referred to, xPy if and only if x is strictly preferred to y:

$$(x)(y)(xPy \rightarrow -yPx). \tag{2}$$

If we interpret 'P' as 'is father of' we obtain a true interpretation of (2) for the domain of human beings:

$$(x)(y)(x \text{ is father of } y \rightarrow -(y \text{ is father of } x)).$$

An example of an interpretation of a formula having free variables also needs to be considered. The technique of interpretation of such variables is simple: all free occurrences of such a variable are replaced by the name of some individual in the domain of individuals with respect to which the interpretation is made. Thus, again drawing upon the theory of preference, consider:

$$xPy \rightarrow (z)(xIz \rightarrow zPy), \tag{3}$$

which says:

> If x is preferred to y, then for all objects z if x and z are indifferent in preference (i.e., equal in value) then z also is preferred to y.

A true arithmetical interpretation of (3) is easily found: Interpret the free variable 'x' as '2', 'y' as '1', 'P' as '$>$', 'I' as '$=$', and we obtain the arithmetical truth:

$$2 > 1 \rightarrow (z)(2 = z \rightarrow z > 1). \tag{4}$$

* Throughout this book we use the words 'axiom' and 'postulate' as synonyms. The vague distinction which is sometimes made between axioms and postulates in elementary geometry is abandoned here.

To simplify the definition of interpretation we shall exclude definite descriptions like 'the husband of x' or 'the smallest prime number greater than y'. We then have as terms, variables, proper names, and expressions built up from variables and names by use of operation symbols, that is, expressions like '$2 + 3$', '$x + y$', '$x + 5$'. To complete our list of intuitive examples of interpretations we need to indicate how proper names and operation symbols are interpreted.

For proper names the answer is simple, they are handled just like free variables: all occurrences of a proper name in a formula are replaced by the name of some individual in the domain of individuals with respect to which the interpretation is made. Consider, for instance, the formula:

(5) x is a city & x is larger than San Francisco.

An arithmetical interpretation of (5) is easily found. Interpret 'is a city' as 'is a positive integer', the free variable 'x' as '1', 'is larger than' as '$>$' and the proper name 'San Francisco' as '2'. We then have as an interpretation of (5):

(6) 1 is a positive integer & $1 > 2$.

Sentence (6) is, of course, a false interpretation. If we existentially quantify the free variable 'x' in (5), we have:

(7) $(\exists x)$(x is a city & x is larger than San Francisco).

As a true interpretation of (7), we then have:

(8) $(\exists x)$(x is a positive integer & $x > 2$).

But it is also easy to find a false interpretation of (7), namely:

(9) $(\exists x)$(x is a positive integer & $x < 1$).

In (9) 'San Francisco' is replaced by '1' rather than '2' in the interpretation and 'is larger than' by '$<$' rather than '$>$'.

Operation symbols are handled like predicates: we substitute in the interpretation an operation symbol defined for the individuals in the domain of interpretation. The operation symbols which occur most frequently in ordinary language are probably those for arithmetical addition and multiplication. In this chapter, we shall scarcely consider operation symbols, but in order to make the definition of interpretation complete, we mention them cursorily now. As one example of an interpretation, we may interpret the symbol for addition as the symbol for multiplication. In this case the domain of individuals is the same for the formula and the interpretation given. Thus, as an interpretation of:

$$(x)(y)(x + y = y + x)$$

we have:

$$(x)(y)(xy = yx).$$

As an interpretation of:

$$(\exists y)(x + y > x)$$

we have:

$$(\exists y)(2y > 2).$$

In this last case the free variable 'x' is assigned the individual constant '2' in the interpretation.*

Without going into exact detail it should be clear that a one-place predicate is interpreted by a one-place predicate, a binary operation symbol like '+' by a binary operation symbol, and so forth. However, there is some flexibility in classifying predicates as one-place, two-place, etc. For instance, we might replace the predicate 'is human' by the arithmetical predicate 'is greater than one', which in ordinary notation looks like a two-place predicate, or we might even replace 'is human' by '$(\exists y)$(... *is greater than* y)'. It is not necessary for our purposes exactly to characterize what predicates may be replaced by what others in an interpretation, but the basic idea is obvious: the predicate of the interpretation should have exactly the same number of places to fill with free variables as the predicate being interpreted. (Not that in the interpretation these places are filled with free variables. It is in fact a characteristic of an interpretation that it has no free variables and is thus either true or false.)

The above remarks and examples should make the general definition of an interpretation easy to comprehend.

> *Sentence* **P** *is an interpretation of formula* **Q** *with respect to the domain of individuals* D *if and only if* **P** *can be obtained from* **Q** *by substituting predicates and operation symbols defined for the individuals in the domain* D *for the predicates and operation symbols respectively of* **Q** *and by substituting proper names of individuals in* D *for proper names (i.e., individual constants) and free variables of* **Q**.†

Using this notion of interpretation we may now define *universal validity*, *logical consequence* or *logical implication*, and *consistency*.

> *A formula is universally valid if and only if every interpretation of it in every non-empty domain of individuals is true.*

The intuitive idea behind this definition is that universally valid formulas should be true in every possible world. If their truth hinges upon con-

* In mathematical contexts, proper names like '1' and '2' are usually called *individual constants*, and predicates are called *relation symbols*.

† A completely precise definition requires the set-theoretical notions developed in Part II of this book. This definition is slightly extended in § 4.3 to take care of the technical device used for handling existential quantifiers.

tingent facts about the actual world, they are not genuinely universally valid.

A formula Q *logically follows from a formula* P *if and only if in every non-empty domain of individuals every interpretation which makes* P *true also makes* Q *true.*

An equivalent definition is simply:

Q *logically follows from* P *if and only if the conditional* P $\rightarrow$ Q *is universally valid.*

A second intuitive idea behind these definitions is that one formula logically follows from another just by virtue of their respective logical forms. By considering all interpretations of a formula we effectively abstract from everything but its bare logical structure. It should be remarked that in making this abstraction we deny the universal validity of certain sentences of ordinary language which would seem to be true in every possible world. A typical example is 'All bachelors are unmarried'. The logical analysis of such sentences is a subtle and complicated matter which we shall avoid in this book. Fortunately the systematic deductive development of any branch of mathematics or theoretical science can proceed without explicit recourse to sentences whose truth follows simply from the meanings of the predicates used.*

The definition of consistency is easily anticipated from Chapter 2.

A formula is consistent if and only if it has at least one true interpretation in some non-empty domain of individuals.

(In dealing with a set of premises, we may take their conjunction to obtain a single formula.)

In the three definitions just given, only *non-empty* domains of individuals have been considered. Probably the strongest single argument for this restriction is that we want to consider the inference from:

$$(x)Hx$$

to:

$$(\exists x)Hx$$

as logically valid. And correspondingly, we want to have the formula:

$$(x)Hx \;\&\; -(\exists x)Hx$$

inconsistent. If the empty domain is included, neither of these wants will be satisfied.

* Almost the whole of Chapter 12 is relevant to this point: only a meager theory of meaning is needed to apply the axiomatic method in a given branch of science.

For explicitness, it is also desirable to define the notion of an *invalid argument.*

> *An argument is invalid if and only if there is an interpretation in some non-empty domain which makes its premises true and its conclusion false.*

We may apply this definition to show explicitly that the flagging restriction on universal generalization, which was introduced in the last section, is necessary. Our notation for derivations is such that a line of a derivation should logically follow from the set of premises corresponding to the set of numbers on the left of the line. Now given the premise 'Hx', suppose we could derive '$(x)Hx$'. It is a simple matter to find an arithmetical interpretation for which 'Hx' is true and '$(x)Hx$' is false. Interpret the free variable 'x' as '1' and 'H' as the arithmetical predicate 'is less than 2'. Obviously 1 is less than 2, whence the premise is true in this interpretation, but certainly it is false that every positive integer is less than 2, whence the conclusion is false in this interpretation, and the argument is invalid. (No confusion should result from the fact that in this example the 'x' in the premise is free, whereas both occurrences of 'x' in the conclusion are bound.)

To find an interpretation which shows that an argument is logically invalid is the same thing as finding a *counterexample* to the argument. The ability to construct appropriate counterexamples for invalid arguments is nearly as important as the ability to construct logical derivations for valid arguments.

In Chapter 2 we saw that there was a mechanical method (by use of truth tables) for testing the truth-functional validity or invalidity of an argument. Such a mechanical method is often called a *decision procedure.* In one sense the existence of a decision procedure for truth-functional arguments trivializes the subject. Fortunately or unfortunately, no such trivialization of the logic of quantification is possible. It was rigorously proved in 1936 by the contemporary American logician Alonzo Church that there is no decision procedure, that is, no mechanical test, for the validity of arbitrary formulas in first-order predicate logic.* Since all of mathematics may be formalized within first-order predicate logic,† the existence of such a decision procedure would have startling consequences: a machine could be built to answer any mathematical problem or to decide on the validity or invalidity of any mathematical argument. But Church's theorem ruins at a stroke all such daydreams of students of logic and mathematics. Not

* First-order predicate logic is the logic of the sentential connectives and quantifiers for individual variables, that is, the logic of the formulas defined in Chapter 3. "First-order" refers to the fact that no quantification of predicates is permitted.

† The standard developments of axiomatic set theory have as one of their aims to establish this fact in substantive detail.

only is there no known decision procedure: his theorem establishes that there never will be any.*

On the other hand, it can be shown that the rules of inference given in this chapter together with those in Chapter 2 are *complete*, meaning that if **Q** logically follows from **P** then **Q** is derivable from **P** by means of our rules of inference. The first completeness proof for a set of rules of inference for first-order predicate logic was given by Kurt Gödel in 1930.

Although our basic logic is complete, most mathematical theories of any complexity are not complete, in the sense that it is not possible to give a list of axioms or postulates of the theory from which all other true assertions of the theory may be derived. In particular, it was shown in 1931 by Gödel that the elementary theory of positive integers, and a fortiori any theory including elementary number theory, is incomplete in the sense just stated. Gödel's theorem on the incompleteness of elementary number theory is probably the most important theorem in the literature of modern logic.

The last three paragraphs have attempted to indicate in a rough way the kind of important notions and results which cluster around the concept of logical validity. But these paragraphs are superficial and constitute a digression from our main enterprise.†

Before returning to our rules of inference, one further useful application of the method of interpretation should be mentioned, namely, its use to show that one formula is logically independent of a set of other formulas. The application is easy: by use of an interpretation it is shown that the given formula is not a logical consequence of the given set of formulas. The most common application of this method is to the problem of showing that a set of axioms for some theory are mutually independent. The technique is to give for each axiom an interpretation such that the given axiom is false in this interpretation and the remaining axioms are true in the interpretation. If the given axiom could then be derived from the remaining, a violation of our basic definition of logical consequence would result.

To illustrate this method of proving independence of axioms, we may again consider the two axioms for weak preference:

I. $(x)(y)(z)(xQy \; \& \; yQz \rightarrow xQz)$
II. $(x)(y)(xQy \vee yQx)$.

* Decision procedures do exist for various special domains of mathematics. Probably the most important positive result is Alfred Tarski's decision procedure for the elementary algebra of real numbers, first published in 1948. Given any sentence about real numbers built up from individual variables or individual constants denoting particular real numbers, the relation '<', the operation symbols of addition and multiplication, the equality sign, the sentential connectives and quantifiers, Tarski provides a mechanical test for deciding on the truth or falsity of the sentence.

† Details concerning all of these theorems are to be found in S. C. Kleene's comprehensive treatise *Introduction to Metamathematics*, New York, 1952.

Axiom I is proved independent by the following arithmetical interpretation:

$$xQy \quad \text{if and only if} \quad x \leq y + 1.$$

For any two positive integers x and y, it is true that

$$x \leq y + 1 \quad \text{or} \quad y \leq x + 1,$$

and thus Axiom II is true in this interpretation; but Axiom I does not hold for all integers x, y, and z, for although

$$3 \leq 2 + 1,$$

$$2 \leq 1 + 1,$$

it is false that

$$3 \leq 1 + 1.$$

That is, if in this interpretation of Axiom I we apply universal specification, putting '3' for 'x', '2' for 'y', and '1' for 'z', we obtain a false sentence. The arithmetical interpretation which establishes the independence of Axiom II has already been mentioned: interpret the predicate 'Q' as 'is greater than'.

Because the method of interpretation provides the only general method of proving arguments invalid, premises consistent, or postulates independent, a list of explicit rules for giving interpretations will be useful. No ideas not already mentioned are put forth in these rules; their aim is to call attention to the most common mistakes in applying the method of interpretation.

The first rule merely suggests a standard format to use in applications.

RULE I. *In applying the method of interpretation always clearly:*

(i) *state what set of objects is the domain of interpretation;*

(ii) *state in the form of equivalences or identities the interpretation of predicates, proper names, operation symbols, and free variables;*

(iii) *write down the interpretation of each original sentence and state if the interpretation of the sentence is true or false.*

Thus consider the argument:

Some men are liars.
Adams is a man.
Therefore, Adams is a liar.

We may symbolize this:

$$(\exists x)(Mx \,\&\, Lx)$$
$$Ma$$
$$\overline{\qquad\qquad}$$
$$La$$

We show this argument is invalid by giving the following interpretation.

(i) Domain of interpretation = set of positive integers.
(ii) Interpretation of predicates and proper name:

$$Mx \leftrightarrow x \text{ is a positive integer}$$
$$Lx \leftrightarrow x \text{ is an even integer}$$
$$a = 1.$$

(iii) Interpretation of original sentences:

T (1) $(\exists x)(x$ is a positive integer & x is an even integer)
T (2) 1 is a positive integer
F (3) Therefore, 1 is an even integer.

From the most familiar facts of arithmetic we see at once that (1) and (2) are true while (3) is false, which establishes that the argument is invalid.

RULE II. *Bound variables of the original sentence must remain untouched in the interpretation of the sentence. In particular, all quantifiers are unchanged.*

To illustrate violation of this rule consider the argument:

There are liars.
Therefore, there are thieves.

We symbolize this:

$$\frac{(\exists x)(Lx)}{(\exists x)(Tx)}.$$

We use the following interpretation to establish the obvious invalidity of this argument.

(i) Domain of interpretation = $\{1, 3\}$.
(ii) Interpretation of predicates:

$$Lx \leftrightarrow x = 1$$
$$Tx \leftrightarrow x = 2$$

(iii) Interpretation of original sentences:

$$\frac{\text{T} \quad (1) \ (\exists x)(x = 1)}{\text{F} \quad (2) \ (\exists x)(x = 2)}.$$

Clearly, for the domain $\{1, 3\}$ sentence (1) is true and (2) is false. Perhaps because the domain is a finite set many students initially try to offer something like:

$$(1') \ 1 = 1$$

as an interpretation of (1), for instance. But (1′) flatly violates Rule II for (1′) does not contain an existential quantifier using the variable 'x' as the original sentence does. It would also be a mistake to offer as an interpretation:

$$\frac{(1'')\ x = 1}{(2'')\ x = 2}.$$

Here the quantifiers are incorrectly omitted. Furthermore, it has already been noted that an interpretation should contain no free variables since it must be a sentence which is either true or false. This last point may be stated as the next rule.

Rule III. *An interpretation of a sentence or formula must contain no free variables.*

The second and third rules have as a consequence that it is improper to universally or existentially specify in giving an interpretation. Specification should never be confused with interpretation. However, in establishing the falsity of an interpreted sentence having a universal quantifier, application of universal specification may be useful, and in establishing the truth of an interpreted sentence having an existential quantifier the most direct method may be to exhibit an object (in the domain of interpretation) which satisfies the existentially quantified formula. Thus to show that sentence (1) above (i.e., '$(\exists x)(x = 1)$') is true, it is sufficient to find one object in the domain satisfying the formula '$x = 1$'. We may indicate this selection by a formula like (1′). Correspondingly, suppose that we have as the interpretation of the conclusion of an invalid argument the sentence:

$$(3) \qquad (x)(x > 1 \rightarrow x > 3)$$

where the domain of interpretation is the set of positive integers. In order for (3) to be true it must hold for any specified instance of a positive integer, whence to show it is false we need exhibit only one positive integer for which it does not hold. In this case we take $x = 2$, that is, we specify '2' for 'x' and obtain the false sentence:

$$(4) \qquad 2 > 1 \rightarrow 2 > 3.$$

It needs to be emphasized again in the case of both (1′) and (4) that they are not interpretations, but are merely auxiliary sentences for establishing the truth or falsity, as the case may be, of the interpretations (1) and (3) respectively.

The fourth and fifth rules are concerned with the kind of predicates which may be used as interpretations of a given predicate.

RULE IV. *An interpretation of a predicate must use the same number of distinct variables as the original predicate.*

Suppose, for instance, that we wanted to show the following premises are consistent:

$$(\exists x)(y)(xQy)$$
$$(x)(y)(xQy \rightarrow yQx)$$
$$(x)(y)(z)(xQy \,\&\, yQz \rightarrow xQz).$$

In order to show that these three premises are consistent, we need to find an interpretation for which they are all true. (Although these premises use the predicate 'Q' of weak preference we analyze their consistency independent of the two axioms for preference previously given.) An interpretation with a single element in its domain is appropriate here.

(i) Domain of interpretation = $\{1\}$.
(ii) Interpretation of predicate:

$$xQy \leftrightarrow x = y.$$

(iii) Interpretation of original sentences:

T (1) $(\exists x)(y)(x = y)$
T (2) $(x)(y)(x = y \rightarrow y = x)$
T (3) $(x)(y)(z)(x = y \,\&\, y = z \rightarrow x = z)$.

For the domain $\{1\}$ the only universal specification instances of (2) and (3) are, respectively:

(2′) $1 = 1 \rightarrow 1 = 1$

(3′) $1 = 1 \,\&\, 1 = 1 \rightarrow 1 = 1$

which are obviously true. In the case of (1) we must take $x = 1$, and obtain:

(4′) $(y)(1 = y)$,

which is true for the only possible specification of 'y'. (We note once again that (2′), (3′) and (4′) are *not* themselves interpretations of the given premises.)

It would have been a violation of Rule IV to take as an interpretation of 'Q':

(5) $xQy \leftrightarrow x = x$,

for the formula '$x = x$' only uses one variable, whereas 'xQy' uses two.

Naturally the interpretation of the premises obtained from (5) would be such that the resulting three interpretations would all be true, but the form of the interpretations would be incorrect.*

RULE V. *An interpretation of a predicate is the same in all occurrences regardless of what variables are used with the predicate.*

Thus, given the premise:

$$(\exists x)(Hx \mathbin{\&} (y)(Hy \rightarrow Lxy))$$

it would be a mistake to interpret:

$$Hx \leftrightarrow x > 1$$
$$Hy \leftrightarrow y > 3.$$

By violating Rule V it is easy to give a spurious demonstration that a valid argument is invalid. For instance, consider the valid argument:

$$\frac{(\exists x)Fx}{(\exists y)Fy}.$$

Let the set of positive integers be the domain and interpret (incorrectly): †

$$Fx \leftrightarrow x = 1$$
$$Fy \leftrightarrow y \neq y.$$

Since the sentence '$(\exists x)(x = 1)$' is true and the sentence '$(\exists y)(y \neq y)$' is false we would wrongly conclude on the basis of this mistaken interpretation that the argument is invalid.

Furthermore, in giving an interpretation of a predicate, we do not have to write down equivalences for all the different variables which occur with this predicate. One instance is sufficient to indicate the interpretation of the predicate; appropriate substitution of variables is then made in interpreting the given formulas. Thus in showing that the three premises following Rule IV are consistent we gave as the single interpretation of 'Q':

$$xQy \leftrightarrow x = y,$$

* Actually Rule IV may be weakened to: An interpretation of a predicate must use *no more* distinct variables than does the original predicate (see related discussion for definitions, p. 157). On the other hand, when the interpretation of a predicate uses bound variables, as for instance: $xQy \leftrightarrow (\exists z)(x \geq z \mathbin{\&} z \geq y)$, then Rule IV must be modified to say that an interpretation of a predicate must use no more distinct *free* variables than does the original predicate. Furthermore, a restriction on the bound variables of the interpretation must be adhered to—namely, a quantifier of the interpretation must not capture variables in the original formula. However, none of the applications considered in the sequel require bound variables in the interpretation of a predicate.

† The predicate '$\neq$' is read 'not identical with'. We could symbolize '$y \neq y$' by '$-(y = y)$'.

and then in interpreting premises (2) and (3) as (2′) and (3′) respectively we made the obvious changes of variables called for; that is, we replaced

'yQx' by '$y = x$'
'yQz' by '$y = z$'
'xQz' by '$x = z$'.

RULE VI. *The interpretation of proper names and free variables must designate objects in the domain of interpretation.*

Consider, for example, three possible postulates of preference where 'a' is the name used for some object, say, an original first edition of Kant's *Critique of Pure Reason:*

(1) $(\exists x)(xPa)$

(2) $(x)(y)(xPy \rightarrow -(yPx))$

(3) $(x)(y)(z)(xPy \,\&\, yPz \rightarrow xPz)$.

We want to show by the method of interpretation that Postulate (1) is independent of the other two postulates. As has been previously explained, to establish such independence amounts to the same thing as showing that (1) is not a logical consequence of (2) and (3). Consequently we want to find an interpretation for which (1) is false, and (2) and (3) are true. Suppose we take as the domain of interpretation the set of positive integers. Then it is a violation of Rule VI to interpret 'a' as some object, say, -2, which is not in this domain. A correct interpretation is the following:

(i) Domain of interpretation = set of positive integers
(ii) Interpretation of predicate and proper name:

$$xPy \leftrightarrow x < y$$

$$a = 1$$

(iii) Interpretation of postulates:

F (1′) $(\exists x)(x < 1)$
T (2′) $(x)(y)(x < y \rightarrow -(y < x))$
T (3′) $(x)(y)(z)(x < y \,\&\, y < z \rightarrow x < z)$.

Since the interpretation of 'a' as '-2' also yields a false interpretation of (1) for this domain:

(1″) $(\exists x)(x < -2)$,

it might be thought that the restriction expressed by Rule VI is really not necessary. However, violation of it can lead to a fallacious demonstration that a valid argument is invalid, or that a dependent postulate is independent. For example, consider the two postulates:

$$Fa$$

$$(x)Fx.$$

It is clear that the first can immediately be derived from the second by universal specification. However, if we take as our domain of interpretation the set of positive integers, interpret 'F' by:

$$Fx \leftrightarrow x \geq 1$$

and interpret 'a' as '-3', then we obtain the fallacious result that the first postulate is independent of the second, since '$-3 \geq 1$' is false whereas '$(x)(x \geq 1)$' is true for the domain selected.

Closely connected with Rule VI is a corresponding rule for operation symbols, which will mainly be of use in subsequent chapters.

> RULE VII. *The interpretation of operation symbols must be such that the interpretation of any term using the operation symbol refers to an element in the domain of interpretation.*

Thus, given the sentence:

$$(x)(y)(x + y = y + x)$$

it would be a mistake to fix upon the set $\{1\}$ as the domain of interpretation and to interpret the operation symbol '$+$' as arithmetical addition, for then the term '$1 + 1$' designates an object not in the domain of interpretation. A technical way of describing the requirement laid down by Rule VII is that the interpretation of an operation symbol must have the *closure property* with respect to the domain of interpretation.

For a second example, suppose we are given the following two postulates on a binary operation symbol '$\circ$':

(1) $$(x)(y)(x \circ y = y \circ x)$$

(2) $$(\exists x)(y)(y \circ x = y),$$

and we want to prove that the first postulate is independent of the second. A common error in connection with an example of this sort is to take as a domain of interpretation the set $\{1, 2\}$ and interpret the operation symbol as division:

(3) $$x \circ y = x/y.$$

Then (2) seems true since we may take $x = 1$ and for any number y

$$y/1 = y.$$

And (1) seems false since universally specifying '1' for 'x' and '2' for 'y' we obtain the obvious falsehood:

$$\frac{1}{2} = \frac{2}{1}.$$

However, the specification made in (1) shows that the domain $\{1, 2\}$ and the interpretation (3) together violate Rule VII, for the number 1/2 is not in the set $\{1, 2\}$. To indicate why the restriction imposed by Rule VII is not capricious, but essential, we need to anticipate some of the ideas expounded at the beginning of the next chapter. There we take as a truth of logic depending on no premises the identity $\mathfrak{t} = \mathfrak{t}$ where $\mathfrak{t}$ is any term. Thus in the present context if we admit '1/2' as a term, then the sentence:

$$\frac{1}{2} = \frac{1}{2} \tag{4}$$

is true, but from (4) we may immediately derive the sentence:

$$(\exists x)(\tfrac{1}{2} = x)$$

which is obviously false when $\{1, 2\}$ is the domain of interpretation.

In connection with a number of exercises in subsequent sections it may be useful to exhibit two different interpretations, one with a finite domain and one with an infinite domain of interpretation, which will prove (1) independent of (2). If we fix upon division as our interpretation of the operation symbol '$\circ$', we need an infinite domain: the set of positive rational numbers. (A positive rational number is a number which is the ratio of two positive integers.) For this domain and interpretation of '$\circ$' Rule VII is satisfied; moreover, the resulting interpretation of (1) is false and that of (2) is true. As a second interpretation, let the domain of interpretation be the set $\{1, 2\}$, and interpret '$\circ$' by the means of the identity:

$$x \circ y = x. \tag{5}$$

When the domain is finite, an interpretation like (5) is sometimes given by means of a table or *matrix* rather than by an identity. Thus we could replace (5) by the matrix:

$\circ$	1	2
1	1	1
2	2	2

This table or matrix is used in the following manner. To find what element $1 \circ 2$ is, say, we look at the entry occurring in the first row and second column and find:

$$1 \circ 2 = 1.$$

We note, on the other hand, that

$$2 \circ 1 = 2.$$

Thus under this interpretation Postulate (1) is false, since the specification of '1' for 'x' and '2' for 'y' yields the absurdity that $1 = 2$. Yet Postulate (2) is, as desired, true in this interpretation as may be seen by taking $x = 1$, since the sentence:

$$(y)(y \circ 1 = y)$$

is true here.

In concluding this section we may summarize again the three most important applications of the method of interpretation:

Prove arguments invalid
Prove premises consistent
Prove axioms independent.

EXERCISES

1. Give arithmetical interpretations to prove that the following formulas are not universally valid.

(a) $(x)(Hx \,\&\, Ax \rightarrow Mx)$
(b) $(\exists x)Hx$
(c) $(\exists x)(Hx \vee -Ax)$
(d) $(x)(Hx \,\&\, -Mx)$

2. Give interpretations (arithmetical or other) to prove that the following arguments are invalid. (Notice that all conclusions are true but not valid.)

(a) All men are animals. All men are mortal. Therefore, all animals are mortal. (Ax, Hx, Mx)
(b) New York is north of Washington. Boston is north of New York. Therefore, Boston is north of Washington. (Nxy, n, w, b)
(c) Some sailors are ignorant. Some Americans are ignorant. Therefore, some sailors are American. (Ax, Ix, Sx)
(d) All men are animals. Some animals are short-lived. Therefore, some men are short-lived. (Sx)

3. Give interpretations to prove that the following sets of premises are consistent. The first three examples deal with the theory of preference, but are independent of previous assumptions about preference and indifference.*

(a) $(\exists x)(y)(xQy)$
$(x)(y)(\exists z)(xQz \,\&\, zQy)$
(b) $(x)(y)(xPy \rightarrow -yPx)$
$(x)(y)(xIy)$
(c) $(x)(\exists y)(yPx)$
$(x)(y)(xPy \rightarrow -yPx)$
$(x)(y)(z)(xPy \,\&\, yPz \rightarrow xPz)$
(d) All unicorns are animals. No unicorns are animals. (Ux)

* The order properties dealt with in these exercises on the theory of preference are of importance in many domains of science. There is an extensive discussion of ordering relations in Chapter 10.

4. Prove by the method of interpretation that the postulates are mutually independent in each of the following. Hint: in using arithmetical interpretations it is often convenient to restrict the domain of individuals to some finite set of integers like the first two or first ten. The examples are again drawn from the theory of preference.

(a) $(x)(y)(xPy \rightarrow -yPx)$
$(x)(y)(z)(xPy \,\&\, yPz \rightarrow xPz)$
(b) $(x)(y)(z)(xIy \,\&\, yIz \rightarrow xIz)$
$(x)(y)(z)(xPy \,\&\, yPz \rightarrow xPz)$
(c) $(x)(\exists y)(xPy)$
$(x)(y)(z)(xPy \,\&\, yPz \rightarrow xPz)$
(d) $(x)(y)(xIy \rightarrow yIx)$
$(x)(y)(xPy \rightarrow -(xIy \vee yPx))$

5. Give interpretations to prove that the following sets of postulates dealing with a binary operation symbol 'o' are consistent. Be careful not to violate Rule VII.

(a) $(x)(y)(x \circ y = y \circ x)$
$(\exists z)(x)(y)(x \circ y = z)$
(b) $(x)(\exists y)(x \circ y = 1)$
$(x)(\exists z)(x \circ z = 0)$
$(x)(\exists w)(x \circ 1 = w)$
(c) $(\exists x)(\exists y)(x \circ y \neq y \circ x)$
$(x)(\exists y)(x \circ y = 0)$
$(x)(\exists z)(z \circ x = 0)$

§ **4.3 Restricted Inferences with Existential Quantifiers.*** In considering how to formulate the rules of inference governing existential quantifiers, probably the most natural idea is to proceed as we did for universal quantifiers and introduce a rule of *existential specification* permitting us simply to drop an existential quantifier. Thus from:

(1) $(\exists x)Hx$

we would infer:

(2) $Hx.$

However, a little reflection shows that this rule, if used without restriction, will produce invalid inferences. For example, if we interpret 'H' as 'greater than 1' and substitute the name '1' for the free variable 'x' in (2), in this arithmetical interpretation from the truth that there is a number greater than 1, we infer the false sentence that 1 is greater than 1. Moreover, if we universally generalize on (2), we are able to infer '$(x)Hx$' from '$(\exists x)Hx$', which is clearly invalid.

* The restriction imposed in this section is that existentially quantified formulas must contain just one individual variable.

Several methods for avoiding such invalid inferences have been developed.* The technical device used here, although apparently new in the literature of first-order predicate logic, is close to an approach that is used continuously in intuitive proofs in mathematics. The device is to replace an existential quantifier by a "temporary" constant; that is, some symbol is introduced as a name for the purposes immediately at hand. The justification is that the existential quantifier guarantees that some individual can be taken to be represented by the new individual constant. For example, suppose we want to derive 'There is a mortal', i.e., '$(\exists x)Mx$' from the premises '$(x)(Hx \rightarrow Mx)$' and '$(\exists x)Hx$'. We might proceed as follows:

{1}	(1) $(x)(Hx \rightarrow Mx)$	P
{2}	(2) $(\exists x)Hx$	P
{2}	(3) $H_{\text{John Doe}}$	Ambiguous name from (2) by existential specification
{1}	(4) $H_{\text{John Doe}} \rightarrow M_{\text{John Doe}}$	1 *US*
{1, 2}	(5) $M_{\text{John Doe}}$	3, 4 T
{1, 2}	(6) $(\exists x)Mx$	Eliminate ambiguous name from (5) by existential generalization

The use of the ambiguous name 'John Doe' in the above derivation permits us to eliminate the existential quantifier and apply sentential methods in the standard way. The ambiguous name 'John Doe' is appropriate in arguments concerning human beings. In mathematical contexts, an ambiguous name is often introduced by adding a star to a variable: thus, 'x^*' or 'z^*'. Naturally we have no definite individual in mind when we use 'John Doe', and it may properly be claimed that 'John Doe' is not a genuine proper name; that is why we use the terminology 'ambiguous name'. The existential premise '$(\exists x)Hx$' guarantees there is some individual to whom we may attach the name 'John Doe'.

To fix on a uniform practice regarding ambiguous names, we use lower case Greek letters for this purpose:

$$\alpha, \beta, \gamma, \delta, \alpha_1, \beta_1, \gamma_1, \delta_1, \ldots$$

As the rough derivation above indicates, we have two rules governing

* See, for example, W. V. Quine, *Methods of Logic*, New York, 1950; and I. Copi, *Symbolic Logic*, New York, 1954. In this connection it should be noted that the notion of *flagging* in Quine's system is quite different from that developed here in the first section of this chapter

existential quantifiers and ambiguous names:

EXISTENTIAL SPECIFICATION (*ES*). *The assertion that there is something satisfying a given condition implies the assertion that this given condition is satisfied by some namable individual.*

EXISTENTIAL GENERALIZATION (*EG*). *The assertion that a condition is satisfied by a named individual implies the assertion that this condition is satisfied by some individual.*

Using a Greek letter for an ambiguous name and introducing the abbreviations *ES* and *EG*, we may rewrite the above derivation:

EXAMPLE 4.

{1}	(1) $(x)(Hx \rightarrow Mx)$	P
{2}	(2) $(\exists x)Hx$	P
{2}	(3) $H\alpha$	2 *ES*
{1}	(4) $H\alpha \rightarrow M\alpha$	1 *US*
{1, 2}	(5) $M\alpha$	3, 4 T
{1, 2}	(6) $(\exists x)Mx$	5 *EG*

Before formally stating the rules of existential specification and generalization, it should be pointed out why a certain restriction has to be imposed. Given the true premises '$(\exists x)Hx$' and '$(\exists x)-Hx$', we may derive the false conclusion '$(\exists x)(Hx \,\&\, -Hx)$' if the use of ambiguous names is not restricted:

{1}	(1) $(\exists x)Hx$	P
{2}	(2) $(\exists x)-Hx$	P
{1}	(3) $H\alpha$	1 *ES*
{2}	(4) $-H\alpha$	2 *ES* (fallaciously)
{1, 2}	(5) $H\alpha \,\&\, -H\alpha$	3, 4 T
{1, 2}	(6) $(\exists x)(Hx \,\&\, -Hx)$	5 *EG*

The difficulty arises from the fact that ambiguous names, like all names, cannot be used indiscriminately. The person who calls a loved one by the name of a *former* loved one is quickly made aware of this. Having in line (3) chosen 'α' as the name of some human being postulated in line (1), we cannot in line (4) also use 'α' as the name of something which is not human as postulated in line (2). Such a happy-go-lucky naming process is bound to lead to error, just as we could infer a false conclusion from true facts about two individuals named 'Fred Smith' if we did not somehow devise a notational device for distinguishing which Fred Smith was being referred to in any given statement. The restriction which we impose to stop such invalid arguments is to require that when we introduce by existential specification an ambiguous name in a derivation, that name has not previously been used in the derivation. This is the simplest, though not the weakest, restriction we could state. It does require immediate application of existential specification before universal specification. Thus in Example 4 above it is necessary for line (3) to precede line (4). If line (4) had been first, 'α' would have been already introduced and it could not then be brought in by existential specification.

The formal statement of the two rules is:

RULE OF EXISTENTIAL SPECIFICATION: *ES*. *If a formula* **S** *results from a formula* **R** *by substituting for every free occurrence of a variable* **v** *in* **R** *an ambiguous name which has not previously been used in the derivation, then* **S** *is derivable from* $(\exists \mathsf{v})\mathsf{R}$.

RULE OF EXISTENTIAL GENERALIZATION: *EG*. *If a formula* **S** *results from a formula* **R** *by substituting a variable* **v** *for every occurrence in* **R** *of some ambiguous (or proper) name, then* $(\exists \mathsf{v})\mathsf{S}$ *is derivable from* **R**.

Another example of applying these two rules is given by the following syllogism.

EXAMPLE 5. *All mammals are animals. Some mammals are two-legged. Therefore, some animals are two-legged.*

{1}	(1) $(x)(Mx \rightarrow Ax)$	P
{2}	(2) $(\exists x)(Mx \,\&\, Tx)$	P
{2}	(3) $M\alpha \,\&\, T\alpha$	2 *ES*
{1}	(4) $M\alpha \rightarrow A\alpha$	1 *US*
{1, 2}	(5) $A\alpha \,\&\, T\alpha$	3, 4 T
{1, 2}	(6) $(\exists x)(Ax \,\&\, Tx)$	5 *EG*

Following up the remarks just made, notice that it is necessary to have line (3) before line (4). If line (4) had been written down first, then according to *ES* some other ambiguous name would be needed for line (3). The appropriate strategy is: whenever possible, drop existential quantifiers *before* universal quantifiers.

The next example illustrates how quantifiers inside a formula are handled.

EXAMPLE 6. *All of Aristotle's followers like all of Aquinas' followers. None of Aristotle's followers like any philosophical idealist. Moreover, Aristotle does have followers. Therefore, none of Aquinas' followers are philosophical idealists.*

{1}	(1) $(x)(Ax \rightarrow (y)(Qy \rightarrow Lxy))$	P
{2}	(2) $(x)(Ax \rightarrow (z)(Iz \rightarrow -Lxz))$	P
{3}	(3) $(\exists x)(Ax)$	P
{3}	(4) $A\alpha$	3 *ES*
{1}	(5) $A\alpha \rightarrow (y)(Qy \rightarrow L\alpha y)$	1 *US*

{2}	(6) $A\alpha \to (z)(Iz \to -L\alpha z)$	2 *US*
{1, 2}	(7) $(y)(Qy \to L\alpha y)$	4, 5 T
{2, 3}	(8) $(z)(Iz \to -L\alpha z)$	4, 6 T
{1, 3}	(9) $Qy \to L\alpha y$	7 *US*
{2, 3}	(10) $Iy \to -L\alpha y$	8 *US*
{1, 2, 3}	(11) $Qy \to -Iy$	9, 10 T
{1, 2, 3}	(12) $(y)(Qy \to -Iy)$	11 *UG*

Line (2) is deliberately written using 'z' instead of 'y' in order to provide another illustration of how *US* is used to change variables. The most important remark to be made in connection with this last example is: a quantifier may only be dropped when it stands at the beginning of a formula and its scope is the whole formula. It is a major error to apply either *US* or *ES* when this condition is not satisfied. Thus in Example 6 it would have been a mistake to apply *US* to line (5) in order to drop the quantifier '(y)', for this quantifier does not stand at the beginning of the formula. A similar mistake would have been made by applying *US* to line (6) to eliminate the quantifier '(z)'. In the context of the present example such an application of *US* would not have yielded a false conclusion, but in other contexts it would, which is why it must be avoided.*

It needs to be remarked that although the rule of existential generalization is primarily used to eliminate ambiguous names in favor of existential quantifiers, it is occasionally necessary to be able to apply an existential quantifier to a free variable. The intuitive justification of this latter application is that an assertion about an arbitrary thing implies an assertion about *some* thing. Formally, such an application of *EG* is justified if we consider an ambiguous name which does not occur at all in the formula in question. For example, '$(\exists x)Fx$' is derivable from 'Fx' by *EG*. We substitute 'x' for *every* occurrence of 'α', say, in 'Fx'. Since there are no occurrences of 'α' in 'Fx', the substitution just results in what we started with. Some applications of this gambit are given in the section on theorems of logic in Chapter 5.

If we interpret ambiguous names in the same way that we interpret proper names and free variables, then not every line of a derivation is a logical consequence of the conjunction of the premises on which it depends. For example, from the premise '$(\exists x)Fx$' we derive by existential specification '$F\alpha$', but obviously if 'α' is interpreted just like a proper name then '$F\alpha$' is not a logical consequence of '$(\exists x)Fx$'. Yet this interpretation is the most natural one, and the simplest procedure is to weaken the requirement that every line of a derivation be a logical consequence of the premises on which it depends. What we may prove is that if a formula in a derivation contains no ambiguous names and neither do its premises,

* See Exercises 6 and 7 of § 4.6.

then it is a logical consequence of its premises. And this state of affairs is in fact intuitively satisfactory, for in a valid argument of use in any discipline we begin with premises and end with a conclusion which contains no ambiguous names. The point of ambiguous names is to provide a smooth-running method for inferring conclusions, and the real test of their adequacy is the soundness of the conclusion derived, not the status of the intermediate lines.

Moreover, the logical status of the relation between premises and conclusion, at least one of which contains an ambiguous name, may be characterized in a rather simple way, in terms of what we shall call the *unambiguous closure* of a formula. The idea is to eliminate ambiguous names by existential quantifiers but only in a certain order in relation to universal quantifiers as determined by the subscripts on the ambiguous names.* For example, the unambiguous closure, or for brevity the U-closure, of '$H\alpha x$' is '$(\exists y)(x)Hyx$'. The U-closure of '$H\alpha_x x$' is '$(\exists y)Hyx$'. Roughly speaking, the U-closure of a formula **S** is the formula obtained from **S** in accordance with the following prescription: if **S** has no ambiguous names it is its own U-closure; if **S** has at least one ambiguous name, first universally quantify all free variables of **S** which do not appear as subscripts, then replace the ambiguous name with the most subscripts by an appropriate existentially quantified variable (if there are two or more ambiguous names with the maximum number of subscripts, replace the one which has the earliest occurrence in **S**); repeat this procedure until all ambiguous names have been eliminated.†

The following result may then be established concerning the logical relation between premises and derived conclusion: the U-closure of the implication whose antecedent is the conjunction of the premises and whose consequent is the conclusion is universally valid. Put another way, every line of a derivation is a consequence in the following sense of the premises corresponding to the set of numbers on the left: the U-closure of the implication consisting of the conjunction of the premises as antecedent and given line as consequent is universally valid, and thus true in every interpretation in a non-empty domain of individuals.

* The intuitive interpretation of subscripts, which are used to indicate dependency relationships, is explained at the beginning of ¶ 4.5 on p. 89.

† Appropriate conventions on alphabetical ordering of variables are easily given in order to make the U-closure of a formula unique.

EXERCISES

Construct (if possible) a derivation corresponding to the following arguments. If a conclusion does not follow, give an arithmetical interpretation which will prove that it does not.

1. Some foolish people drink whiskey. Some students do not drink whiskey. Therefore, some students are not foolish. (Fx, Wx, Sx)

2. All boxers are strong. Some policemen are strong. Therefore, some policemen are boxers. (Bx, Sx, Px)

3. Some scientific subjects are not interesting, but all scientific subjects are edifying. Therefore, some edifying things are not interesting. (Sx, Ix, Ex)

4. No intelligent person who drinks to excess also eats to excess. Some prudent persons eat to excess. Therefore, some prudent persons are not intelligent. (Ix, Dx, Ex, Px)

5. No red-haired women use hair tonic, but some red-haired women use perfume. Therefore, some women use perfume and not hair tonic. (Rx, Wx, Hx, Px)

6. Some soldiers are heroes. Some soldiers are not brave. Therefore, some heroes are not brave. (Sx, Hx, Bx)

7. No nominee lives in California. Some people who live in California are eligible candidates. Therefore, some eligible candidates are not nominees. (Nx, Lx, Cx)

8. Every member of the policy committee is a Republican. Some members of the tax committee are not Republicans. Therefore, some members of the tax committee are not members of the policy committee. (Px, Rx, Tx)

9. Some members of the tax committee are on the policy committee. Some members of the policy committee are Democrats. Therefore, some members of the tax committee are Democrats. (Tx, Px, Dx)

10. Every member of the policy committee is either a Democrat or a Republican. Some members of the policy committee are wealthy. Adams is not a Democrat, but he is wealthy. Therefore, if Adams is a member of the policy committee, he is a Republican. (Px, Dx, Rx, Wx, a)

11. If anyone likes Aquinas, then he does not like Kant. Everyone either likes Kant or likes Russell. Someone does not like Russell. Therefore, someone does not like Aquinas. (Ax, Kx, Rx)

12. All Texans speak to anyone whom they know intimately. No Texan speaks to anyone who is not a Southerner. Therefore, Texans know only Southerners intimately. (Tx, Sxy, Kxy, Ux)

13. Anyone who is guilty of larceny is not guilty of obtaining by false pretenses. Each prisoner in the line-up is guilty of either larceny or obtaining by false pretenses. Some prisoners in the line-up are guilty of obtaining by false pretenses. Therefore, some prisoners in the line-up are not guilty of larceny. (Lx, Ox, Px)

14. Some persons who buy goods in good faith get a good title. No person who buys goods for a trifling sum gets a good title. Therefore, no person who buys goods for a trifling sum is a person who buys goods in good faith. (Bx, Tx, Sx)

15. In the psychology of perception and in epistemology as well, a relation between objects of being alike in color is often discussed. Thus two shades of red are alike; however, as ordinarily conceived, this relation of being alike is not transitive, that is, color a is like b, b is like c, but a is not like c. So the problem naturally arises of defining a transitive relation of being *exactly* alike between colors. Let 'L' be the predicate standing for the first relation of being alike, and let 'E' be the

predicate 'is exactly alike'. We then introduce the definition:

$$(x)(y)(xEy \leftrightarrow (z)(xLz \leftrightarrow yLz) \mathbin{\&} xLy \mathbin{\&} yLx).$$

Using this definition as a single premise, prove that 'E' is symmetric and transitive, that is, prove:

(a) $(x)(y)(xEy \rightarrow yEx)$
(b) $(x)(y)(z)(xEy \mathbin{\&} yEz \rightarrow xEz)$.*

16. With reference to the preceding exercise, give an interpretation of 'L' and 'E' to prove that the sentence '$(x)(xEx)$' is not a logical consequence of the equivalence defining 'E' in terms of 'L'.

§ 4.4 Interchange of Quantifiers. From considerations of ordinary usage it is clear that the sentence:

(1) Every mathematician admires Archimedes

is equivalent to:

(2) It is not the case that there is a mathematician who does not admire Archimedes.

Yet (1) uses a universal quantifier and (2) uses an existential quantifier. Symbolization of (1) yields:

(3) $$(x)(Mx \rightarrow Ax),$$

and symbolization of (2) yields:

(4) $$-(\exists x)(Mx \mathbin{\&} -Ax).$$

Moreover, since '$Mx \mathbin{\&} -Ax$' is tautologically equivalent to '$-(Mx \rightarrow Ax)$', (4) is equivalent to:

(5) $$-(\exists x)-(Mx \rightarrow Ax).$$

The close relation between (3) and (5) suggests the rule:

Q1. *If* **v** *is any variable and if formula* **S** *results from* **R** *by replacing at least one occurrence of the universal quantifier* (**v**) *by* $-(\exists$**v**$)-$, *then* **S** *is derivable from* **R**, *and conversely.*

In the above example, **v** is, of course, the variable 'x'. Using the methods of § 5.3 below, we may derive Rule Q1 from the previous rules given, but for our present purposes we may accept it as sufficiently well-grounded intuitively to need no further justification. To assert something of every x is just the same thing as to deny there is an x not satisfying this assertion.

* This exercise was suggested by Leo Simons. For an application of these notions in epistemology, see Hilary Putnam, "Reds, Greens and Logical Analysis," *Philosophical Review*, Vol. 65 (1956) pp. 206–217.

A similar rule is suggested by the following example. Clearly the sentence:

(6) There is a two-headed calf

is equivalent to:

(7) It is not the case that no calves are two-headed.

We may symbolize (6) by:

(8) $(\exists x)(Cx \mathbin{\&} Tx)$,

and (7) by:

(9) $-(x)(Cx \rightarrow -Tx)$.

Since '$Cx \rightarrow -Tx$' is tautologically equivalent to '$-(Cx \mathbin{\&} Tx)$', (9) is equivalent to:

(10) $-(x)-(Cx \mathbin{\&} Tx)$.

The relation between (8) and (10) suggests the rule:

Q2. *If* v *is any variable and if formula* S *results from* R *by replacing at least one occurrence of the existential quantifier* (∃v) *by* −(v)−, *then* S *is derivable from* R, *and conversely.*

The meaning of Q2 is that to assert there is an x having some property is equivalent to asserting that not every x is without this property.

In the discussion of Q1 and Q2 we have made use of certain tautological equivalences in an intuitively unobjectionable manner. A rule of inference justifying this use is the following.

RULE FOR TAUTOLOGICAL EQUIVALENCES: T.E. *If formula* P *occurs as part of formula* R, *if formula* Q *is tautologically equivalent to* P, *and if formula* S *results from* R *by replacing at least one occurrence of* P *in* R *by* Q, *then* S *is derivable from* R, *and conversely.*

Thus Q1 and T.E. formally justify the inference of (4) from (3); and Q2 and T.E. together countenance the inference of (9) from (8). The three rules introduced are used in the following inference.

EXAMPLE 7. *If there is a federal court which will sustain the decision then every member of the bar is wrong. However, some members of the bar are not wrong. Therefore, no federal court will sustain the decision.*

{1}	(1) $(\exists x)(Fx \mathbin{\&} Sx) \rightarrow$ $(y)(My \rightarrow Wy)$	P
{2}	(2) $(\exists y)(My \mathbin{\&} -Wy)$	P
{2}	(3) $-(y)-(My \mathbin{\&} -Wy)$	2 Q2

{2}	(4) $-(y)(My \rightarrow Wy)$	3 T.E.
{1, 2}	(5) $-(\exists x)(Fx \& Sx)$	1, 4 T
{1, 2}	(6) $--(x)-(Fx \& Sx)$	5 Q2
{1, 2}	(7) $(x)-(Fx \& Sx)$	6 T
{1, 2}	(8) $(x)(Fx \rightarrow -Sx)$	7 T.E.

In obtaining line (6) from (5), the existential quantifier '$(\exists x)$' was replaced by '$-(x)-$', although the existential quantifier did not stand at the beginning of the formula in (5). Such contextual interchange is permitted by Q1 and Q2. This differentiates them from *US* and *ES*, which apply only when the quantifier stands in front of the whole formula.

A number of further rules concerning quantifiers are stated in § 5.3.

EXERCISES

1. If every member of the bar is wrong, then there is a federal court which will sustain the decision. However, no federal court will sustain the decision. Therefore, some members of the bar are not wrong. (Mx, Wx, Fx, Sx)

2. If every witness is telling the truth, then Bluenose will be found guilty or a hung jury will result. But a hung jury will not result. Therefore, either Bluenose will be found guilty or some witness is not telling the truth. (Wx, Tx, B, H)

3. If Bluenose is guilty then no witness is lying unless he is fearful. There is a witness who is fearful. Therefore, Bluenose is not guilty. (B, Wx, Lx, Fx)

§ 4.5 General Inferences. By limiting the kind of inferences considered, we were able to hold the number of restrictions on the rules governing quantifiers to two: one on *UG* (flagging restriction) and one on *ES* (introduce a new ambiguous name with each application). To extend our rules to general inferences and not violate Criterion I, that is, not permit the inference of a false conclusion from true premises, we must add five further restrictions: one on *US*, a second one on *UG*, and three on *EG*. Also the statement and use of *ES* must be complicated slightly further.

These five restrictions involve but one notational innovation. We introduce the use of variables as *subscripts* on ambiguous names to keep *ES* from leading us to a contradiction. A simple example will illustrate the difficulty and our particular way of resolving it. It is a truth of arithmetic that there is no largest number, that is,

(1) $$(x)(\exists y)(x < y).$$

We apply *US* and derive:

(2) $$(\exists y)(x < y).$$

If we apply *ES* as stated in § 4.3, we next derive:

(3) $$x < \alpha,$$

and then by *EG* we obtain the false assertion that

(4) $$(\exists x)(x < x).$$

We block this fallacious inference by requiring that when an ambiguous name is introduced by *ES* it include as subscripts all the free variables occurring in the original formula. Thus in (2), 'x' is a free variable. Hence we now derive not (3) but:

(3′) $$x < \alpha_x.$$

Finally we restrict the use of *EG*.

> FIRST NEW RESTRICTION ON *EG*. *We may not apply an existential quantifier to a given formula using a variable which occurs as a subscript in the formula.*

Hence we cannot infer (4) from (3′), since 'x' occurs as a subscript in (3′).

It should be emphasized that this use of subscripts is not an arbitrary technical device, but is very close to ordinary mathematical methods. In (3) the determination of α depends on the number x. We make this dependence of α explicit in (3′) by writing 'α_x'. Since α_x specifically depends on x, the variable 'x' cannot be quantified as long as ambiguous names on which it occurs as a subscript are on the scene.

We repeat in capsule form the procedure for applying subscripts.

> *When an ambiguous name is introduced by ES it must include as subscripts all the free variables occurring in the formula to which ES is applied.**

It should be remarked that if we have a line like (3′), we may apply an existential quantifier using any variable other than 'x'. Thus it is perfectly correct to derive from (3′):

$$(\exists z)(x < z).$$

The restriction on *EG* must also be extended to *UG*, as the following line of inference shows. Let us go back to (1), which asserts that there is no largest number. As before, we derive (2) by *US:*

$$(\exists y)(x < y),$$

and we now apply the newly modified rule of existential specification to obtain (3′):

$$x < \alpha_x.$$

So far so good, but to this last result, we now apply universal generalization:

(5) $$(x)(x < \alpha_x),$$

* It should be explicitly noted that any occurrence of a variable as a subscript is a free occurrence. Thus from '$(\exists y)(\alpha_x < y)$' we derive '$\alpha_x < \beta_x$', not '$\alpha_x < \beta$'.

and to this we apply *EG* to obtain the false sentence:

$$(6) \qquad (\exists y)(x)(x < y).$$

This fallacious argument is blocked by prohibiting the inference of (5) from (3′). That is, we restrict *UG* like *EG*.

NEW RESTRICTION ON *UG*. *We may not apply a universal quantifier to a given formula using a variable which occurs as a subscript in the formula.*

This last example justifying the restriction on *UG* may be written as a fallacious derivation to make explicit how the new restriction operates in a derivation.

FALLACIOUS EXAMPLE FOR SUBSCRIPT RESTRICTION ON *UG*

{1}	(1) $(x)(\exists y)(x < y)$	P
{1}	(2) $(\exists y)(x < y)$	1 *US*
{1}	(3) $x < \alpha_x$	2 *ES*
{1}	(4) $(x)(x < \alpha_x)$	3 *UG* (fallaciously)
{1}	(5) $(\exists y)(x)(x < y)$	4 *EG*

The variable 'x' occurs as a subscript in line (3), and it is therefore incorrect to apply *UG* to derive (4).

The justification of the new restriction on universal specification is simple to illustrate. Again we begin with the true premise that there is no largest number:

$$(x)(\exists y)(x < y),$$

and we apply *US*, replacing 'x' by 'y' to obtain the false sentence:

$$(\exists y)(y < y).$$

The fallacious derivation is as follows:

{1}	(1) $(x)(\exists y)(x < y)$	P
{1}	(2) $(\exists y)(y < y)$	1 *US* (fallaciously)

The kind of restriction on *US* which is required is obvious from this example: do not substitute a variable which becomes bound. Thus in line (2) above 'y' replaces 'x' and becomes bound, which is the source of the invalid inference.

NEW RESTRICTION ON *US*. *Do not substitute a term containing a variable which becomes bound by a quantifier in the original formula.*

The restriction is stated for terms, since not only substitution of a variable might lead to trouble, but also substitution of a term which is not a variable. For example, it would also have been fallacious above to substitute the term '$x + y$' for 'x', since 'y' becomes bound in (2) whether it is part of a term or standing by itself. We would have obtained:

$$(\exists y)(x + y < y).$$

Universally generalizing on this, we get:

$$(x)(\exists y)(x + y < y),$$

and by universal specification we then derive the false sentence:

$$(\exists y)(0 + y < y).$$

Our next restriction, the second new one on existential generalization, is just like the new restriction on universal specification: Do not substitute a variable which is captured by a quantifier already present in the formula. The following example shows the necessity of this restriction on *EG*. We begin with the true sentence of arithmetic:

$$(1) \qquad (\exists x)(y)(x + y = y).$$

To see that (1) is true, take 'x' as '0'. We apply existential specification to obtain:

$$(2) \qquad (y)(\alpha + y = y).$$

To (2) we apply *EG* and get into trouble:

$$(3) \qquad (\exists y)(y)(y + y = y).$$

Finally to (3) we apply *ES* and get the false sentence:

$$(4) \qquad (y)(y + y = y).$$

Since in line (2) 'α' occurs within the scope of the quantifier '(y)', it is a mistake to replace 'α' by 'y' and add the existential quantifier '$(\exists y)$' in line (3). Notice that we obtain line (4) from (3) by *ES* trivially, since there are no free variables in '$(y)(y < y)$' to replace by an ambiguous name. Formulas like (3) satisfy the definition of formulas given in Chapter 3, but the rules of inference are so framed that the outside quantifier is essentially redundant. When an occurrence of a variable falls within the scope of two quantifiers, the inside quantifier always governs. Summarizing, we have:

SECOND NEW RESTRICTION ON *EG*. *Do not replace an ambiguous name by a variable which becomes bound by a quantifier in the original formula.*

Our final restriction is the third one on *EG*. It is concerned with existentially generalizing on a flagged variable and an ambiguous (or proper) name simultaneously. For difficulties to arise, the name must actually occur at least once in the formula being generalized on. An arithmetical example using only the predicate 'O', where 'Ox' means that x is an odd integer, will illustrate the difficulty.

{1}	(1) $(\exists x)-Ox$	P
{2}	(2) Ox	xP
{1}	(3) $-O\alpha$	1 *ES*
{1, 2}	(4) $Ox \& -O\alpha$	x 2, 3 T
{1, 2}	(5) $(\exists x)(Ox \& -Ox)$	4 *EG* (fallaciously)

Line (1) is a truth of arithmetic, and if we interpret the free variable 'x' in (2) as '1', then (2) is true, but (5) is patently false. Line (4) should not have been existentially generalized in terms of 'x' to yield (5), since 'α' and flagged 'x' occur together. This final restriction is not often violated in practice.* Its formal statement is:

THIRD NEW RESTRICTION ON *EG: Do not use a variable flagged in a formula to eliminate an actual occurrence of a name from the formula.*

There is no harm in existentially generalizing on a flagged variable in a formula when this generalization does not actually eliminate some occurrence of a name. Thus, from the premise 'Fx' we may infer '$(\exists x)Fx$' even though 'x' is flagged in the premise. If we think of eliminating 'α' in this generalization, 'α' does not actually occur in 'Fx'. Its elimination is vacuous and there is no difficulty.

We conclude this section with several examples illustrating the use of the rules in final form, that is, with subscripts when necessary on ambiguous names and with the five new restrictions satisfied.

The nineteenth-century British logician De Morgan maintained that the classical syllogistic logic was too weak to derive that all heads of horses are heads of animals from the premise that all horses are animals. The desired derivation illustrates the use of subscripts. We use 'H' for 'is a head of' and 'P' for 'is a horse', and we then translate the desired conclusion thus:

$$(x)[(\exists y)(Py \& Hxy) \rightarrow (\exists y)(Ay \& Hxy)].$$

EXAMPLE 8. *All horses are animals. Therefore, all heads of horses are heads of animals.*

{1}	(1) $(x)(Px \rightarrow Ax)$	P
{2}	(2) $(\exists y)(Py \& Hxy)$	xP
{2}	(3) $P\alpha_x \& Hx\alpha_x$	x2 *ES*
{1}	(4) $P\alpha_x \rightarrow A\alpha_x$	1 *US*
{1, 2}	(5) $A\alpha_x \& Hx\alpha_x$	x3, 4 T
{1, 2}	(6) $(\exists y)(Ay \& Hxy)$	x5 *EG*

* Strictly speaking, this restriction was needed in § 4.3.

{1}	(7) $(\exists y)(Py \& Hxy) \rightarrow (\exists y)(Ay \& Hxy)$	2, 6 C.P.
{1}	(8) $(x)[(\exists y)(Py \& Hxy) \rightarrow (\exists y)(Ay \& Hxy)]$	7 *UG*

In line (2) we introduce the antecedent of the conclusion as a working premise. Note that 'x' is free in this premise. In line (3) we use *ES* to introduce the ambiguous name 'α_x'. Intuitively, α_x is the horse of which x is the head, the subscript 'x' being used to indicate the dependence.

The following example deals with the fourteenth-century nominalistic philosopher William of Ockham and the great seventeenth-century philosopher Thomas Hobbes.

EXAMPLE 9. *None of Ockham's followers like any realist. Any of Ockham's followers likes at least one of Hobbes' followers. Moreover, Ockham does have followers. Therefore, some of Hobbes' followers are not realists.*

{1}	(1) $(x)(Ox \rightarrow (y)(Ry \rightarrow -Lxy))$	P
{2}	(2) $(x)(Ox \rightarrow (\exists y)(Hy \& Lxy))$	P
{3}	(3) $(\exists x)(Ox)$	P
{3}	(4) $O\alpha$	3 *ES*
{1}	(5) $O\alpha \rightarrow (y)(Ry \rightarrow -L\alpha y)$	1 *US*
{2}	(6) $O\alpha \rightarrow (\exists y)(Hy \& L\alpha y)$	2 *US*
{2, 3}	(7) $(\exists y)(Hy \& L\alpha y)$	4, 6 T
{2, 3}	(8) $H\beta \& L\alpha\beta$	7 *ES*
{1, 3}	(9) $(y)(Ry \rightarrow -L\alpha y)$	4, 5 T
{1, 3}	(10) $R\beta \rightarrow -L\alpha\beta$	9 *US*
{1, 2, 3}	(11) $-R\beta$	8, 10 T
{1, 2, 3}	(12) $H\beta \& -R\beta$	8, 11 T
{1, 2, 3}	(13) $(\exists x)(Hx \& -Rx)$	12 *EG*

Although two ambiguous names are needed in this example, neither requires subscripts, since there are no free variables in either line (3) or line (7).

The following simple derivation shows that there is no difficulty in handling ambiguous names with more than one subscript. The example deals with points on a line: xPy when x precedes y on the line; $B(x, y, z)$ when y is between x and z. The facts stated in the premises are obvious, as is the conclusion.

EXAMPLE 10. *For every x and y if x precedes y then it is not the case that y precedes x. For every x, y, and z, if x precedes y and y precedes z then x precedes z. For every x and y if x precedes y then x is not equal to y. For every x, y, and z if y is between x and z then either x precedes y and y precedes z or z precedes y and y precedes x. For every x and z if*

$x \neq z$ then there is a y such that y is between x and z. Therefore, for every x and z if x precedes z then there is a y such that x precedes y and y precedes z.

{1}	(1)	$(x)(y)(xPy \rightarrow -yPx)$	P
{2}	(2)	$(x)(y)(z)(xPy \,\&\, yPz \rightarrow xPz)$	P
{3}	(3)	$(x)(y)(xPy \rightarrow x \neq y)$	P
{4}	(4)	$(x)(y)(z)(B(x, y, z) \rightarrow (xPy \,\&\, yPz) \vee (zPy \,\&\, yPx))$	P
{5}	(5)	$(x)(z)(x \neq z \rightarrow (\exists y)B(x, y, z))$	P
{6}	(6)	xPz	x, z P
{3}	(7)	$xPz \rightarrow x \neq z$	3 *US* x/x, z/y
{5}	(8)	$x \neq z \rightarrow (\exists y)B(x, y, z)$	5 *US* x/x, z/z
{3, 5, 6}	(9)	$(\exists y)B(x, y, z)$	x, z 6, 7, 8 T
{3, 5, 6}	(10)	$B(x, \alpha_{xz}, z)$	x, z 9 *ES*
{4}	(11)	$B(x, \alpha_{xz}, z) \rightarrow (xP\alpha_{xz} \,\&\, \alpha_{xz}Pz) \vee (zP\alpha_{xz} \,\&\, \alpha_{xz}Px)$	4 *US* x/x, α_{xz}/y, z/z
{3, 4, 5, 6}	(12)	$(xP\alpha_{xz} \,\&\, \alpha_{xz}Pz) \vee (zP\alpha_{xz} \,\&\, \alpha_{xz}Px)$	x, z 10, 11 T
{1}	(13)	$xPz \rightarrow -zPx$	1 *US* x/x, z/y
{1, 6}	(14)	$-zPx$	x, z 6, 13 T
{2}	(15)	$zP\alpha_{xz} \,\&\, \alpha_{xz}Px \rightarrow zPx$	2 *US* z/x, α_{xz}/y, x/z
{1, 2, 6}	(16)	$-(zP\alpha_{xz} \,\&\, \alpha_{xz}Px)$	x, z 14, 15 T
{1, 2, 3, 4, 5, 6}	(17)	$xP\alpha_{xz} \,\&\, \alpha_{xz}Pz$	x, z 12, 16 T
{1, 2, 3, 4, 5, 6}	(18)	$(\exists y)(xPy \,\&\, yPz)$	x, z 17 *EG*
{1, 2, 3, 4, 5}	(19)	$xPz \rightarrow (\exists y)(xPy \,\&\, yPz)$	6, 18 C.P.
{1, 2, 3, 4, 5}	(20)	$(x)(z)(xPz \rightarrow (\exists y)(xPy \,\&\, yPz)$	19 *UG*

Since 'x' and 'z' are both free in line (9), they occur as subscripts in line (10) and subsequently. In lines (7), (8), (11), (13) and (15) some new notation is introduced at the right to indicate a multiple use of *US*. For example, in deriving (7) from (3) both universal quantifiers in (3) were dropped. According to a strict reading of *US* such a double application is not permitted, but it is clear that a derived rule permitting such inferences can easily be established. The new notation used at the right is self-explanatory. In line (11), for instance, 'x/x, α_{xz}/y, z/z' indicates that in applying *US* to line (4), 'x' and 'z' were trivially substituted for themselves, and 'α_{xz}' was substituted for 'y'. Whenever we make such a multiple application of *US* the substitution will be indicated at the right.

Since line (6) is a premise, variables 'x' and 'z' are flagged in this line, and in every subsequent line depending on this premise. The fact that in line (10) these two variables appear both as flagged and as subscripts is a logical accident. There is no direct systematic connection between the two, except that both operations apply to free variables.

When derivations run to as many lines as Example 10, a useful working strategy is to sketch out on scratch paper a line of inference that seems to be sound before attempting to write down a detailed formal derivation. Ways of shortening or eliminating the routine steps in a derivation are discussed in Chapter 5 and in Chapter 7 *in extenso.*

EXERCISES

Construct (if possible) a derivation corresponding to the arguments in Exercises 1–8. If the argument is invalid, prove it by the method of interpretation.

1. Some of Aristotle's followers like all of Aquinas' followers. None of Aristotle's followers like any idealist. Therefore, none of Aquinas' followers are idealists. (Ax, Lxy, Qx, Ix)
2. If the team wins, then someone in the backfield is a good tailback. Adams is a good tailback. Therefore, if Adams is in the backfield, the team wins. (W, Bx, Tx, a)
3. None of the paintings is valuable, except the battle pieces. All the battle pieces are painted in oils. Some of the paintings are not painted in oil. Some paintings are not framed. Therefore, none of the paintings not painted in oils is valuable. (Px, Vx, Bx, Ox, Fx)
4. Some psychologists admire Freud. Some psychologists like no one who admires Freud. Therefore, some psychologists are not liked by all psychologists. (Px, Fx, Lxy)
5. If Round Robin won the race, then some people who were at the track were happy. If everyone who bet on the race lost money, then none who were at the track were happy. Therefore, if Round Robin won the race, then someone who bet on the race did not lose money. (R, Tx, Hx, Bx, Lx)
6. Kilroy was here. Therefore, someone was here. (Hx, k)
7. All good critics like every poet mentioned in the lecture. No good critic likes Edgar Guest, although Edgar Guest is a poet. Therefore, Edgar Guest was not mentioned in the lecture. (Cx, Lxy, Mx, Px, g)
8. Every philosophical empiricist admires Hume. Some philosophical idealists like no one who admires Hume. Therefore, some philosophical idealists like no philosophical empiricist. (Ex, Hx, Ix, Lxy)
9. The theory of empirical measurements affords some of the simplest examples of non-trivial scientific theories. This exercise deals with a set of axioms concerned with the measurement of mass. Intuitively the domain of individuals is a large set of physical objects which we may place on one of the two pans of an equal-arm balance. As in the case of the theory of preference, 'Q' is a two-place predicate with properties like '$\leq$'. If two objects stand in the relation Q, that is, if xQy,

then the balance is either in horizontal equilibrium (the objects are equal in mass) or the pan on which x is, is higher than y's pan (x has less mass than y). Thus by a qualitative observation we can always decide if two physical objects are in the relation Q when each is placed on a pan of the balance. We also have an operation $\star$ for combining objects to form new objects. Thus if $x \star y\, Q\, z$ then x and y are on one pan of the balance, z on the other; and the experimenter observes—

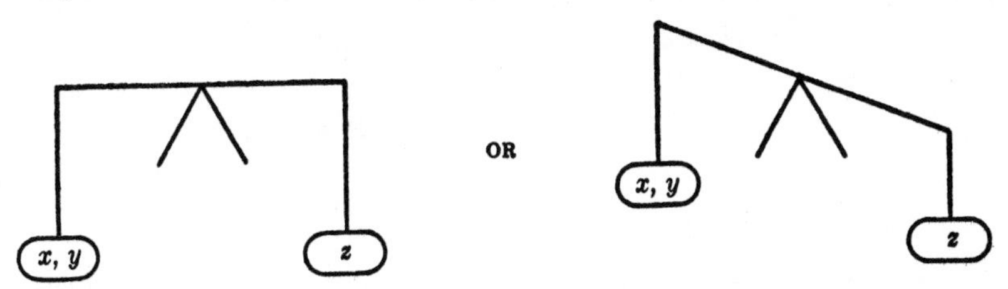

that is, the combination of x and y is equal to or less in mass than z. From a logical standpoint, '$\star$' is an operation symbol which we use to form terms, like '$x \star y$' '$z \star x$'. It is important to realize that '$(x \star y) \star z$' is not the same as '$x \star (y \star z)$'; in other words it is logically possible that the order of grouping or combining, as indicated by the parentheses, might make a difference in mass. (It is a consequence of the axioms that it does not, but this has to be proved.)

We want to state five axioms which we might expect to be satisfied, and then to state eight theorems which may be derived from these axioms by the rules of inference we have developed. To continue this list of theorems in order finally to prove that we may properly introduce numbers to measure mass, one additional axiom (a so-called Archimedean axiom) is needed. We do not give this last axiom here because of its more advanced mathematical character.*

The five axioms are:

(1) $(x)(y)(z)(xQy \;\&\; yQz \rightarrow xQz)$

(2) $(x)(y)(z)[(x \star y) \star z\, Q\, x \star (y \star z)]$

(3) $(x)(y)(z)(xQy \rightarrow x \star z\, Q\, z \star y)$

(4) $(x)(y)(-xQy \rightarrow (\exists z)(x\, Q\, y \star z \;\&\; y \star z\, Q\, x))$

(5) $(x)(y)(-x \star y\, Q\, x)$

The first axiom says that the relation of equal to or less than in mass is transitive. The second axiom countenances rearrangements of objects. The third axiom asserts that if x is equal to or less than y in mass then combining x with z and z with y will not change the inequality. The fourth axiom says that if x is strictly heavier than y then an object z can be found which combined with y exactly balances x. The fifth axiom says every object has positive mass; that is, combine any object x with any object y and the combination is heavier than x. Treating the five axioms as premises, formally derive the following eight theorems as conclusions.

(To avoid one long derivation, it is suggested that in the derivation of all but the first theorem, you insert when needed a previous theorem, justifying it on the right by the theorem number and listing on the left the axioms on which it depends. For example, you will want to use Theorem 1 in proving Theorem 2. The insertion

* The full set of axioms is to be found in my article, "A Set of Independent Axioms for Extensive Quantities," *Portugaliae Mathematica*, Vol. 10 (1951) pp. 163–172.

in a derivation of such previously established results is discussed systematically in the next chapter.)

THEOREM 1. $(x)(xQx)$
THEOREM 2. $(x)(y)(x \star y \; Q \; y \star x)$
THEOREM 3. $(x)(y)(u)(v)(xQy \; \& \; uQv \rightarrow x \star u \; Q \; y \star v)$
THEOREM 4. $(x)(y)(z)[x \star (y \star z) \; Q \; (x \star y) \star z]$

(This derivation is long.)

THEOREM 5. $(x)(y)(xQy \lor yQx)$

(Hint: Use an indirect proof and Axiom (4).)

THEOREM 6. $(x)(y)(z)(x \star z \; Q \; y \star z \rightarrow xQy)$
THEOREM 7. $(x)(y)(z)(u)(y \star z \; Q \; u \; \& \; xQy \rightarrow x \star z \; Q \; u)$
THEOREM 8. $(x)(y)(z)(u)(u \; Q \; x \star z \; \& \; xQy \rightarrow u \; Q \; y \star z)$

10. Consider the five axioms given in the preceding exercise. The following arithmetical interpretation proves that Axiom (1) is independent of the remaining four:

(i) The domain of individuals is the set of positive integers.
(ii) $xQy \leftrightarrow x \leq y + 1$.
(iii) $x \star y = x + y + 2$.

It is easily verified that Axiom (1) is false under this interpretation and the remaining four axioms are true. Find arithmetical interpretations which prove that the other four axioms are also each independent. As a hint in constructing an appropriate interpretation, a satisfactory domain of individuals is given for each axiom:

For Axiom (2): The set of all positive rational numbers (a rational number is the ratio of two integers; finding the proper interpretation for this axiom is a little difficult).

For Axiom (3): The set of all positive rational numbers.
For Axiom (4): The set of positive integers with the exception of one.
For Axiom (5): The set consisting of just the number one.

§ 4.6 Summary of Rules of Inference. The point of this section is to summarize in tabular form the seven basic rules of inference which have been introduced in Chapter 2 and the present chapter.

Rather than repeat the complex substitution phraseology used in the initial statement of the rules, we use a convenient abbreviated notation. If S(v) is any formula in which the variable v is free, then S(t) is the formula which results from S(v) by substituting the term t for every free occurrence of v in S. Thus if v is 'x', S(v) is 'Fx', and t is 'y', then S(t) is 'Fy'. As a second example, if

$$\mathsf{v} = \text{`}z\text{'}$$
$$\mathsf{S(v)} = \text{`}(\exists z)(x + z = 0) \; \& \; y + z > z + 2\text{'}$$
$$\mathsf{t} = \text{`}x + y\text{'}$$

then

$$\mathsf{S(t)} = \text{`}(\exists z)(x + z = 0) \; \& \; y + (x + y) > (x + y) + 2\text{'}.$$

Similarly, if $\mathsf{S}(\boldsymbol{\nu})$ is any formula in which the ambiguous name $\boldsymbol{\nu}$ occurs, then $\mathsf{S}(\mathsf{v})$ is obtained by substitution from $\mathsf{S}(\boldsymbol{\nu})$ if and only if every occurrence of $\boldsymbol{\nu}$ in $\mathsf{S}(\boldsymbol{\nu})$ is replaced by the variable v. Thus, if

$$\boldsymbol{\nu} = \text{‘}\alpha\text{’}$$
$$\mathsf{S}(\boldsymbol{\nu}) = \text{‘}\alpha + x = \beta \;\&\; \alpha > 0\text{’}$$
$$\mathsf{v} = \text{‘}x\text{’}$$

then

$$\mathsf{S}(\mathsf{v}) = \text{‘}x + x = \beta \;\&\; x > 0\text{’}.$$

BASIC RULES OF INFERENCE

Abbreviation	Rule	Restriction
P	Introduction of premises	None
T	Use of tautologies	None
C.P.	Conditional Proof	None
US	Universal Specification: From $(\mathsf{v})\mathsf{S}$ derive $\mathsf{S}(\mathsf{t})$	No free occurrence of v within scope of quantifier using variable of t.
UG	Universal Generalization: From S derive $(\mathsf{v})\mathsf{S}$	(1) v not flagged (2) v not a subscript
ES	Existential Specification: From $(\exists\mathsf{v})\mathsf{S}(\mathsf{v})$ derive $\mathsf{S}(\boldsymbol{\nu})$	Ambiguous name $\boldsymbol{\nu}$ not previously used
EG	Existential Generalization: From $\mathsf{S}(\boldsymbol{\nu})$ derive $(\exists\mathsf{v})\mathsf{S}(\mathsf{v})$	(1) v not a subscript (2) No occurrence of name $\boldsymbol{\nu}$ within scope of quantifier using v (3) v not flagged if $\boldsymbol{\nu}$ actually occurs in $\mathsf{S}(\boldsymbol{\nu})$

FLAGGING: A variable free in a premise is flagged and remains flagged in any line in which it is free and which depends on the premise.

SUBSCRIPTS: When an ambiguous name is introduced by existential specification it must have as subscripts all the free variables occurring in the formula to which *ES* is applied.

EXERCISES

What error is committed at what point in the fallacious derivations 1–5?

1. Given any thing there exists another thing distinct from it. Therefore, there exists something which is distinct from itself.

$\{1\}$ (1) $(x)(\exists y)(x \neq y)$
$\{1\}$ (2) $(\exists y)(y \neq y)$

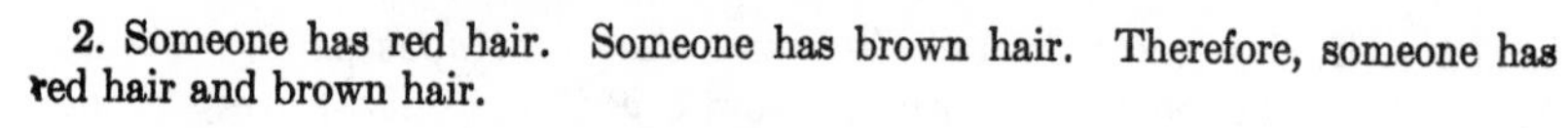

2. Someone has red hair. Someone has brown hair. Therefore, someone has red hair and brown hair.

{1}	(1) $(\exists x)Rx$
{2}	(2) $(\exists x)Bx$
{1}	(3) $R\alpha$
{2}	(4) $B\alpha$
{1, 2}	(5) $R\alpha \& B\alpha$
{1, 2}	(6) $(\exists x)(Rx \& Bx)$

3. Some men are over six feet tall. Therefore, all men are over six feet tall.

{1}	(1) $(\exists x)Ox$
{1}	(2) $O\alpha$
{1}	(3) $(x)Ox$

4. Every man is just as tall as himself. Therefore, there is a man who is just as tall as every man.

{1}	(1) $(x)Txx$
{1}	(2) $T\alpha x$
{1}	(3) $(x)T\alpha x$
{1}	(4) $(\exists y)(x)Tyx$

5. Everything is identical with itself. Therefore, everything is identical with everything else.

{1}	(1) $(x)(x = x)$
{1}	(2) $x = y$
{1}	(3) $(y)(x = y)$
{1}	(4) $(x)(y)(x = y)$

6. Construct a fallacious derivation which violates the restriction that a universal quantifier may be dropped by *US* only when the quantifier stands in front of the formula and its scope is the whole formula.

7. Construct a fallacious derivation which violates the restriction that an existential quantifier may be dropped by *ES* only when the quantifier stands in front of the formula and its scope is the whole formula.

CHAPTER 5

FURTHER RULES OF INFERENCE

§ 5.1 **Logic of Identity.** In everyday language we often put a form of the verb 'to be' between two terms in order to indicate that they designate the same entity. Thus we write:

(1) Elizabeth II is the present Queen of England,

to mean that Elizabeth II is the same person as the present Queen of England. Or (here using the past tense) we write:

(2) James Madison was the fourth President of the United States,

to mean that James Madison and the fourth President of the United States are one and the same person.

Everyday language puts the verb 'to be' to a great many other uses, however. For example, we sometimes use this verb to indicate that a certain entity has a certain property. For instance, when we write:

(3) Salt is white,

we certainly do not mean that salt is the same thing as white—this doesn't make sense—or even that salt is the same thing as whiteness—which is false—but that salt possesses the property of being white.

Although we may contrive to muddle through somehow with the ambiguous verb 'to be' in everyday English, in philosophical discussions the ambiguity of this word can lead to obscurity and endless confusion. Thus in more exact discourse it is convenient to divide up the various meanings of 'to be' among various words and symbols which are specially introduced for this purpose. In particular, we use the words 'is identical with', or the symbol '$=$', for the first meaning mentioned above. (In Chapter 9 we shall introduce other words and symbols for other meanings of 'to be'.) Instead of (1) we write:

Elizabeth II is identical with the present Queen of England, or simply:

Elizabeth II = the present Queen of England.

And instead of (2) we write:

James Madison is identical with the fourth President of the United States,

or simply:

(4) James Madison = the fourth President of the United States.

But we do not replace the 'is' of (3) by 'is identical with' or '='; since the 'is' in (3), as we remarked above, has quite a different meaning from the 'is' in (1), or the 'was' in (2).

We shall sometimes read the symbol '=' also as 'equals'. But it should be emphasized that some people use 'equals' in a somewhat broader sense: thus in elementary geometry two line segments AB and CD are sometimes called 'equal' if they have the same length. We shall never speak in this way, however. If we were to write:

$$AB = CD,$$

or to say that AB equals CD, we should mean that the segment AB and the segment CD are one and the same. We could, of course, express the fact that AB and CD have the same length by writing:

The length of AB = the length of CD.

It is clear from the meaning given to '=' that everything is identical with itself: i.e., if x is anything whatsoever, then $x = x$. This principle is sometimes called the *law of identity*.

Moreover, for any x and y, if $x = y$, then $y = x$. This fact is sometimes expressed by saying that the relation of identity is *symmetric*.

We notice also that if $x = y$ and $y = z$, then $x = z$. This fact is expressed by saying that the relation of identity is *transitive*.

If $x = y$, then whatever is true of x is also true of y, and whatever is true of y is also true of x. Thus from (4), together with the fact that James Madison was born in Port Conway, Virginia, one can conclude that the fourth President of the United States was born in Port Conway, Virginia. This principle, often called the *principle of extensionality*, is sometimes expressed by saying that equals may be substituted for equals.*

* This is not literally true, for there also exist so-called non-extensional contexts into which substitutions cannot be made. Consider, for example, the following three statements:

(5) The morning star is identical with the evening star;

(6) John Smith knows that the morning star is the same as Venus;

(7) John Smith knows that the evening star is the same as Venus.

If every property of x is also a property of y, then $x = y$; for x has the property of being identical with x, and hence if every property of x is a property of y, then y has the property of being identical with x, so that $y = x$, and hence $x = y$. This principle is sometimes called *Leibniz's law*, or the *principle of the identity of indiscernibles.*

We have not as yet introduced any special rules which permit us to draw *logical inferences* involving identities. We could, of course, take the view that the relation of identity has no more special status than any other relation, and that when derivations involve identities, special premises, such as the principle of extensionality, shall be introduced as needed. It is more convenient, however, because of the ubiquitous character of identities—we use them in every systematic domain of investigation—to introduce a special rule of inference for them.

The rule is divided into two parts. The first part simply converts the principle of extensionality into a principle of inference. As the example concerning James Madison suggests, we want the rule to apply not just to variables, but to terms in general. An algebraic example will further emphasize this point. Suppose we have as a line of a derivation:

$$x + y = 2, \tag{1}$$

and we know that

$$x = y + 3. \tag{2}$$

We want to be able to substitute the term '$y + 3$' for the variable 'x' in equation (1). Without a rule permitting such substitution it would be difficult to solve the simplest algebraic problems.

To avoid lengthy restrictions regarding the presence of quantifiers we restrict applicability of the rule to *open* formulas, that is, formulas which have no quantifiers. The following fallacious inference shows the need for restriction in some direction.

{1}	(1) $x = y$	x, y P
{2}	(2) $(\exists x)-(x = y)$	y P
{1, 2}	(3) $(\exists x)-(x = x)$	1, 2 (fallacious unrestricted rule for identities)

Here (5) is true, and (7) results from (6) by substituting 'evening star' for 'morning star'; but it can very well happen that John Smith's knowledge of astronomy happens to be such that (6) is true, while (7) is false. However, in mathematics and its applications to the empirical sciences, we do not need to make use of non-extensional contexts, just as we do not need non-truth-functional sentential connectives. The point is that by rejecting non-extensional contexts and non-truth-functional connectives we can keep our basic logic relatively simple and yet not give up any capacity for expression which is essential to the vast superstructure of mathematics.

Line (3) is false, and it is easy to find an interpretation such that (1) and (2) are true: Let the free variable 'x' be 'James Madison' and 'y' be 'the fourth President of the United States'. The source of the difficulty here is that when 'x' is substituted in (2) for 'y' on the basis of (1), 'x' is captured by the existential quantifier already present in (2).

The second part of the rule for identities simply asserts that the law of identity is independent of any premises, that is, for any term $\mathfrak{t}$, the assertion $\mathfrak{t} = \mathfrak{t}$ is derivable from the *empty set* of premises. In a derivation we indicate that a line depends on no premises by writing a capital Greek lambda, 'Λ', at the left. (This letter is often used to designate the empty set, that is, the set which has no members.) The formal statement of the rule follows.

> RULE GOVERNING IDENTITIES: I. *If* S *is an open formula, from* S *and* $\mathfrak{t}_1 = \mathfrak{t}_2$, *or from* S *and* $\mathfrak{t}_2 = \mathfrak{t}_1$ *we may derive* T, *provided that* T *results from* S *by replacing one or more occurrences of* $\mathfrak{t}_1$ *in* S *by* $\mathfrak{t}_2$. *Moreover, the identity* $\mathfrak{t} = \mathfrak{t}$ *is derivable from the empty set of premises.*

The first part of the rule, which is more widely used in logical inference than the second part, is applied in the following example.

EXAMPLE 1. *The man who committed the crime was in the apartment. Now if anyone was in the apartment, he was in town. If anyone was in Mexico he was not in town. In point of fact, Barnes was in Mexico. Therefore, Barnes is not the man who committed the crime.*

{1}	(1) Ac	P
{2}	(2) $(x)(Ax \rightarrow Tx)$	P
{3}	(3) $(x)(Mx \rightarrow -Tx)$	P
{4}	(4) Mb	P
{3}	(5) $Mb \rightarrow -Tb$	3 *US*
{3, 4}	(6) $-Tb$	4, 5 T
{2}	(7) $Ab \rightarrow Tb$	2 *US*
{2, 3, 4}	(8) $-Ab$	6, 7 T
{9}	(9) $b = c$	P
{2, 3, 4, 9}	(10) $-Ac$	8, 9 I (rule governing identities)
{1, 2, 3, 4, 9}	(11) $Ac \& -Ac$	1, 10 T
{1, 2, 3, 4}	(12) $b \neq c$	9, 11 R.A.A.

Use of the newly introduced rule for identities is exemplified in line (10). This derivation illustrates a method of attack which is usually sound: if we want to establish the negation of an identity, it is natural to use an indirect proof. Note that the application of the rule governing identities usually involves two previous lines ((8) and (9) in the example), namely, the line asserting an identity and the line asserting a formula in which a replacement of a term is made by use of the identity. The resulting line naturally depends on any premises on which these two lines depended. Thus line (10) depends on the premises introduced in lines (2), (3), (4), and (9).

In mathematical contexts identities most commonly occur in conjunction with operation symbols. Formulas like:

$$(x)(y)(\exists z)(x + y = z)$$
$$x > 0 \rightarrow (\exists y)(x + y = 0)$$

are typical. The binary operation symbols of addition and multiplication are most familiar, but the binary operation symbol of subtraction and the *unary* operation symbol for finding the negative of a number also occur frequently and have meanings familiar to all readers of this book:

$$x - y = 1,$$
$$x + (-x) = 0.$$

To illustrate how the rules of inferences for identities are used in connection with operation symbols, we may derive some consequences of one of the simplest but one of the more important sets of axioms in modern mathematics, namely, the axioms for a *group*. The axioms use three non-logical symbols: the binary operation symbol '$\circ$', the unary operation symbol '$^{-1}$', and the individual constant 'e'. The most familiar interpretation is to take as the domain of interpretation the set of integers (positive, negative, and zero), to interpret '$\circ$' as '$+$', to interpret '$^{-1}$' as the operation symbol for negating a number, and to interpret 'e' as the name of zero. The three axioms for a group are:

AXIOM (1). $(x)(y)(z)(x \circ (y \circ z) = (x \circ y) \circ z)$.
AXIOM (2). $(x)(x \circ e = x)$.
AXIOM (3). $(x)(x \circ x^{-1} = e)$.

Thus in the arithmetical interpretation just given, the three axioms have as interpretations the following three true sentences:

(1′) $(x)(y)(z)(x + (y + z) = (x + y) + z)$.
(2′) $(x)(x + 0 = x)$.
(3′) $(x)(x + (-x) = 0)$.

In usual mathematical terminology, Axiom (1) says that the operation $\circ$ is *associative;* Axiom (2) says that e is a right-hand *identity* element with respect to the operation $\circ$; and Axiom (3) says that each element of the group has an *inverse element* with respect to the operation $\circ$. Thus zero is the identity element for addition of integers, and the negative of a number is its inverse with respect to addition.

We now want to derive from the axioms what is known as the *right-hand cancellation law:* $(x)(y)(z)(x \circ z = y \circ z \rightarrow x = y)$. Note that all three axioms are used in the derivation.

{1}	(1) $(x)(y)(z)(x \circ (y \circ z) = (x \circ y) \circ z)$	P (Ax. (1))
{2}	(2) $(x)(x \circ e = x)$	P (Ax. (2))
{3}	(3) $(x)(x \circ x^{-1} = e)$	P (Ax. (3))
Λ	(4) $(x \circ z) \circ z^{-1} = (x \circ z) \circ z^{-1}$	I
{5}	(5) $x \circ z = y \circ z$	x, y, z P
{5}	(6) $(x \circ z) \circ z^{-1} = (y \circ z) \circ z^{-1}$	x, y, z 4, 5 I
{1}	(7) $x \circ (z \circ z^{-1}) = (x \circ z) \circ z^{-1}$	1 *US* $x/x, z/y, z^{-1}/z$
{1}	(8) $y \circ (z \circ z^{-1}) = (y \circ z) \circ z^{-1}$	1 *US* $y/x, z/y, z^{-1}/z$
{1, 5}	(9) $x \circ (z \circ z^{-1}) = (y \circ z) \circ z^{-1}$	x, y, z 6, 7 I
{1, 5}	(10) $x \circ (z \circ z^{-1}) = y \circ (z \circ z^{-1})$	x, y, z 8, 9 I
{3}	(11) $z \circ z^{-1} = e$	3 *US*
{1, 3, 5}	(12) $x \circ e = y \circ e$	x, y 10, 11 I
{2}	(13) $x \circ e = x$	2 *US*
{1, 2, 3, 5}	(14) $x = y \circ e$	x, y 12, 13 I
{2}	(15) $y \circ e = y$	2 *US*
{1, 2, 3, 5}	(16) $x = y$	x, y 14, 15 I
{1, 2, 3}	(17) $x \circ z = y \circ z \rightarrow x = y$	5, 16 C.P.
{1, 2, 3}	(18) $(x)(y)(z)(x \circ z = y \circ z \rightarrow x = y)$	17 *UG*

The use of the rule governing identities seven times in this derivation indicates how often it is appealed to in dealing with operation symbols. In Chapter 7 methods for reducing the length of derivations are developed, but until then all derivations should be written out in full. The above example makes obvious the need for distinguishing the routine from the non-routine steps. Probably *the* crucial step in deriving the right-hand cancellation law is realizing what substitution in Axiom (1) is appropriate (line (7)). For with this insight goes the perception that $z \circ z^{-1} = e$ by virtue of Axiom (3). The rest of the derivation is just a matter of eliminating the identity element by use of Axiom (2).

We return to the axioms for groups in § 5.2.*

* For a clear and detailed but elementary discussion of various sets of axioms for groups, see Alfred Tarski, *Introduction to Logic*, New York, 1941.

Since we have made the logic of identity part of our basic first-order predicate logic (often called with this extension: *first-order predicate logic with identity*), we need to remark on the extension of the notion of interpretation to formulas involving identities. The extension is simple: the identity predicate like the other logical constants and unlike all other predicates is held fixed in its original meaning in all interpretations. Thus, given the formula:

$$(1) \qquad x = 0 \;\&\; y = 0 \rightarrow x + y = 0,$$

it would be a mistake to interpret '=' as, say, 'is greater than', just as it would be a mistake to interpret '→' as '&'. This means that in formulas like (1) the identity sign is treated as a universal logical symbol and not like '+' as a symbol of arithmetic only.*

On the other hand, in giving interpretations of non-logical predicates it is sometimes convenient to interpret a predicate as the logical predicate of identity. In fact, this was already done in one interpretation for the predicate 'I' of indifference given in Chapter 4.

EXERCISES

1. The verb 'to be' is used in the sense of identity in which of the following sentences?

(a) Women are wonderful.
(b) Washington was an American.
(c) Washington was the first President of the United States.
(d) Washington was the last husband of Martha Washington.
(e) Jefferson was not the second President of the United States.
(f) Men are not angels.
(g) Simone de Beauvoir is not a great writer.
(h) Stephen Dedalus was not a lover of Molly Bloom.
(i) Stendhal was not the author of NORTHANGER ABBEY.

2. Some members of the swimming team have not lost a race. Jones is on the swimming team; furthermore he is the fastest man on the team. Therefore, Jones has not lost a race. (Mx, Lx, j, f)

3. Is the following set of premises consistent? Adams is the man who signed the contract, and the man who signed the contract is liable. However, Adams is bankrupt, and if anyone is bankrupt then he is not liable. (Lx, Bx, a, c)

4. Is the following set of premises consistent? Horatio is the bravest man in the county, but not the strongest. Yet none but the brave are strong. (h, b, s, Bx, Sx)

5. This exercise is related to the axioms for a group but not directly dependent on them. Given that the operation $\circ$ is commutative and satisfies the right can-

* The standard reference on the logic of identity is D. Hilbert and P. Bernays, *Grundlagen der Mathematik*, Vol. 1, pp. 164–209. It may be shown that first-order predicate logic with identity is complete in the sense defined in § 4.2.

cellation law, prove that it satisfies the left cancellation law, that is, from:

(1) $(x)(y)(x \circ y = y \circ x)$,
(2) $(x)(y)(z)(x \circ z = y \circ z \rightarrow x = y)$,

derive:

$$(x)(y)(z)(z \circ x = z \circ y \rightarrow x = y).$$

6. Given the following three axioms on a binary operation $\circ$, either prove that the axioms are inconsistent or give an interpretation to show that they are consistent. (This exercise in no way depends on the axioms for groups.)

AXIOM 1. $(x)(y)(x \circ y = y \circ x)$.
AXIOM 2. $(x)(y)(x \circ y = y)$.
AXIOM 3. $(x)(\exists y)(x \neq y)$.

7. Prove by the method of interpretation that the first axiom for groups is independent of the other two. (Warning: do not violate Rule VII, §4.2.)

8. Prove by the method of interpretation that the third axiom for groups is independent of the first two. (Warning: do not violate Rule VII, §4.2.)

§ 5.2 Theorems of Logic. So far we have mainly been concerned with the derivation of conclusions from given premises. We want now to give some brief consideration to formulas which are derivable from the empty set of premises. Such formulas are called *theorems of logic*, since they are true independent of the truth or falsity of any particular factual premises. Referring to the notions of § 4.2 it may be shown that a formula is a theorem of logic if and only if it is universally valid. Since a theorem of logic depends on no premises, the strategy of proof is slightly different from that for deriving conclusions from given premises. If the formula we are attempting to establish as a theorem is an implication, the natural thing is to assume the antecedent of the implication as an initial premise. In many cases a second premise may be obtained by assuming the negation of the consequent of the implication and embarking upon an indirect proof. This approach may be illustrated by proving the theorem '$(\exists x)Fx \rightarrow -(x)-Fx$'.

{1}	(1) $(\exists x)Fx$	P
{2}	(2) $(x)-Fx$	P
{1}	(3) $F\alpha$	1 *ES*
{2}	(4) $-F\alpha$	2 *US*
{1, 2}	(5) $F\alpha \;\&\; -F\alpha$	3, 4 T
{1}	(6) $-(x)-Fx$	2, 5 R.A.A.
Λ	(7) $(\exists x)Fx \rightarrow -(x)-Fx$	1, 6 C.P.

As the first premise we assume the antecedent '$(\exists x)Fx$', and as the second premise '$(x)-Fx$' which is the negation of the consequent. The theorem is established by showing in line (7) that it depends on the empty set of premises, in other words, on no premise at all.* Thus, (7) is true regardless of what interpretations we give the predicate 'F'.

It is sometimes necessary to adopt a more subtle strategy, as in the proof that '$-(x)-Fx \rightarrow (\exists x)Fx$' is a theorem.

{1}	(1) Fx	x P
{1}	(2) $(\exists x)Fx$	1 *EG*
Λ	(3) $Fx \rightarrow (\exists x)Fx$	1, 2 C.P.
{4}	(4) $-(\exists x)Fx$	P
{4}	(5) $-Fx$	3, 4 T
{4}	(6) $(x)-Fx$	5 *UG*
Λ	(7) $-(\exists x)Fx \rightarrow (x)-Fx$	4, 6 C.P.
Λ	(8) $-(x)-Fx \rightarrow (\exists x)Fx$	7 T

In this case it would not work to assume the antecedent '$-(x)-Fx$' since the beginning negation sign prevents us from doing anything with the formula. In line (4) we assume as a premise the negation of the conclusion, but in this case not for the purposes of an indirect proof. The main trick in this proof is to apply *EG* to 'Fx' to obtain (2). As already remarked, such applications of *EG*, where no ambiguous names occur, are infrequent but occasionally crucial.

The two theorems just proved are closely related to Rule Q2 introduced in § 4.4; the interesting thing to note is that Q2 is not used in their proofs. These two theorems are in fact the basis for showing that Q2 may be derived from the sentential rules and the four rules introduced in Chapter 4.

Since a theorem of logic already established may be useful in proving a new theorem, it is desirable to have a derived rule which permits us to use the previously established theorem directly in the derivation establishing the new theorem. Since the use of previously established results need not be restricted to theorems of logic, we state the derived rule in a general form.

As might be expected, a restriction regarding ambiguous names is required, for the intended referent of an ambiguous name changes from derivation to derivation. Thus consider the derivation:

{1}	(1) $(\exists x)Fx$	P
{1}	(2) $F\alpha$	1 *ES*
Λ	(3) $(\exists x)Fx \rightarrow F\alpha$	1, 2 C.P.

Since line (3) is a theorem of logic we may feel we can introduce it at any point in some other derivation without creating difficulties. But the fal-

* It was explained in the previous section that in this book the symbol 'Λ' designates the empty set. Note its use in line (7).

lacious inference from '$(\exists x)Fx$' to '$(x)Fx$' is easily constructed by using line (3) above and a similar derivation concerning '$-Fx$'.

{1}	(1) $(\exists x)-Fx$	P
{1}	(2) $-F\alpha$	1 *ES*
Λ	(3) $(\exists x)-Fx \rightarrow -F\alpha$	1, 2 C.P.
Λ	(4) $F\alpha \rightarrow -(\exists x)-Fx$	3 T
Λ	(5) $(\exists x)Fx \rightarrow F\alpha$	Previous theorem (fallaciously)
Λ	(6) $(\exists x)Fx \rightarrow -(\exists x)-Fx$	4, 5 T
Λ	(7) $(\exists x)Fx \rightarrow (x)Fx$	6 Q1

The difficulty comes from the appearance of the ambiguous name 'α' in line (5) by virtue of introducing the preceding theorem. If the two derivations had been combined into one, line (7) could not have been reached, for in line (2) of the second derivation 'β' or some other ambiguous name other than 'α' would have had to be introduced, and the derivation of (6) would be blocked. It was pointed out when ambiguous names were first discussed that they were for immediate contextual use. This example shows in particular that they cannot be carried from derivation to derivation.

Derived Rule for Introducing Previous Results. *If formula* S *is derivable from formula* T *and a set of premises* $\mathscr{P}$, *and if* T *is derivable from* $\mathscr{P}$, *then* S *is derivable from* $\mathscr{P}$ *alone, provided* T *contains no ambiguous names.*

If S is a theorem of logic then, of course, $\mathscr{P} = \Lambda$. We shall not give a detailed proof of this derived rule but it is obvious how it goes. By hypothesis S is derivable from T and $\mathscr{P}$, and T is derivable from $\mathscr{P}$. To show that S is derivable from $\mathscr{P}$ alone, we simply construct a new derivation for S such that the first part of the new derivation is just a derivation of T from $\mathscr{P}$.

To illustrate the labor-saving nature of this derived rule, let us derive the theorem '$(\exists x)Fx \leftrightarrow -(x)-Fx$'.

Λ	(1) $(\exists x)Fx \rightarrow -(x)-Fx$	Previous theorem of logic
Λ	(2) $-(x)-Fx \rightarrow (\exists x)Fx$	Previous theorem of logic
Λ	(3) $(\exists x)Fx \leftrightarrow -(x)-Fx$	1, 2 T

Without the rule, derivations of (1) and (2) would have had to be included, which would have greatly increased the number of lines. The proof of this theorem illustrates a typical and standard strategy when proving a theorem whose main sentential connective is 'if and only if'; we break the proof into the derivation of two implications.

The theorems of logic whose proofs are given as exercises all exemplify useful logical principles, some of which will be discussed in the next section.

The use of this derived rule for introducing previous results which are not theorems of logic may be illustrated by returning to the axioms for groups given in the previous section. There we proved that the axioms for groups imply the right-hand cancellation law, which we may label:

THEOREM 1 OF GROUP THEORY. $(x)(y)(z)(x \circ z = y \circ z \rightarrow x = y)$.

And we may use this first theorem in proving a second theorem concerning the commutativity of the identity element.

THEOREM 2 OF GROUP THEORY. $(x)(x \circ e = e \circ x)$.

DERIVATION

{1}	(1) $(x)(y)(z)(x \circ (y \circ z) = (x \circ y) \circ z)$	P (Ax. 1)
{2}	(2) $(x)(x \circ e = x)$	P (Ax. 2)
{3}	(3) $(x)(x \circ x^{-1} = e)$	P (Ax. 3)
{1}	(4) $e \circ (x \circ x^{-1}) = (e \circ x) \circ x^{-1}$	1 *US* e/x, x/y, x^{-1}/z
{3}	(5) $x \circ x^{-1} = e$	3 *US*
{1, 3}	(6) $e \circ e = (e \circ x) \circ x^{-1}$	4, 5 I
{2}	(7) $e \circ e = e$	2 *US*
{1, 2, 3}	(8) $e = (e \circ x) \circ x^{-1}$	6, 7 I
{1, 2, 3}	(9) $x \circ x^{-1} = (e \circ x) \circ x^{-1}$	5, 8 I
{1, 2, 3}	(10) $(x)(y)(z)(x \circ z = y \circ z \rightarrow x = y)$	Th. 1
{1, 2, 3}	(11) $x \circ x^{-1} = (e \circ x) \circ x^{-1} \rightarrow x = e \circ x$	10 *US* x/x, $(e \circ x)/y$, x^{-1}/z
{1, 2, 3}	(12) $x = e \circ x$	9, 11 T
{2}	(13) $x \circ e = x$	2 *US*
{1, 2, 3}	(14) $x \circ e = e \circ x$	12, 13 I
{1, 2, 3}	(15) $(x)(x \circ e = e \circ x)$	14 *UG*

Note that Theorem 1 is used to justify the introduction of line (10). Obviously it would have been tedious and useless to rederive Theorem 1 at this point. The rule for introducing previous results permits its introduction without such repetition. Some of the exercises are concerned with further elementary theorems of group theory and repeated applications of the rule for introducing previous results. In working these exercises, a rough line of attack should be sketched on scratch paper before a formal derivation is attempted. In a first try at getting a line of argument going, it is best not to be too concerned with logical rigor but to experiment freely with a variety of applications of the axioms and the theorems already proved. In the derivation of Theorem 2 the two crucial insights are concerned with what substitution to make in Axiom (1), line (4), and how to make use of Theorem 1, line (11).

EXERCISES

1. Prove that the following formulas are theorems. You may use preceding formulas in the derivation of later ones.

(a) $(x)Fx \rightarrow -(\exists x)-Fx$ (Do not use Q1 in proof)
(b) $-(\exists x)-Fx \rightarrow (x)Fx$ (Do not use Q1 in proof)
(c) $(x)(y)Gxy \rightarrow (y)(x)Gxy$
(d) $(\exists x)(\exists y)Gxy \rightarrow (\exists y)(\exists x)Gxy$
(e) $(\exists x)(y)Gxy \rightarrow (y)(\exists x)Gxy$
(f) $(x)(Fx \,\&\, Hx) \rightarrow (x)Fx \,\&\, (x)Hx$
(g) $(x)Fx \,\&\, (x)Hx \rightarrow (x)(Fx \,\&\, Hx)$
(h) $(x)Fx \vee (x)Hx \rightarrow (x)(Fx \vee Hx)$
(i) $(x)(Fx \rightarrow Hx) \rightarrow [(\exists x)Fx \rightarrow (\exists x)Hx]$
(j) $(\exists x)(Fx \,\&\, Hx) \rightarrow (\exists x)Fx \,\&\, (\exists x)Hx$
(k) $Hy \,\&\, (\exists x)Fx \rightarrow (\exists x)(Hy \,\&\, Fx)$
(l) $(\exists x)(Fx \vee Hx) \leftrightarrow (\exists x)Fx \vee (\exists x)Hx$
(m) $(x)(Hy \vee Fx) \leftrightarrow Hy \vee (x)Fx$
(n) $(\exists x)(Hy \vee Fx) \leftrightarrow Hy \vee (\exists x)Fx$
(o) $(x)(Hy \,\&\, Fx) \leftrightarrow Hy \,\&\, (x)Fx$
(p) $(\exists x)(Hy \,\&\, Fx) \leftrightarrow Hy \,\&\, (\exists x)Fx$
(q) $(x)(Hy \rightarrow Fx) \leftrightarrow [Hy \rightarrow (x)Fx]$
(r) $(\exists x)(Hy \rightarrow Fx) \leftrightarrow [Hy \rightarrow (\exists x)Fx]$
(s) $(x)(Fx \rightarrow Hy) \leftrightarrow [(\exists x)Fx \rightarrow Hy]$
(t) $(\exists x)(Fx \rightarrow Hy) \leftrightarrow [(x)Fx \rightarrow Hy]$

2. For each of the following formulas prove by a derivation or an interpretation, as the case may be, if the formula is (1) a theorem of logic, (2) the negation of a theorem of logic, or (3) neither (1) nor (2).

(a) $(x)(y)(z)((Fxy \,\&\, Fyz) \rightarrow Fxz)$
(b) $(x)(y)(Fx \vee -Fy)$
(c) $(x)(\exists y)(Fx \,\&\, Gy) \rightarrow (\exists y)(x)(Fx \,\&\, Gy)$
(d) $(x)(Fx \leftrightarrow -Fx) \rightarrow (\exists y)Fy$
(e) $(x)(\exists y)(Fx \,\&\, -Fx \,\&\, -Fy)$
(f) $(\exists x)(Fx \vee -Fx)$
(g) $(x)Fx \vee (\exists y)-Fy$

3. Prove that the following five formulas concerning the relation of identity are theorems of logic.

(a) $(x)(y)(x = y \rightarrow y = x)$
(b) $(x)(y)(z)[(x = y \,\&\, y = z) \rightarrow x = z]$
(c) $(x)(y)(z)[(x = z \,\&\, y = z) \rightarrow x = y]$
(d) $(x)(\exists y)(x = y)$
(e) $(x)[Fx \leftrightarrow (\exists y)(x = y \,\&\, Fy)]$

4. This exercise deals with elementary group theory. Using the axioms for groups (p. 105) and the two theorems already established, provide formal derivations for the following theorems.

THEOREM 3 (UNIQUENESS OF IDENTITY).

$$(y)[(x)(x \circ y = x) \rightarrow y = e]$$

THEOREM 4 (COMMUTATIVITY OF INVERSES).

$$(x)(x \circ x^{-1} = x^{-1} \circ x)$$

THEOREM 5 (LEFT-HAND CANCELLATION LAW).

$$(x)(y)(z)(z \circ x = z \circ y \rightarrow x = y)$$

THEOREM 6 (UNIQUENESS OF INVERSE).

$$(x)(y)(x \circ y = e \rightarrow y = x^{-1})$$

THEOREM 7. $(x)((x^{-1})^{-1} = x)$

THEOREM 8. $(x)(y)(\exists z)(x = y \circ z)$

THEOREM 9. $(x)(z)(\exists y)(x = y \circ z)$

5. The last two theorems of the preceding exercise may be combined with the associativity axiom to give a new set of axioms for group theory based only on the single non-logical symbol '$\circ$'; that is, in this formulation the axioms are:

AXIOM 1. $(x)(y)(z)(x \circ (y \circ z) = (x \circ y) \circ z)$
AXIOM 2. $(x)(y)(\exists z)(x = y \circ z)$
AXIOM 3. $(x)(z)(\exists y)(x = y \circ z)$

To indicate how the development goes, prove the following four theorems on the basis of these three new axioms:

THEOREM 1 (EXISTENCE OF RIGHT-HAND IDENTITY ELEMENT).

$$(\exists y)(x)(x \circ y = x)$$

THEOREM 2 (EXISTENCE OF LEFT-HAND IDENTITY ELEMENT).

$$(\exists y)(x)(y \circ x = x)$$

THEOREM 3 (UNIQUENESS OF RIGHT-HAND IDENTITY ELEMENT).

$$(y)(z)[(x)(x \circ y = x \;\&\; x \circ z = x) \rightarrow y = z]$$

THEOREM 4 (UNIQUENESS OF LEFT-HAND IDENTITY ELEMENT).

$$(y)(z)[(x)(y \circ x = x \;\&\; z \circ x = x) \rightarrow y = z]$$

§ 5.3 Derived Rules of Inference. In § 2.4 we introduced a derived rule for indirect proofs; it was remarked that the three rules of § 4.4 could be derived; and a derived rule concerning the use of theorems was stated in the last section. We now want to examine in a more general way the character of such rules. We may approach this problem by asking how derived rules of inference differ from original rules such as the rule for universal specification or the rule of conditional proof. There are at least three essential points of difference.

(i) Derived rules of inference are dispensable; from the theoretical standpoint they are a mere convenience. Any inference which uses a derived rule can always be replaced by one which does not. In the case of the derived rule for indirect proofs, for instance, it is clear from the discussion in § 2.4 that it is a trivial and simple matter to eliminate its use.

(ii) Derived rules of inference must be *proved* by appeal to the original rules. In this context a proof primarily consists of giving an explicit set of directions for eliminating the use of the rule on any particular occasion. The attitude is that a derived rule may be used in inference if one can write down for a doubting Thomas a set of explicit directions for a formal derivation which does not use the derived rule and yet is able to come to the same conclusion. The proof of Rule R.A.A. gave such a set of directions. In that proof the essential idea is that we can eliminate appeal to the rule by making use of the Law of Absurdity.

(iii) The original rules are used to characterize in an *effective* or *finitistic* manner the intuitive notion of logical inference. As we saw in § 4.2 the notion of a valid argument or a universally valid sentence is defined in terms of the notion of a true interpretation of a formula and independent of *any* rules of inference. However, it is ordinarily impractical to prove that an argument is valid by showing that every interpretation which makes the premises true also makes the conclusion true. Since there is an infinite number of interpretations of any argument, it is impossible literally to examine each interpretation and determine its truth or falsity for the premises and conclusions. Thus we are naturally led to the idea of developing a finite number of rules of inference which we can use to establish the validity of arguments. The important two properties the original rules should have are those of soundness and completeness (already discussed in Chapters 2 and 4). The derived rules, being theoretically redundant, do not enter directly into a proof of the soundness or completeness of the basic rules of inference, although the proof of a new derived rule should show that the new rule cannot convert a sound system of inference into an unsound one.

So much for the difference between original and derived rules of inference. To every theorem of logic there is a corresponding derived rule. Since there is an unlimited number of theorems, there is an unlimited number of derived rules of inference. However, the number of such rules which are used repeatedly in inference is relatively small. We shall limit the remainder of this section to the consideration of some further rules governing the manipulation of quantifiers, and an extension of the rule concerning tautological equivalences (which was stated in § 4.4) to a rule for interchanging *logically* equivalent formulas.

If we wanted to establish Q1 as a derived rule, it would be reasonable to begin by proving:

(A) From $(\mathsf{v})\mathsf{S}$ we may derive $-(\exists \mathsf{v})-\mathsf{S}$.

PROOF: Introducing $(\mathsf{v})\mathsf{S}(\mathsf{v})$ and $(\exists \mathsf{v})-\mathsf{S}(\mathsf{v})$ as premises, we derive a contradiction by (i) using *ES* to obtain $-\mathsf{S}(\boldsymbol{\nu})$ from $(\exists \mathsf{v})-\mathsf{S}(\mathsf{v})$, where $\boldsymbol{\nu}$ is an appropriate ambiguous name, and (ii) using *US* to obtain $\mathsf{S}(\boldsymbol{\nu})$ from $(\mathsf{v})\mathsf{S}(\mathsf{v})$. The contradiction $\mathsf{S}(\boldsymbol{\nu})\ \&\ -\mathsf{S}(\boldsymbol{\nu})$ establishes the rule by indirect proof.

We may use Rule (A) to give a two-line derivation of '$-(\exists x)-(x + 0 = x)$' from '$(x)(x + 0 = x)$'.

{1}	(1) $(x)(x + 0 = x)$	P
{1}	(2) $-(\exists x)-(x + 0 = x)$	1, (A)

If challenged on the validity of this derivation, we may use the proof of Rule (A) to tell us how to eliminate the appeal to (A) in line (2); we get then in terms of the original rules the longer derivation:

{1}	(1) $(x)(x + 0 = x)$	P
{2}	(2) $(\exists x)-(x + 0 = x)$	P
{2}	(3) $-(\alpha + 0 = \alpha)$	2 *ES*
{1}	(4) $\alpha + 0 = \alpha$	1 *US*
{1, 2}	(5) $(\alpha + 0 = \alpha)\ \&\ -(\alpha + 0 = \alpha)$	3, 4 T
{1}	(6) $-(\exists x)-(x + 0 = x)$	2, 5 R.A.A.

The function of derived rules is primarily to eliminate unnecessary repetition of recurring patterns of inference. Once the general method of attack is clear, there is no point in duplicating the argument corresponding to lines (2)–(5) when we want to pass from a universal statement to the negation of an existential one.

The most important properties of quantifiers are summarized in the following rule; the first half of Part 1a is what has just been discussed as Rule (A).

DERIVED RULE GOVERNING QUANTIFIERS: Q

(1) *If* S *is any formula, then:*

(a) *from* $(\mathsf{v})\mathsf{S}$ *we may derive* $-(\exists \mathsf{v})-\mathsf{S}$, *and conversely;*
(b) *from* $(\mathsf{v})-\mathsf{S}$ *we may derive* $-(\exists \mathsf{v})\mathsf{S}$, *and conversely;*
(c) *from* $(\exists \mathsf{v})\mathsf{S}$ *we may derive* $-(\mathsf{v})-\mathsf{S}$, *and conversely;*
(d) *from* $(\exists \mathsf{v})-\mathsf{S}$ *we may derive* $-(\mathsf{v})\mathsf{S}$, *and conversely;*
(e) *from* $(\mathsf{v})(\mathsf{w})\mathsf{S}$ *we may derive* $(\mathsf{w})(\mathsf{v})\mathsf{S}$;
(f) *from* $(\exists \mathsf{v})(\exists \mathsf{w})\mathsf{S}$ *we may derive* $(\exists \mathsf{w})(\exists \mathsf{v})\mathsf{S}$;
(g) *from* $(\exists \mathsf{v})(\mathsf{w})\mathsf{S}$ *we may derive* $(\mathsf{w})(\exists \mathsf{v})\mathsf{S}$.

(2) *If* S *and* T *are any formulas, then:*

(a) *from* (v)(S & T) *we may derive* (v)S & (v)T, *and conversely;*
(b) *from* (∃v)(S ∨ T) *we may derive* (∃v)S ∨ (∃v)T, *and conversely;*
(c) *from* (v)(S → T) *we may derive* (∃v)S → (∃v)T;
(d) *from* (∃v)(S & T) *we may derive* (∃v)S & (∃v)T;
(e) *from* (v)S ∨ (v)T *we may derive* (v)(S ∨ T).

(3) *If* S *and* T *are any formulas and* v *is not free in* S, *then:*

(a) *from* (v)(S ∨ T) *we may derive* S ∨ (v)T, *and conversely;*
(b) *from* (∃v)(S ∨ T) *we may derive* S ∨ (∃v)T, *and conversely;*
(c) *from* (v)(S & T) *we may derive* S & (v)T, *and conversely;*
(d) *from* (∃v)(S & T) *we may derive* S & (∃v)T, *and conversely;*
(e) *from* (v)(S → T) *we may derive* S → (v)T, *and conversely;*
(f) *from* (∃v)(S → T) *we may derive* S → (∃v)T, *and conversely;*
(g) *from* (v)(T → S) *we may derive* (∃v)T → S, *and conversely;*
(h) *from* (∃v)(T → S) *we may derive* (v)T → S, *and conversely.*

The proof of the first half of Part 1a was given above. The remainder of the proof of the various parts of this rule is left to the student. The proof of nearly every part parallels the derivation of some one or two theorems in Exercise 1 of the previous section. For example, (1a) of the rule corresponds to (a) and (b) of the Exercise; (3a) of the rule corresponds to (m) of the Exercise. The proof of (1b) and (1d) is trivial once (1a) and (1c) respectively are established.

You will scarcely absorb at once all parts of the above rule, but you will find it useful to have in one place for ready reference a rule as complete as this one. Since students occasionally have difficulty in correctly interpreting the restriction in Part (3) of the rule that v not be free in S, it may be useful to consider one or two examples. Given the formula '$(x)(Mx \rightarrow Ax)$' may we use (3e) to derive '$Mx \rightarrow (x)Ax$'? No; since 'x' is free in 'Mx', the restriction is not satisfied. In this example, S is 'Mx' and T is 'Ax'. The fact that 'x' is bound in the whole formula '$(x)(Mx \rightarrow Ax)$' is irrelevant; the question is always: Is 'x' free in S considered by itself? As another example, suppose we are given '$(\exists y)[(x)(Mx \rightarrow Ax) \vee Ay]$'. May we use (3*b*) to derive '$(x)(Mx \rightarrow Ax) \vee (\exists y)Ay$'? Yes; for here S is '$(x)(Mx \rightarrow Ax)$', v is the variable '$y$', and clearly '$y$' is not free in '$(x)(Mx \rightarrow Ax)$'. In this example, T is, of course, 'Ay'.

The uses of the derived rule governing quantifiers are widened by extending the rule permitting replacement of a formula by a tautologically equivalent one to replacement by a *logically* equivalent formula. We have not yet precisely characterized the notion of logical equivalence, but the appropriate definition should be obvious from § 4.2. Two formulas are *logically equivalent* if and only if each logically implies the other. And this is the same as saying that the biconditional formed from them is universally valid. From the soundness and completeness of the basic rules of inference it follows that two formulas are logically equivalent if and only if each is derivable from the other by the rules of inference. Or, put another way, if and only if the biconditional formed from them is a theorem of logic.

Suppose, for instance, we wanted to show that the formula:

$$(x)(\exists y)(xQy \rightarrow (z)(\exists u)(yQz \rightarrow uQx)) \tag{1}$$

implies and is implied by:

$$(x)(\exists y)(z)(xQy \,\&\, yQz \rightarrow (\exists u)(uQx)). \tag{2}$$

We first observe that it follows from Rule Q3*e* above that the formula:

$$xQy \rightarrow (z)(\exists u)(yQz \rightarrow uQx) \tag{3}$$

is logically equivalent to:

$$(z)(xQy \rightarrow (\exists u)(yQz \rightarrow uQx)). \tag{4}$$

Moreover, from Q3*f* we observe that the formula:

$$(\exists u)(yQz \rightarrow uQx) \tag{5}$$

is logically equivalent to:

$$yQz \rightarrow (\exists u)(uQx). \tag{6}$$

Replacing then (5) by (6) in (4), we obtain as logically equivalent to (4) and thus as logically equivalent to (3):

$$(z)(xQy \rightarrow (yQz \rightarrow (\exists u)uQx)), \tag{7}$$

but the formula:

$$xQy \rightarrow (yQz \rightarrow (\exists u)uQx)$$

is tautologically equivalent (by importation of 'yQz') to:

$$xQy \,\&\, yQz \rightarrow (\exists u)uQx, \tag{8}$$

whence (7), and thus (3), is logically equivalent to:

$$(z)(xQy \,\&\, yQz \rightarrow (\exists u)uQx). \tag{9}$$

Replacing (3) by (9) in (1) we immediately obtain (2) as desired. Note that Q3*e* and Q3*f* could not be applied directly to (1), since the quantifiers to which we applied them are not standing in front of the formula. The rule permitting mutual replacement of logically equivalent formulas was needed. Naturally, the equivalence of (1) and (2) could have been shown by a derivation making no use of derived rules, but this involves the tedious operations of dropping and re-adding all quantifiers—the sort of thing derived rules are designed to help avoid.

The rule for mutual replacement of logically equivalent formulas must, like the rule for introducing previous results, be restricted with respect to ambiguous names. The construction of a fallacious inference justifying the restriction is left as an exercise.

> RULE FOR LOGICALLY EQUIVALENT FORMULAS: L.E. *Let* P *and* Q *be logically equivalent formulas which contain no ambiguous names. If* P *occurs as part of formula* R *and if formula* S *results from* R *by replacing at least one occurrence of* P *in* R *by* Q, *then* S *is derivable from* R, *and conversely; that is,* R *and* S *are logically equivalent.*

The intuitive soundness of the rule should be obvious. Since a precise proof of this rule involves use of the principle of mathematical induction on the length of R, details are omitted here, but the intuitive idea may be stated. If R is simply P, the proof is trivial, for the logical equivalence of R and S is the same thing as the logical equivalence of P and Q. We now suppose the theorem holds for formulas of length not greater than n, and show that it holds for any formula R of length $n + 1$. This latter argument breaks down into cases. R must be of the form: $-R_1$, R_1 & R_2, $R_1 \vee R_2$, $R_1 \rightarrow R_2$, $R_1 \leftrightarrow R_2$, $(v)R_1$, or $(\exists v)R_1$, where R_1 and R_2 are both of length less than $n + 1$. The problem is then to show that for each of the cases the theorem holds for R on the inductive hypothesis that it holds for R_1 and R_2; the proof for each case is straightforward and simple. Suppose, for instance, R is R_1 & R_2. Let S_1 be the result of replacing any number of occurrences of P in R_1 by Q, and let S_2 result in a similar way from R_2. Then by the inductive hypothesis, we have as logically equivalent, and thus as theorems of logic,

$$R_1 \leftrightarrow S_1 \tag{1}$$

$$R_2 \leftrightarrow S_2, \tag{2}$$

and (1) and (2) tautologically imply

$$R_1 \,\&\, R_2 \leftrightarrow S_1 \,\&\, S_2;$$

that is, R and S are logically equivalent as desired, since R is R_1 & R_2, and S is S_1 & S_2.

With respect to the remarks made in § 4.4 about the derived character of Rules Q1, Q2, and T.E., it is now clear that all three of them follow at once from the rule for quantifiers stated in this section and the rule of replacement for logically equivalent formulas.

We have now completed our development of the rules of logical inference. It is characteristic of the elementary treatment contained in Chapters 1–5 that we have proved nothing elaborate or sophisticated *about* logic. The aim of these chapters has been to teach you how to *use* exact and explicit rules of inferences. The remaining chapters (with the exception of the one immediately to follow) are concerned with applying the concepts and methods of logic already introduced to the broader context of mathematics. In other words, we shall be concerned with applying our logical tools to improving our understanding of mathematics and the character of theoretical science in general.

EXERCISES

1. In each of the following cases, construct a fallacious derivation to show how the adoption of the particular rule in question would lead from true premises to a false conclusion.

   (a) From $(\mathsf{v})(\exists \mathsf{w})\mathsf{S}$ we may derive $(\exists \mathsf{w})(\mathsf{v})\mathsf{S}$.
   (b) From $(\mathsf{v})(\mathsf{S} \rightarrow \mathsf{T})$ we may derive $(\exists \mathsf{v})\mathsf{S} \rightarrow (\mathsf{v})\mathsf{T}$.
   (c) From $(\exists \mathsf{v})\mathsf{S} \,\&\, (\exists \mathsf{v})\mathsf{T}$ we may derive $(\exists \mathsf{v})(\mathsf{S} \,\&\, \mathsf{T})$.
   (d) From $\mathsf{S} \,\&\, (\exists \mathsf{v})\mathsf{T}$ we may derive $(\exists \mathsf{v})(\mathsf{S} \,\&\, \mathsf{T})$. (HINT: Consider the case where v is free in S).
   (e) From $(\mathsf{v})(\mathsf{S} \vee \mathsf{T})$ we may derive $(\mathsf{v})\mathsf{S} \vee (\mathsf{v})\mathsf{T}$.
   (f) From $(\mathsf{v})(\mathsf{S} \rightarrow \mathsf{T})$ we may derive $(\exists \mathsf{v})\mathsf{S} \rightarrow \mathsf{T}$.
   (g) From $(\mathsf{v})\mathsf{S} \rightarrow (\mathsf{v})\mathsf{T}$ we may derive $(\mathsf{v})(\mathsf{S} \rightarrow \mathsf{T})$.

2. Prove the following theorems of logic.

   (a) $(x)(Fx \vee -Fx)$
   (b) $Gy \rightarrow (x)(Gy \vee Fx)$
   (c) $(x)(Fx \leftrightarrow Gx) \rightarrow ((x)Fx \leftrightarrow (x)Gx)$
   (d) $-(x)(y)(\exists z)Fxyz \leftrightarrow (\exists x)(\exists y)(z)-Fxyz$

3. Prove the various parts of the derived rule governing quantifiers.

4. In which of the following is a correct application made of the indicated part of the rule governing quantifiers? Note that usually the crucial question is deciding if restrictions on free variables have been satisfied.

   (a) By (3a) from '$Fy \vee (x)Fx$' we may derive '$(x)(Fy \vee Fx)$'.
   (b) By (3b) from '$Hxy \vee (\exists y)Fy$' we may derive '$(\exists y)(Hxy \vee Fy)$'.
   (c) By (3c) from '$(x)(\exists y)(Fy \,\&\, Hxy)$' we may derive '$(\exists y)(Fy \,\&\, (x)Hxy)$'.
   (d) By (3d) from '$(\exists x)((x)Fx \,\&\, Hxy)$' we may derive '$(x)Fx \,\&\, (\exists x)Hxy$'.
   (e) By (3e) from '$(x)(y \neq 0 \rightarrow (x \neq 0 \rightarrow x \cdot y \neq 0)$' we may derive '$y \neq 0 \rightarrow (x)(x \neq 0 \rightarrow x \cdot y \neq 0)$'.
   (f) By (3f) from '$(\exists y)(x > 1 \,\&\, y > 1 \rightarrow x \cdot y > 1)$' we may derive '$(x > 1 \,\&\, y > 1) \rightarrow (\exists y)(x \cdot y > 1)$'.
   (g) By (3g) from '$(z)(x + z = y + z \rightarrow x = y)$' we may derive '$(\exists z)(x + z = y + z) \rightarrow x = y$'.
   (h) By (3h) from '$(\exists y)(y > 0 \rightarrow x > 0)$' we may derive '$(y)(y > 0) \rightarrow x > 0$'.

5. In each of the following inferences, use is made of the rule for replacing logically equivalent formulas (L.E.). In each case, state the particular logical

equivalence being used. The formula on the right is derived from the one on the left, and conversely.

(a) $(x)(y)(\exists z)(x > y \rightarrow x > z \,\&\, z > y)$

$(x)(y)(x > y \rightarrow (\exists z)(x > z \,\&\, z > y))$

(b) $(\exists x)(y)(y \geq 0 \rightarrow y > x)$ $\quad (\exists x)(y)(y > x \vee -(y \geq 0))$

(c) $(\exists x)(y)-(y > x)$ $\quad (\exists x)-(\exists y)(y > x)$

(d) $(\exists x)-(y)(x > 0 \rightarrow y - 1 < 2)$ $\quad (\exists x)(\exists y)(x > 0 \,\&\, -(y - 1 < 2))$

6. Prove the following derived rule (SUBSTITUTION FOR FREE VARIABLES): *From* S(v) *we may derive* S(t), *provided* (i) v *is not flagged,* (ii) v *is not a subscript, and* (iii) *no free occurrence of* v *is within the scope of a quantifier using a variable of* t.

7. Construct a fallacious inference which will justify the restriction concerning ambiguous names in the rule for replacing logically equivalent formulas.

CHAPTER 6

POSTSCRIPT ON USE AND MENTION *

§ 6.1 Names and Things Named. Ordinarily in using language there is no possibility of confusing a thing and its name. We use names to talk about things; and, so it would seem, only an idiot could mix up William Shakespeare, say, and his name. However, when we want to *mention* names or expressions in general, and not merely use them, we do not have to be idiots to become confused. That is, certain special problems arise when the things named are themselves linguistic expressions. The standard method of naming expressions is to use single or double quotation marks; in previous chapters we have used single quotes.

Consider the following sentences:

(1) California is a state.
(2) California has ten letters.
(3) 'California' is a state.
(4) 'California' has ten letters.
(5) 'California' is a name of California.
(6) "California" is a name of a name of California.

On the basis of the remarks in the first paragraph it should be clear that sentences (1), (4), (5), and (6) are true, while (2) and (3) are false. In sentences (3), (4), and (5) the word 'California' is being mentioned rather than used. In such cases it is somewhat clarifying to read the phrase 'the word' before the mention of the word. Thus rephrased, (3) would read:

(3′) The word 'California' is a state.

A more subtle test of our understanding of these matters may be obtained by baptizing the word 'California' with a personal name. Let

(7) Jeremiah = 'California'.

* This chapter may be omitted without any loss of continuity.

On the basis of (7) let us see which of the following are true:

(8) Jeremiah is a name.
(9) Jeremiah has ten letters.
(10) Jeremiah has eight letters.
(11) 'Jeremiah' has eight letters.
(12) 'Jeremiah' is a name of a state.

It should be evident that (8), (9), and (11) are true, while (10) and (12) are false. If in (12) the word 'Jeremiah' had been used rather than mentioned, a true sentence would have been obtained:

(12′) Jeremiah is a name of a state.

Sentence (12) could also have been rendered true by inserting another 'name of' clause:

(12″) 'Jeremiah' is a name of a name of a state.

The view accepted by most philosophers is that not every expression of a language names something. An extreme example is the left-parenthesis. Nearly everyone would agree that this expression is the name of no object. More perplexing problems arise when predicates are considered. For example, what kind of entity, if any, does the predicate 'is a President of the United States in the nineteenth century' name? We shall not enter into this controversy here, nor are any controversial assumptions concerning such questions necessary for our subsequent developments. The distinction of serious concern in this book is that between expressions and their names. The left-parenthesis has a name even though it designates nothing itself.

EXERCISES

1. Which of the following sentences are true?

(a) Kant was a German philosopher.
(b) 'Kant' has four letters.
(c) Kant is a name of Kant.
(d) Kant is a name of 'Kant'.
(e) Newton has a longer last name than Kant.
(f) Newton is a longer name than Kant.

2. If, baptizing, we let

Mary = 'Marilyn',

which of the following sentences are true?

(a) Mary is a longer name than 'Mary'.
(b) 'Mary' has seven letters.
(c) Mary is a common girl's name.
(d) 'Mary' is a common girl's name.

(e) There is no person named Mary.
(f) Mary is a name of Marilyn.
(g) 'Mary' is a name of Marilyn.
(h) Mary is a name of 'Marilyn'.
(i) 'Mary' is a name of 'Marilyn'.

§ 6.2 Problems of Sentential Variables. In Chapter 1 we were faced with the problem of using variables for which sentences or names of sentences may be substituted. For several reasons the latter alternative was chosen: for the letters 'P', 'Q', 'R', etc. of Chapter 1 we substitute names of sentences. When one has in mind substituting sentences rather than names of sentences for the variables, it is customary to use lower-case letters 'p', 'q', 'r', etc. On the basis of the conventions stated, it is appropriate to call the letters 'P', 'Q', and 'R' *sentential variables*, and the letters 'p', 'q', and 'r' *propositional variables*. Now the word 'proposition' has a tarnished history in recent philosophy. We understand here that a proposition is an entity named by a sentence. If you do not believe or want to believe that sentences name anything, do not use propositional variables.

The remarks just made about propositional variables would be regarded as suspect in some philosophical strongholds, but the basis of them seems sound to the author and may be explained by an appeal to the previous discussions of variables and terms in general in Chapters 3 and 4. The standard viewpoint of formal logic is:

(I) *When a variable is replaced by a term which itself contains no variables, then the substituted term must name some entity.*

This thesis, i.e., (I), is so fundamental to the development of a sound, smooth-running logic of inference that it is to be abridged only for very profound reasons. (A specific application of this thesis is to be found in Chapter 8 in the discussion of what to do about the arithmetical problem of dividing by zero.) Given the true sentence of arithmetic:

$$(x)(y)(x + y = y + x),$$

if we infer by universal specification:

$$3 + 5 = 5 + 3,$$

we are then committed by (I) to the view that '$3 + 5$' and '$5 + 3$' are terms designating a certain number. The letters 'x' and 'y' are called *numerical variables* because they may be replaced by terms designating numbers. Similarly, from the truth of logic that for any propositions p and q

$$p \rightarrow (p \vee q),$$

we may infer:

$$1 = 1 \rightarrow (1 = 1 \vee 2 = 3).$$

Since the variable '*p*' is replaced by the sentence '1 = 1', according to (I), this sentence must be the name of some entity. Thus the need for propositions when variables are used for which we substitute sentences and not names of sentences.*

Skepticism about the existence of such entities as propositions aside, variables which are replaced by names of sentences rather than sentences have certain technical advantages. The simple assertion:

(1) Every sentence P is either true or false

is not easy to state using propositional rather than sentential variables. We might begin by trying:

(2) Every proposition *p* is either true or false.

But if we apply universal specification to (2), replacing '*p*' by 'Geronimo is dead', we obtain a grammatically meaningless expression—namely, a sentence which appears to have two main verbs:

(3) Geronimo is dead is either true or false.

As a second try we might rephrase (2):

(4) For every proposition *p*, '*p*' is either true or false.

If we did not balk at substituting for '*p*' when it is inside quotes, we could obtain from (4) the sensible sentence:

(5) 'Geronimo is dead' is either true or false.

However, there are good reasons for balking at this substitution. Consider, for instance, the sentence:

(6) For every *p*, the letter '*p*' is the sixteenth letter of the alphabet.

This sentence would, I believe, ordinarily be said to be true, and the quantifier 'For every *p*' would be regarded as redundant and not binding the occurrence of '*p*' inside the quotes. If we regard (6) as true and consider the quantifier as binding the occurrence of '*p*' in quotes, we obtain by the substitution used in obtaining (5) from (4) the false assertion:

(7) 'Geronimo is dead' is the sixteenth letter of the alphabet.

The rule blocking (7) may be summarized as follows:

(II) *Quantifiers standing outside of quotes cannot bind variables occurring inside quotes.*

Rule (II), like (I), is to be abandoned only for profound reasons.

* It is perhaps worth remarking that in Quine's *Methods of Logic* the letters '*p*', '*q*', '*r*' are not used but mentioned. Hence they are not used as variables by Quine and what has been said above does not entail that he has tacitly committed himself to propositions.

Without claiming to have exhausted all alternatives in translating (1) into a sentence using propositional variables, I believe we may rightly claim that the alternatives considered do argue for sentential as opposed to propositional variables.

EXERCISES

1. In each of the following sentences, decide if the letters 'A', 'B', and 'C' should be propositional or sentential variables.

(a) For every A and B, A implies the disjunction of A and B.
(b) For every A, B, and C, if A only if B and B only if C then A only if C.
(c) For every A and B the implication whose antecedent is the conjunction of A and B and whose consequent is the disjunction of A and B is a tautology.

2. Point out the confusion between propositional and sentential variables in the following sentence:

If p implies q and q implies p, then p if and only if q.

§ 6.3 Juxtaposition of Names. In the preceding section we defended the choice of sentential variables. But we did not mention a certain difficulty which arises in their use. Consider the sentence:

(1) For all sentences P and Q, P → (P ∨ Q) is true.

Applying universal specification to (1) by replacing the sentential variables by names of some sentences, we obtain:

(2) 'Geronimo is dead' → ('Geronimo is dead' ∨ 'Crazy Horse was two-faced') is true.

The problem is that the subject of (2) is not a sentence; in fact, as it stands (2) is not meaningful since it has no proper grammatical subject. We would expect to find the name of a sentence preceding the phrase 'is true', but instead we find 'is true' preceded by a conglomeration of names of sentences, signs for sentential connectives, and parentheses.

To remedy this situation, we need to introduce two conventions, which are fundamental to the usage in this book.

CONVENTION (I). *The following logical signs are used as names of themselves:* −, &, ∨, →, ↔, (,), ∃, =.

Convention (I) countenances such unusual appearing identities as:

& = '&',

→ = '→'.

Using (I), we may replace (2) by (3):

(3) 'Geronimo is dead' '→' '(' 'Geronimo is dead' '∨' 'Crazy Horse was two-faced' ')' is true.

The second convention will permit the conversion of (3) into a proper sentence.

CONVENTION (II). PRINCIPLE OF JUXTAPOSITION. *If* N_1 *is a name of the expression* E_1 *and* N_2 *is a name of the expression* E_2, *then the expression consisting of* N_1 *followed immediately by* N_2 *is a name of the expression consisting of* E_1 *followed by* E_2.

The following identities exemplify the principle:

(4)
$$\begin{cases} \text{`}Fx\text{' `}\rightarrow\text{'} = \text{`}Fx \rightarrow\text{'}, \\ \text{`}-\text{' `(' `Geronimo is dead' `)'} = \text{`}-\text{(' `Geronimo is dead)'} \\ \qquad = \text{`}-\text{(Geronimo is dead)'}. \end{cases}$$

Or using (I), we could write in place of (4):

$$\text{`}Fx\text{'} \rightarrow\ = \text{`}Fx \rightarrow\text{'}$$

$$-\text{(`Geronimo is dead')} = \text{`}-\text{(Geronimo is dead)'}.$$

Using (II), we may rectify (3) to obtain the true sentence:

(5) 'Geronimo is dead → (Geronimo is dead ∨ Crazy Horse was two-faced)' is true.

The extension of (II) to the juxtaposition of more than two names is obvious and has been used in the above examples.*

Conventions (I) and (II) have been tacitly used throughout the previous chapters.† Their use has not been restricted to sentential variables; it is also required to make exact, literal sense of the statements concerning inferences with quantifiers in Chapters 4 and 5. For example, consider the statement of the rule of universal generalization:

> From **S** we may derive (**v**)**S**, provided **v** is not flagged in **S** and does not occur as a subscript in **S**.

Here '**S**' is a variable which we replace by the name of a formula, i.e., '**S**' is a sentential variable—with the notion of sentence extended to the more general one of formula. The letter '**v**', on the other hand, is a variable

* In mathematical language juxtaposition of names of expressions is used to denote a binary operation on expressions, just as juxtaposition of numerical variables is used to denote the binary operation of multiplication. Sometimes it is convenient to have a symbol for multiplication, and similarly we could if desired introduce Tarski's symbol '⁀' of concatenation to denote the appropriate operation on pairs of expressions. We would then write, for instance,

$$\text{`}Fx\text{'}{}^\frown\text{`}\rightarrow\text{'} = \text{`}Fx \rightarrow\text{'}$$

† We have also used the convention of printing in boldface all variables which are replaced by names of expressions.

which we replace by names of individual variables such as 'x' and 'y'. Thus, as an instance of *UG* we have:

(6) From '$(x + x > x)$' we may derive ('x')'$(x + x > x)$'.

Using (I) and (II) we rectify (6) into the sound statement:

(7) From '$x + x > x$' we may derive '$(x)(x + x > x)$'.

Certain minor inconsistencies or extensions of usage, which are not covered by (I) and (II), have been permitted in the preceding chapters. Explicit attention to some of the violations is directed by the exercises given below. In this connection it perhaps needs to be said that in Chapters 4 and 5 we used neither variables which are replaced by predicates nor variables which are replaced by names of predicates. When such variables are wanted, a choice similar to that between sentential and propositional variables needs to be made. It is customary to say that a variable which is replaced by a predicate takes as its values *propositional functions*, that is, predicates denote propositional functions; a variable which is replaced by a name of a predicate takes as its values, *sentential functions*, but sentential functions are of course nothing but predicates. Many philosophers find it even harder to believe that there are entities like propositional functions than that there are propositions. No commitment to either propositions or propositional functions has been made in this book. The letters 'F', 'G', 'H', and the like used in Chapters 4 and 5 are not variables but predicate constants. If we want to be explicit about their introduction, we may provide equivalences for each one:

$$Hx \leftrightarrow x \text{ is human,}$$

and so on.

EXERCISES

1. Exactly what extension of Convention (I) is required to cover the use of sentential variables in displayed formulas in § 1.3 giving idiomatic equivalents of 'if ... then ...': P only if Q, etc.

2. What is the difference concerning use and mention of formulas between the way derivations are written in Chapter 2 and in Chapter 4?

3. Since juxtaposition of names denotes a binary operation on expressions, we may ask:

   (a) Is the operation denoted by juxtaposition of names associative?
   (b) Does this operation satisfy the right-hand and left-hand cancellation laws? (For explanation of this terminology see the discussion and exercises on group theory in the preceding chapter.)
   (c) Does this operation satisfy the second or third axioms for groups introduced in Chapter 5?

CHAPTER 7

TRANSITION FROM FORMAL TO INFORMAL PROOFS

§ 7.1 General Considerations. It is not customary in mathematics and the empirical sciences to present derivations of conclusions from premises in a manner as formal as that developed in Chapters 4 and 5. On the other hand, beginning students of the sciences are often puzzled by the criteria which govern the acceptance and rejection of proposed mathematical proofs when the proofs are set forth in an informal style. For example, if in an informal proof a student makes a mistake in substituting variables it is often hard to convince him that he is wrong—unless a well-defined body of formal rules is at hand to clarify the mistake.

The purpose of this chapter is to help you make the transition from formal derivations to informal proofs in the usual mathematical sense. You will no doubt experience some uneasiness in passing from the neat, precise domain of formal logic, where there is a rule for everything, to the more complicated, less precise world of ordinary mathematics and the empirical sciences.

In an informal proof enough of the argument is stated to permit anyone conversant with the subject to follow the line of thought with a relatively high degree of clarity and ease. It is presumably intuitively transparent how to fill in the logical lacunae in the proof. In many respects the standards of intelligibility for informal proofs are similar to those for informal conversation. Thus if someone asks if you are driving home today, suppose you answer, "No, because my car is in the garage being repaired." You do not then proceed to state all the other obvious premises which, together with the assertion that your car is in the garage, will logically imply the conclusion that you are not driving home today. Analogously, in giving an informal proof, we try to cover the essential, unfamiliar, unobvious steps and omit the trivial and routine inferences. However, it is a commonplace of exact philosophy that the concepts of being essential, being unfamiliar, or being trivial are not precise and are not easily made precise. The very vagueness of the criteria governing informal proofs is a primary justification for a precise definition of a formal proof. In cases of controversy or doubt concerning the validity of an informal proof it is exceedingly useful to have available a clear, exact standard to which appeal may be made.

Our approach to this problem of transition shall be to introduce a number of axioms which express basic facts about numbers, and then to derive both formally and informally some intuitively familiar conclusions. Logical inferences connected with the subject matter of arithmetic have been selected for three reasons to illustrate methods of informal proof. In the first place, the material is intuitively familiar to everyone from childhood. In the second place, examples of inferences in this domain have a genuine mathematical content in the sense that the inferences are not intuitively trivial. Many of the exercises in Chapter 4 do not have mathematical content in this sense, for it is often intuitively obvious that the conclusions logically follow from the premises given in the exercises. In comparison, it is by no means obvious that the full arithmetic of addition, multiplication, and subtraction can be derived from the fifteen basic axioms introduced in the next section. In the third place, only a few more axioms need to be added to the ones to be introduced to provide a systematic basis for the logical development of the differential and integral calculus, the branch of mathematics which has to date been the most important in the empirical sciences.

§ 7.2 Basic Number Axioms. We hope you will recognize immediately each of the following axioms as a familiar truth of arithmetic. In addition to the general apparatus of logic, the axioms use only the individual constants '0' and '1', the familiar operation symbols '+' and '·' denoting the operations of addition and multiplication respectively, and the relation symbol '<' denoting the relation *less than.**

(1) $(x)(y)(x + y = y + x)$

(2) $(x)(y)(x \cdot y = y \cdot x)$

(3) $(x)(y)(z)[(x + y) + z = x + (y + z)]$

(4) $(x)(y)(z)[(x \cdot y) \cdot z = x \cdot (y \cdot z)]$

(5) $(x)(y)(z)[x \cdot (y + z) = (x \cdot y) + (x \cdot z)]$

(6) $(x)(x + 0 = x)$

(7) $(x)(x \cdot 1 = x)$

(8) $(x)(\exists y)(x + y = 0)$

(9) $(x)(y)[y \neq 0 \rightarrow (\exists z)(x = y \cdot z)]$

(10) $(x)(y)[x < y \rightarrow -(y < x)]$

(11) $(x)(y)(z)[(x < y \;\&\; y < z) \rightarrow x < z]$

(12) $(x)(y)[x \neq y \rightarrow (x < y \vee y < x)]$

(13) $(x)(y)(z)[y < z \rightarrow (x + y < x + z)]$

(14) $(x)(y)(z)[(0 < x \;\&\; y < z) \rightarrow x \cdot y < x \cdot z]$

(15) $0 \neq 1$

* The axioms are essentially those given by Tarski in his *Introduction to Logic.*

Since many of these axioms have familiar verbal descriptions, it is desirable to examine them individually. Axioms 1 and 2 assert that addition and multiplication are both *commutative* operations, that is, if we add or multiply two numbers, it does not matter in what order we consider them. For example, $2 + 3 = 3 + 2$, and $4 \cdot 3 = 3 \cdot 4$. Axioms 3 and 4 assert that addition and multiplication are both *associative* operations, that is, the result of adding or multiplying three numbers is independent of the way in which they are grouped. Axiom 5 says that multiplication is *distributive* with respect to addition; it is this distributive property which permits us to "multiply out." Axiom 6 says that the addition of zero to any number simply yields that number again. More technically, the content of this axiom is that zero is a *right-hand identity element for addition.* (In general, an element e is a right-hand identity element for an operation $\circ$ if for every x

$$x \circ e = x.$$

A left-hand identity element is defined similarly. An element which is both a right-hand and a left-hand identity element is called simply an *identity element.*) Axiom 7 says that 1 is a right-hand identity element for multiplication. We shall shortly prove that both 0 and 1 are in fact unique identity elements. Axiom 8 says that for every number x there exists a number y such that $x + y = 0$. We shall later show (what we intuitively know) that it follows from our axioms that y is unique. The element y is the *inverse element* of x (with respect to addition). Axiom 9 says that if x is a number and y is a number not equal to 0 then we can find a number z such that $y \cdot z = x$. As in the case of Axiom 8, it is not difficult to show that this z is unique. The intuitive content of this axiom is that division except by zero is always possible. Axioms 10–12 assert familiar ordering properties of the relation $<$. Axiom 13 asserts that adding the same number to two given numbers preserves the inequality between the given numbers, and Axiom 14 asserts that multiplying two members of an inequality by a positive number leaves the inequality unchanged. The final axiom, Axiom 15, provides that '0' and '1' designate distinct elements.

The axioms we have been discussing are satisfied by the set of all rational numbers (a rational number is a number equal to the ratio of two integers), and also by the set of all real numbers. For the present, we may characterize the set of real numbers as the set of all unlimited decimals. For example, $\sqrt{2}$ is a real number but not a rational number, since it is not equal to the ratio of two integers; on the other hand, every rational number is a real number. The rational number $4/3$, for example, is identical with the unlimited decimal 1.3333 .... In other words, we may pick as our domain of individuals either the set of rational numbers or the more inclusive set of real numbers. As we shall see, Axioms 1–15 are adequate to develop the arithmetic of the fundamental operations of addition, multiplication, and subtraction.

EXERCISES

1. Which of the Axioms 1–15 are not satisfied by

   (a) the positive integers?
   (b) the integers?
   (c) the non-negative rational numbers?

2. An operation $\circ$ is *commutative* if for every x and y

$$x \circ y = y \circ x.$$

Axioms 1 and 2 say that addition and multiplication are commutative operations.

   (a) Is subtraction commutative?
   (b) Is division commutative?

3. An operation $\circ$ is *associative* if for every x, y, and z,

$$(x \circ y) \circ z = x \circ (y \circ z).$$

Axioms 3 and 4 say that addition and multiplication are associative operations.

   (a) Is subtraction associative? If not, give an explicit counterexample.
   (b) Is division associative? If not, give an explicit counterexample.

4. An operation $\star$ is *distributive* with respect to an operation $\circ$ if for every x, y, and z

$$x \star (y \circ z) = (x \star y) \circ (x \star z).$$

Axiom 5 says that multiplication is distributive with respect to addition. When the answer to any of the following questions is negative, give an explicit counterexample.

   (a) Is addition distributive with respect to multiplication?
   (b) Is subtraction distributive with respect to multiplication?
   (c) Is multiplication distributive with respect to subtraction?
   (d) Is subtraction distributive with respect to division?

5. Is there a left-hand or right-hand identity element for the operation of subtraction?

6. Give an interpretation which proves that Axiom 15 is independent of the other fourteen. (Warning: do not violate Rule VII of §4.2.)

7. Give an interpretation which proves that Axiom 8 is independent of the other fourteen. (Warning: do not violate Rule VII of §4.2.)

8. Give an interpretation which proves that Axiom 9 is independent of the other fourteen. (Warning: do not violate Rule VII of §4.2.)

§ 7.3 Comparative Examples of Formal Derivations and Informal Proofs. We turn now to the main business of this chapter: the introduction of informal proofs. We shall use the fifteen axioms introduced in the previous section as our basic premises. The formal derivations will thus begin with line (16), following the rules laid down in previous chapters. Our only condensation will be to make multiple applications of the rule governing identities. In Chapter 5, any formula derivable from the empty set of premises was called a theorem of logic. Here we term any formula derivable from our basic axioms a *theorem of arithmetic.*

It is important to realize that there is an enormous difference between the relatively trivial task of recognizing the truth of the theorems we shall state and the much more difficult enterprise of showing that they are in fact *logical consequences* of the fifteen axioms stated in the previous section. You should not be thrown intellectually off balance by the seemingly trivial content of some of the theorems. You may in such cases be inclined to say to yourself, "This is so obvious it is silly to ask me to prove it." But if you say this you are confusing the obviousness of the truth of the theorem with the obviousness of the proof that the theorem is a logical consequence of the axioms.

Our first theorem states that zero is the unique right-hand identity element for addition. In the formal derivation references to *US* are omitted, since the actual substitutions made are indicated, and the substitution notation itself signifies an application of *US*.

THEOREM 1. $(x)[(y)(y + x = y) \rightarrow x = 0]$

FORMAL DERIVATION

{16}	(16) $(y)(y + x = y)$	x P
{16}	(17) $0 + x = 0$	x 16 $0/y$
{6}	(18) $x + 0 = x$	6 x/x
{1}	(19) $x + 0 = 0 + x$	1 x/x, $0/y$
{1, 6, 16}	(20) $x = 0$	x 17, 18, 19 I
{1, 6}	(21) $(y)(y + x = y) \rightarrow x = 0$	16, 20 C.P.
{1, 6}	(22) $(x)[(y)(y + x = y) \rightarrow x = 0]$	21 *UG*

INFORMAL PROOF. By hypothesis of the theorem, for every y, $y + x = y$. Hence, (putting '0' for 'y') we obtain:

$$0 + x = 0.$$

But Axiom 6 asserts that $x + 0 = x$. Since by virtue of Axiom 1 addition is commutative, that is, $x + 0 = 0 + x$, we infer immediately that $x = 0$. Q.E.D.

As is to be expected from what has been said before, the informal proof mentions explicitly none of the uses of logical rules of inference; only specific axioms are mentioned by name. In other words, the formal rules of derivation are taken for granted. The parenthetical expression in the second sentence of the informal proof is a concession to explicitness not ordinarily made. The particular substitutions made in axioms or previous theorems are not described unless they are complicated or unobvious. Except when an informal proof is unusually long, there is no "summing up" in the routine manner of lines (21) and (22) of the formal derivation. The obvious task set by Theorem 1 is to show that '$x = 0$' follows from the hypothesis of the theorem. When this is done, the informal proof is satisfactorily completed. The 'Q.E.D.' at the end, standing for *quod erat demonstrandum*, is a traditional way of signifying that the proof is finished.

The working rule followed in the informal proofs given in this chapter and recommended for the exercises is:

> *In an informal proof explicitly indicate EVERY use of an axiom or previous theorem of arithmetic. Ordinarily suppress mention of the logical rules of inference used.*

Since the proof of the second theorem, stating that 1 is the unique right-hand identity element for multiplication, is very similar to that of Theorem 1, we leave it as an exercise.

THEOREM 2. $(x)[(y)(y \cdot x = y) \rightarrow x = 1]$.

The next theorem asserts the cancellation law for addition. Its proof should be compared with the corresponding proof for arbitrary groups in Chapter 5. The basic difference is that we have not defined the notion of an inverse element (with respect to addition) and thus cannot follow completely the strategy of the earlier proof. In the proof of the cancellation law for addition an extension of the rule governing identities is conveniently used. This extension we abbreviate: I2.

ADDITIONAL RULE FOR IDENTITIES: I2. *If* $\mathfrak{t}_1$, $\mathfrak{t}_2$ *and* $\mathfrak{t}_3$ *are any terms and* $\circ$ *is any operation symbol then from* $\mathfrak{t}_2 = \mathfrak{t}_3$ *we may derive* $\mathfrak{t}_1 \circ \mathfrak{t}_2 = \mathfrak{t}_1 \circ \mathfrak{t}_3$ *and* $\mathfrak{t}_2 \circ \mathfrak{t}_1 = \mathfrak{t}_3 \circ \mathfrak{t}_1$.

PROOF: We have as a truth of logic

$$(1) \qquad \mathfrak{t}_1 \circ \mathfrak{t}_2 = \mathfrak{t}_1 \circ \mathfrak{t}_2.$$

Assuming now that the formula $\mathfrak{t}_2 = \mathfrak{t}_3$ holds, we apply the original rule for identities and substitute $\mathfrak{t}_3$ for the second occurrence of $\mathfrak{t}_2$ in (1) and obtain

$$\mathfrak{t}_1 \circ \mathfrak{t}_2 = \mathfrak{t}_1 \circ \mathfrak{t}_3.$$

By a similar argument we easily obtain $\mathfrak{t}_2 \circ \mathfrak{t}_1 = \mathfrak{t}_3 \circ \mathfrak{t}_1$.

As a special case of this rule we obtain for addition and multiplication as theorems of logic:

(1) $$(x)(y)(z)(y = z \rightarrow x + y = x + z),$$

(2) $$(x)(y)(z)(y = z \rightarrow x \cdot y = x \cdot z).$$

It is not uncommon to see sentences like (1) and (2) given as axioms of some special domain of mathematics like arithmetic, but this practice rests on a confusion, for these two sentences are truths of logic and it is redundant to add them as new subject matter axioms. The third theorem is for obvious reasons called the *cancellation law* for addition.

THEOREM 3. $(x)(y)(z)(x + y = x + z \rightarrow y = z)$.

FORMAL DERIVATION

{8}	(16) $(\exists y)(x + y = 0)$	8 x/x
{8}	(17) $x + \alpha_x = 0$	16 *ES*
{1}	(18) $x + \alpha_x = \alpha_x + x$	1 x/x, α_x/y
{1, 8}	(19) $\alpha_x + x = 0$	17, 18 I
{1}	(20) $0 + y = y + 0$	1 $0/x$, y/y
{6}	(21) $y + 0 = y$	6 y/x
{1, 6}	(22) $0 + y = y$	20, 21 I
{1, 6, 8}	(23) $(\alpha_x + x) + y = y$	19, 22 I
{3}	(24) $(\alpha_x + x) + y = \alpha_x + (x + y)$	3 α_x/x, x/y, y/z
{6}	(25) $z + 0 = z$	6 z/x
{1}	(26) $0 + z = z + 0$	1 $0/x$, z/y
{1, 6, 8}	(27) $(\alpha_x + x) + z = z$	19, 25, 26 I
{3}	(28) $(\alpha_x + x) + z = \alpha_x + (x + z)$	3 α_x/x, x/y, z/z
{29}	(29) $x + y = x + z$	x, y, z P
{29}	(30) $\alpha_x + (x + y) = \alpha_x + (x + z)$	x, y, z 29 I2
{1, 3, 6, 8, 29}	(31) $y = z$	y, z, 23, 24, 27, 28, 30 I
{1, 3, 6, 8}	(32) $x + y = x + z \rightarrow y = z$	29, 31 C.P.
{1, 3, 6, 8}	(33) $(x)(y)(z)(x + y = x + z \rightarrow y = z)$	32 *UG*

INFORMAL PROOF. By Axiom 8 there is a number, say u, such that $x + u = 0$, and hence by the commutative law for addition:

(1) $$u + x = 0.$$

Using Axiom 6, and the commutative and associative laws for addition, we obtain the following series of identities:

$$
\begin{aligned}
y &= y + 0 && \text{(Axiom 6)}\\
&= 0 + y && \text{(Commutative law)}\\
&= (u + x) + y && \text{(By (1))}\\
&= u + (x + y) && \text{(Associative law)}\\
&= u + (x + z) && \text{(Hypothesis of theorem)}\\
&= (u + x) + z && \text{(Associative law)}\\
&= 0 + z && \text{(By (1))}\\
&= z + 0 && \text{(Commutative law)}\\
&= z && \text{(Axiom 6)} \quad \text{Q.E.D.}
\end{aligned}
$$

The notation at the right in the informal proof should be self-explanatory: for instance, the first identity follows from Axiom 6, and the second identity: $y = 0 + y$ follows from the first by the commutative law for addition (and the logical rule governing identities). In several respects the structure of the informal proof is clearer than the formal derivation; in particular, the main line of the argument is presented without encumbering nuisances such as lines (18), (22), (25)–(28) of the formal derivation—not to mention the redundant summing up in (32) and (33). The notation used at the right exemplifies a general tendency of informal proofs: standard properties, such as commutativity of addition are referred to by name rather than by reference to the appropriate axiom or theorem. Notice that in the informal proof 'u' plays the role of an ambiguous name; it corresponds to 'α_x' in the formal derivation. The phrase 'say u' at the beginning of the informal proof is often used by mathematicians to indicate that a letter which is ordinarily used as a variable is being used for immediate purposes as an ambiguous name.

It is not intended to give the impression that there is exactly one correct style to be employed in writing informal proofs in the sense that an Elizabethan sonnet must be fourteen lines of iambic pentameter grouped into three quatrains and a couplet. Some mathematicians would criticize the informal proof of Theorem 3 for being insufficiently literary because each indicated step of the proof is not expressed in a complete, well-formed sentence, but rather a barbarous notation of incomplete phrases in parentheses is used. Still more mathematicians would criticize the undue length of both the informal proofs so far considered, but this latter criticism may perhaps be rejected on the ground that when we first begin to do informal proofs it is better to be clear and somewhat too prolix than brief but confused.

The fourth theorem states the familiar fact that the multiplication of any number by zero yields zero as the result.

THEOREM 4. $(x)(x \cdot 0 = 0)$.

FORMAL DERIVATION

{6}	(16) $x + 0 = x$	6 x/x
{6}	(17) $(x \cdot x) + 0 = x \cdot x$	6 $(x \cdot x)/x$
{6}	(18) $(x \cdot x) + 0 = x \cdot (x + 0)$	16, 17 I
{5}	(19) $x \cdot (x + 0) = (x \cdot x) + (x \cdot 0)$	5 x/x, x/y, $0/z$
{5, 6}	(20) $(x \cdot x) + (x \cdot 0) = (x \cdot x) + 0$	18, 19 I
{1, 3, 6, 8}	(21) $(x \cdot x) + (x \cdot 0) = (x \cdot x) + 0 \rightarrow (x \cdot 0) = 0$	Th. 3 $(x \cdot x)/x$, $(x \cdot 0)/y$, $0/z$
{1, 3, 5, 6, 8}	(22) $x \cdot 0 = 0$	20, 21 T
{1, 3, 5, 6, 8}	(23) $(x)(x \cdot 0 = 0)$	22 *UG*

INFORMAL PROOF 1. We have the following identities:

$$\begin{aligned} (x \cdot x) + (x \cdot 0) &= x \cdot (x + 0) && \text{(Distributive law)} \\ &= x \cdot x && \text{(Axiom 6)} \\ &= (x \cdot x) + 0 && \text{(Axiom 6 again)} \end{aligned}$$

Hence, putting '$x \cdot x$' for 'x', '$x \cdot 0$' for 'y' and '0' for 'z' in Theorem 3, we obtain at once the desired result: $x \cdot 0 = 0$. Q.E.D.

Since many mathematicians would consider the listing of the substitutions made in Theorem 3 as rather inelegant, we may rewrite this informal proof and eliminate this usage. Notice that Theorem 3 is now referred to by the name of the property it expresses.

INFORMAL PROOF 2. Since by the distribution law and Axiom 6,

$$(x \cdot x) + (x \cdot 0) = x \cdot (x + 0) = x \cdot x = (x \cdot x) + 0,$$

we may use the cancellation law for addition to obtain: $x \cdot 0 = 0$. Q.E.D.

The remarks about informal proofs in this section have been almost entirely stylistic in character. Without any doubt stylistic problems are serious ones for those just beginning to acquire a firm notion of mathematical proof. But perhaps still more important are those problems arising from efforts to develop an efficient *strategy* for finding proofs. There are, of course, no sure-fire methods for developing a good strategy, but we shall try to give some useful hints along the way. In the proof of Theorem 4, for example, the natural thing is to begin by scanning the axioms to pick out those involving zero. Axiom 6 appears the most promising. In view of the methods used to prove the preceding theorem, the next step is to consider whether any application of the commutative, associative, and distributive laws can be made. By the time the distributive law is singled out, some sort of a proof is beginning to shape up. These remarks are after the fact, so to speak; it is an important element of good strategy to be willing to try a number of different approaches. If the approach just described had not worked, then something else should have been tried, involving perhaps some of the axioms not yet used.

Certain general structures recur rather often in proofs, and it is desirable to be explicitly aware of them. One such structure is exemplified in the proofs of Theorems 3 and 4: the proofs primarily consist of a string of identities. A second kind of structure is exemplified by the proof of Theorem 6 of § 7.5: the proof consists mainly of a series of implications. A third sort of structure is provided by the familiar method of indirect proof, which we first use in this chapter in the proof of Theorem 14 of § 7.5.

Theorem 4 is the last theorem for which we give a formal derivation. Consequently in the subsequent statement of theorems (or axioms) we shall usually omit the initial universal quantifiers whose scopes extend over the remainder of the formula—a practice customary in mathematics. In formal derivations we have no rule permitting us to substitute for free variables such as occur in theorems when universal quantifiers are omitted, but we need to apply these theorems to situations involving other variables than those which occur in the initial formulations of the theorems. As might be expected the practice in informal proofs is to substitute for free variables in axioms and theorems whenever necessary. On the other hand, if in the proof of a theorem based on the axioms, we introduce the antecedent of the theorem as a premise (see, for example, Theorem 3) we treat the free variables in this premise as flagged and do not substitute for them.* The intuitive reason for this is obvious. Suppose the antecedent involves free variables 'x', 'y', and 'z'. Clearly it will not do merely to prove the theorem for the special case of $x = y = z$; that is, we would not establish the theorem in its full generality if we substituted 'x' for 'y' and 'z'. The next section is devoted to consideration of such fallacious methods. Further clarification of the use of free variables is also provided there.

EXERCISES

1. Prove both formally and informally Theorem 2.
2. Give formal derivations and corresponding informal proofs for the following:

   (a) $(x)-(x < x)$
   (b) $(x)(y)(x = y \rightarrow -(x < y \vee y < x))$
   (c) $(x)(y)(x < y \rightarrow -(x = y \vee y < x))$
   (d) $(x)(y)(z)(x + y < x + z \rightarrow y < z)$

* For an exact statement of a derived rule governing substitution for free variables see Exercise 6 of § 5.3.

3. Given the additional premise:

(16) $$(x)(y)(x \leq y \leftrightarrow (x = y \vee x < y),$$

prove both formally and informally that

(a) $(x)(x \leq x)$
(b) $(x)(y)[(x \leq y \,\&\, y \leq x) \rightarrow x = y]$
(c) $(x)(y)(z)[(x \leq y \,\&\, y \leq z) \rightarrow x \leq z]$
(d) $(x)(y)(x \leq y \vee y \leq x)$

4. Given the additional premises:

(16) $$2 = 1 + 1$$
(17) $$3 = 2 + 1$$
(18) $$4 = 3 + 1,$$

prove both formally and informally that $2 \cdot 2 = 4$.

§ 7.4 Examples of Fallacious Informal Proofs. In Chapters 4 and 5 a number of fallacious formal derivations were presented to justify various restrictions on the formal rules of inference and to exemplify some of the more common errors. The purpose of this section is to point out some of the more frequent mistakes committed in informal proofs. As is to be expected, mistakes in informal proofs have their analogues in formal derivations. Unfortunately, however, you may find yourself making mistakes in informal proofs that you would not make in formal derivations. Some of the reasons for such a discrepancy are not hard to find. Once explicit and exact adherence to the formal rules of inference is given up, there is a natural tendency to think that now anything is permissible. There is also a feeling of bewilderment concerning exactly what is and what is not now considered an appropriate statement of a piece of reasoning. As some psychologists would put it, in order to make the transition from formal derivations to informal proofs we must develop a tolerance for ambiguity.

There are positive ways to think about this ambiguity. The main rule of thumb is to regard an informal proof as an abbreviation of a formal derivation. There should be no single step in an informal proof which cannot be expanded into a reasonable number of formal steps. It cannot be emphasized too often that whenever you have any doubt about a particular step in an informal proof, the thing to do is to try to expand the step in question into a fragment of a formal derivation. If you cannot make this expansion, if you cannot become fully clear in a formal manner about the validity of the step, then there is good reason to doubt its logical correctness.

Probably the most frequent type of error made in informal proofs is illustrated by the following proof.

THEOREM. *For every x, y, and z, if $x + y < x + z$ then $y < z$.*

FALLACIOUS PROOF. By the hypothesis of the theorem,

(1) $$x + y < x + z.$$

Putting '0' for 'x' in (1) we obtain:

(2) $$0 + y < 0 + z.$$

From Axioms 1 and 6, we have:

(3) $$0 + y = y + 0 = y$$

and

(4) $$0 + z = z + 0 = z.$$

Hence using the identities (3) and (4), we infer from (2) that

$$y < z.$$ Q.E.D.

The mistake is made in inferring (2) from (1). No substitutions may be made for the variables 'x', 'y', and 'z' in (1). Why? Well, from a logical standpoint these variables are flagged, so we cannot universally generalize (1) and then apply universal specification to make the appropriate substitution. Thus if we turn our logical microscope on the proposed inference of (2) from (1), we obtain the following fallacious derivation:

{1}	(1) $x + y < x + z$	x, y, z P
{1}	(2) $(x)(x + y < x + z)$	1 *UG* (fallaciously)
{1}	(3) $0 + y < 0 + z$	2 *US*

Since the error committed in this proof is so common, some further reflections on it will not be amiss. Ordinarily when we want to substitute for free variables in an informal proof, we do not think of universally generalizing and then universally specifying. As was remarked in the last section, a derived rule permitting direct substitution may be used. This rule is given as Exercise 6 in § 5.3, and it reads as follows:

> *From* S(v) *we may derive* S(t) *provided* (i) v *is not flagged,* (ii) v *is not a subscript, and* (iii) *no free occurrence of* v *is within the scope of a quantifier using a variable of* t.

Since the rule simply collapses an application of *UG* followed by an application of *US* into one step, the restrictions on it are just the restrictions on *UG* and *US*. For use in informal proofs, we may restate the rule to take account of almost every case of such substitution that is needed.

> INFORMAL RULE FOR SUBSTITUTING FOR FREE VARIABLES. (i) *You may substitute terms for free variables occurring in axioms or previously proved theorems.* (ii) *Any case not covered by* (i) *should be referred to the formal rule; in particular you cannot substitute for free variables in the hypothesis of the theorem you are trying to prove.*

As remarked in the last section we shall henceforth omit universal quantifiers standing at the beginning of theorems. The above informal rule is explicitly framed to permit substitution in these theorems. For instance, following the convention now in force regarding universal quantifiers, we would state Theorem 3 as:

THEOREM 3. *If* $x + y = x + z$ *then* $y = z$.

We would then be permitted to substitute for the variables 'x', 'y' and 'z' in applications of Theorem 3, as sanctioned by the above informal rule.

In the above example of a fallacious proof we considered a genuine theorem for which a correct proof exists (see Exercise 2d of the previous section). The next example is a fallacious theorem for which no correct proof exists.

FALLACIOUS THEOREM. $x = 1$.

(Thus this fallacious theorem asserts that every number is equal to one. The variable 'x' has been left free in accordance with the convention just stated.)

FALLACIOUS PROOF. Let x be an arbitrary number. By Axiom 2 and Theorem 4, we have:

$$(1) \qquad 0 \cdot x = x \cdot 0 = 0.$$

Hence letting $y = 0$ in Theorem 2 we infer from (1) that

$$x = 1. \qquad \text{Q.E.D.}$$

In logical terms the mistake committed in this proof is in applying universal specification to the quantifier using 'y' in Theorem 2, for the scope of this quantifier is not the whole formula. This kind of error was strenuously warned against in Chapter 4, but it is of the sort you might make in an informal proof without making it in a formal derivation. In order to make use of Theorem 2, one would have to establish that for all y, not just 0, x has the property that

$$y \cdot x = y.$$

The use of the phrase 'letting $y = 0$ in Theorem 2 we infer ...' should not throw you off guard in analyzing the fallacious proof. This phrase is mathematical lingo for: applying *US* to Theorem 2 by substituting '0' for 'y' we infer .... Mathematicians often use the identity sign to indicate an application of *US*. You must learn to distinguish such cases from those which represent a genuine application of the rule governing identities. The context of application usually makes such differentiation rather simple.

Further examples of fallacious informal proofs are given in the exercises below.

EXERCISES

In each of the following fallacious proofs, find the error and state what logical blunder has been committed. Fallacious proofs of both true and false statements are given.

1. THEOREM 3. *If* $x + y = x + z$ *then* $y = z$.

FALLACIOUS PROOF. By the hypothesis of the theorem,

$$x + y = x + z. \tag{1}$$

Letting $x = 0$ in (1) we have:

$$0 + y = 0 + z. \tag{2}$$

From Axioms 1 and 6, we have the identities:

$$0 + y = y + 0 = y$$

and

$$0 + z = z + 0 = z.$$

Whence, using these identities, we obtain from (2) that

$$y = z. \qquad \text{Q.E.D.}$$

2. FALLACIOUS THEOREM. *For every* x *and* y, *if* $x \cdot y = 1$ *then* $x = 1$.

FALLACIOUS PROOF. By hypothesis

$$x \cdot y = 1. \tag{1}$$

Letting $y = 1$ we have then:

$$x \cdot 1 = 1. \tag{2}$$

But by Axiom 7

$$x \cdot 1 = x. \tag{3}$$

It follows at once from (2) and (3) that

$$x = 1. \qquad \text{Q.E.D.}$$

3. FALLACIOUS THEOREM. *If there is a* y *such that* $x \cdot y = 1$ *then* $x = 1$.

FALLACIOUS PROOF. By hypothesis there is a y such that

$$x \cdot y = 1.$$

Letting $y = 1$ we have then:

$$x \cdot 1 = 1.$$

But by Axiom 7

$$x \cdot 1 = x.$$

Hence

$$x = 1. \qquad \text{Q.E.D.}$$

4. Fallacious Theorem. *There is an x such that for every y, $x + y = x$.*

FALLACIOUS PROOF. By Axiom 6

$$x + 0 = x.$$

Let $y = 0$; we then have immediately:

$$x + y = x. \qquad \text{Q.E.D.}$$

§ 7.5 Further Examples of Informal Proofs. A continuation of the systematic development of the arithmetic of the rational and real numbers will afford an opportunity for considering more informal proofs. In this section we shall also define the negative operation and the operation of subtraction. From a formal standpoint a definition is simply an additional premise. The logical character of definitions and the logical problems they generate are examined in detail in the next chapter. The only essential point needed for this section is that a definition introducing a new operation symbol needs to be preceded by a theorem guaranteeing that the definition introduces a uniquely defined operation. The first theorem of this section justifies in this sense the introduction of the negative operation. Note that the variable 'x' is free in accordance with the stipulation previously laid down.

Theorem 5. *There is exactly one y such that $x + y = 0$.*

PROOF. By Axiom 8 there is *at least* one y such that

$$\text{(1)} \qquad x + y = 0.$$

Thus to prove the theorem we need to show there is *at most* one such y. Let y' be a number such that

$$\text{(2)} \qquad x + y' = 0.$$

Then

$$\text{(3)} \qquad x + y = x + y'$$

and by the cancellation law

$$\text{(4)} \qquad y = y'. \qquad \text{Q.E.D.}$$

This proof exemplifies the typical way we break up the proof that there is *exactly* one entity satisfying some condition. First, we prove there is at least one, and then that there is at most one, for to say there is exactly one is just to say there is at least one and at most one. Notice that from (1) and (2) we get (3) by such an obvious application of the rule governing identities that no justification of any sort is given. Also, having inferred (4), we do not add the redundant phrase, 'and thus there is at most one y'', since we started by considering an arbitrary y' with $x + y' = 0$. To say there is at most one y such that $x + y = 0$ is just to say that for every y' if $x + y' = 0$ then $y = y'$. In a formal derivation this would have to be spelled out but it is not needed in an informal proof.

DEFINITION 1. $-x = y$ *if and only if* $x + y = 0$.

The *negative* of x is, of course, $-x$.

To be clear about the use of definitions in proofs, we need only observe that in a formal derivation Definition 1 would be treated as the sixteenth premise:

$$(x)(y)(-x = y \leftrightarrow x + y = 0). \tag{16}$$

As the deductive development of arithmetic (by informal proofs) is continued we use the new operations introduced without explicit reference to their definitions.

We now prove five theorems asserting familiar facts about the negative operation.

THEOREM 6. $-(-x) = x$.

PROOF. By virtue of Theorem 5 there are numbers y and z such that

$$x + y = 0 \tag{1}$$

and

$$(-x) + z = 0. \tag{2}$$

Hence, using Definition 1, we have:

$$-x = y \tag{3}$$

$$-(-x) = z. \tag{4}$$

From (2) and (3) we get:

$$y + z = 0. \tag{5}$$

Using the commutative law for addition we infer from (1) and (5) that

$$y + x = y + z,$$

and hence by the cancellation law for addition

$$x = z;$$

that is, in view of (4),

$$-(-x) = x. \qquad \text{Q.E.D.}$$

The strategy used in this proof is to treat x and its negative in symmetric fashion. The results are then successfully combined in (5). The proof of this theorem poses a problem that is not always easy to solve. There is no hypothesis stated in the theorem, so there is no obvious initial assumption. Further, no string of identities suggests itself, as is the case with Theorem 8. Consequently, we are forced to fall back on the definition of the negative operation and the theorem justifying the definition (i.e., Theorem 5). Our tactical problem is to find our way from this *terra firma* to the desired conclusion. The point here is that in the absence of the other two possibilities mentioned, Theorem 5 and Definition 1 constitute the most reasonable basis for a proof.

THEOREM 7. $x + (-x) = 0$.

PROOF. It is a truth of logic that

$$-x = -x,$$

whence by virtue of Definition 1,

$$x + (-x) = 0. \qquad \text{Q.E.D.}$$

This proof violates the general rule of not referring to principles of logic. Since it involves a somewhat subtle application of the rule governing identities, the violation is justified. It is worth noting that Axioms 3 and 6, together with Theorem 7, are just the three axioms for a group given in Chapter 5, where addition is the group operation and zero is the group identity element.

THEOREM 8. $(-x) \cdot y = -(x \cdot y)$.

PROOF. We have the following identities:

$$\begin{aligned} 0 &= y \cdot 0 && \text{(Theorem 4)} \\ &= y \cdot (x + (-x)) && \text{(Theorem 7)} \\ &= (y \cdot x) + (y \cdot (-x)) && \text{(Distributive law)} \\ &= (x \cdot y) + ((-x) \cdot y) && \text{(Commutative law)} \end{aligned}$$

And thus by Definition 1,

$$(-x) \cdot y = -(x \cdot y). \qquad \text{Q.E.D.}$$

As already remarked, the proof of this theorem follows a familiar pattern. Given a theorem which asserts an identity, the most natural approach is to prove it by a string of identities. Note that the proof consists of a little more than the identities. The last line is needed to apply the result of the chain of identities.

THEOREM 9. $-y < -x$ *if and only if* $x < y$.

PROOF. First, let us assume that $x < y$. Applying Axiom 13, we obtain:

$$-x + x < -x + y,$$

and thus by Theorem 7 and the commutative law for addition

$$0 < -x + y. \tag{1}$$

Using again Axiom 13, we obtain from (1):

$$-y + 0 < -y + (-x + y),$$

whence by Axiom 6

$$-y < -y + (-x + y). \tag{2}$$

Applying now the commutative and associative laws to the right-hand side of (2), we conclude:

$$-y < -x + (-y + y);$$

and since $-y + y = 0$, we get the desired result:

$$-y < -x.$$

Assuming now that $-y < -x$, we infer by an argument exactly similar that $x < y$. Q.E.D.

Theorem 9 is the first one which has been in the form of an equivalence. As already observed in Chapter 5, we first prove that one member of the equivalence implies the other, and then establish the implication in the opposite direction. In this connection we are often able to take liberties in an informal proof that would not be permitted in a formal derivation. Thus in the proof of Theorem 9 we essentially prove only one of the two implications and then simply remark that the proof of the second implication is similar. Naturally if the proof of the second implication involved an argument not used in proving the first, such a remark would not be permissible.

In the proof of the next theorem, which is also an equivalence, rather than dismiss the second implication with a remark, we sketch the proof without justifying the development of the argument.

THEOREM 10. $0 < -x$ *if and only if* $x < 0$.

PROOF. If $x < 0$, then by Axiom 13,

$$-x + x < -x + 0,$$

and hence applying Axiom 6 to the right-hand side, and Theorem 7 and the commutative law to the left-hand side, we infer:

$$0 < -x.$$

Similarly, if $0 < -x$, then

$$x + 0 < x + (-x)$$

and thus

$$x < 0. \qquad \text{Q.E.D.}$$

In connection with the statement of Theorems 9 and 10, there is another style with which you should be familiar. Theorem 9 could be worded:

In order to have $-y < -x$, it is necessary and sufficient to have $x < y$.

The phrase 'necessary and sufficient' plays the same role that 'if and only if' plays in the original statement. 'If $x < y$' is the sufficient condition, and 'only if $x < y$' is the necessary condition. (This idiom was discussed in § 1.3.)

The next theorem justifies the operation of subtraction.

THEOREM 11. *There is exactly one number z such that $x = y + z$.*

PROOF. We first show there is at least one z satisfying the theorem. By virtue of Theorem 7

$$y + (-y) = 0, \tag{1}$$

whence

$$(y + (-y)) + x = 0 + x,$$

and thus by the commutative law and Axiom 6

$$x = (y + (-y)) + x. \tag{2}$$

Applying the associative law to (2) we obtain:

$$x = y + ((-y) + x),$$

and clearly $(-y) + x$ is a satisfactory z.

The proof that there is at most one z is exactly like the similar part of the proof of Theorem 5: consider a z' such that $x = y + z'$; then

$$y + z = y + z',$$

and thus by the law of cancellation for addition

$$z = z'. \qquad \text{Q.E.D.}$$

In this proof it is clear from (2) that $(-y) + x$ is a number satisfying the desired condition; that is, if $z = (-y) + x$, then $x = y + z$. In these various proofs involving only addition and not multiplication we refer to *the* commutative and associative laws; it is understood we mean the laws for the operation of addition. The strategy of this proof is to pick the axiom closest to the theorem (Axiom 8) and work from it. There was no need to use the negative operation in this proof; it is sufficient to have in (1) that there are numbers u and v with $x + u = 0$ and $y + v = 0$. We introduced the operation merely to give some further practice in handling it.

DEFINITION 2. $x - y = z$ *if and only if* $x = y + z$.

The traditional notation of arithmetic has been followed in using the same sign '$-$' in both Definitions 1 and 2, although in Definition 1 it denotes a unary operation (that is, a function of one argument) and in Definition 2, it denotes a binary operation (that is, a function of two arguments). This confusion in the standard notation is more or less justified by the following theorem.

THEOREM 12. $x - y = x + (-y)$.

PROOF. By Theorem 11 there is a unique z such that

$$x = y + z \tag{1}$$

and hence by Definition 2

$$x - y = z.$$

The following identities show that $x + (-y)$ equals z and hence equals $x - y$:

$$\begin{aligned}
x + (-y) &= (-y) + x && \text{(Commutative law)}\\
&= (-y) + (y + z) && \text{(By (1))}\\
&= ((-y) + y) + z && \text{(Associative law)}\\
&= (y + (-y)) + z && \text{(Commutative law)}\\
&= 0 + z && \text{(Theorem 7)}\\
&= z + 0 && \text{(Commutative law)}\\
&= z && \text{(Axiom 6)}
\end{aligned}$$

Q.E.D.

The next three theorems exemplify some typical and useful properties of the relation $<$.

THEOREM 13. *It is not the case that* $x < x$.

PROOF. The theorem follows immediately from Axiom 10 by taking $x = y$. Q.E.D. This proof is so obvious that it can be dismissed in a sentence. If we put 'x' for 'y' in Axiom 10, we obtain the formula:

If $x < x$ then it is not the case that $x < x$,

which is logically equivalent to:

It is not the case that $x < x$.

A brief indirect proof could also be given by supposing there is an x such that $x < x$, and then using Axiom 10 to derive a contradiction. Notice the use of the phrase 'by taking $x = y$' to indicate an application of *US*.

THEOREM 14. *Exactly one of the following holds:* $x = y$, $x < y$, $y < x$.

PROOF. It is clear from Axiom 12 that at least one of the three formulas holds. We show at most one holds by showing that no two of the formulas may hold. First, if $x = y$ and $x < y$, we obtain at once: $x < x$, which contradicts Theorem 13, and obviously exactly the same argument shows that we may not have that both

$$x = y \quad \text{and} \quad y < x.$$

Finally, if both $x < y$ and $y < x$ Axiom 10 is contradicted. Q.E.D.

The proof of Theorem 14 has been our first occasion to use an indirect proof informally. The two applications in the proof are too obvious to require comment.

THEOREM 15. *If* $x < 0$ *and* $y < z$ *then* $x \cdot z < x \cdot y$.

PROOF. From the hypothesis of the theorem and Theorem 10,

$$0 < -x \quad \text{and} \quad y < z.$$

Hence by Axiom 14

$$(-x) \cdot y < (-x) \cdot z,$$

and then by Theorem 8

$$-(x \cdot y) < -(x \cdot z),$$

whence by Theorem 9

$$x \cdot z < x \cdot y \qquad \text{Q.E.D.}$$

This proof exemplifies the standard approach when the theorem to be proved is a conditional sentence: assume the hypothesis and see if the conclusion can be obtained by direct methods.

The next theorem is the cancellation law for multiplication. The proof of this law is the most intricate proof yet given and should be studied rather carefully. Notice that it is an indirect proof; the standard, direct method just mentioned does not work for this theorem.

THEOREM 16. *If* $x \neq 0$ *and* $x \cdot y = x \cdot z$ *then* $y = z$.

PROOF. We give an indirect proof. Suppose $y \neq z$. Then by Axiom 12, $y < z$ or $z < y$. Since the hypothesis of the theorem is symmetric concerning y and z, it will be sufficient to prove the theorem on the assumption that

$$(1) \qquad y < z.$$

By the hypothesis of the theorem, $x \neq 0$; hence, by Axiom 12, $0 < x$ or $x < 0$, which yields two cases to be considered separately. If $0 < x$ then from (1) and Axiom 14, we have:

$$x \cdot y < x \cdot z,$$

which together with Theorem 14 contradicts the hypothesis that $x \cdot y = x \cdot z$. On the other hand, if $x < 0$, then by (1) and Theorem 15

$$x \cdot z < x \cdot y,$$

which together with Theorem 14 again contradicts the hypothesis that $x \cdot y = x \cdot z$. Q.E.D.

In this proof we have at various stages alternative cases to consider. Mathematicians dislike proofs which require subdivision into a large number of cases, but the consideration of two or three cases is frequent and often unavoidable. At the beginning we have the alternatives: $y < z$ or $z < y$. But we do not need to consider these two cases separately since the roles of y and z are exactly the same in the hypothesis. For a proof of one case can be repeated exactly by simply interchanging letters to give a proof of the other case. When we are faced with the alternatives: $0 < x$ or $x < 0$, the situation is different. The proof for one case $(0 < x)$ depends on Axiom 14, and the proof for the other case $(x < 0)$ depends on Theorem 15. When at some stage in a proof a variety of possibilities turn up, the proof cannot progress until each of these possibilities (i.e., alternatives) has been tracked down and satisfactorily disposed of. Of course, in many instances the possibilities can be immediately reduced in number by symmetry considerations, as at the beginning of this proof. The point to remember is that if alternative cases do arise, you must indicate in the proof some method of disposing of them.

At this point we conclude the explicit analysis of informal proofs. The systematic development of arithmetic is carried somewhat further in the exercises accompanying this section.

EXERCISES

1. Formulate Theorem 5 in logical notation. (This exercise is meant to dispel any lingering confusions about the exact meaning of the phrase, 'there is exactly one ...'.)

2. Prove the following assertions concerning the negative operation and the operation of subtraction.

(a) $-x = (-1) \cdot x$
(b) $(-x) \cdot y = x \cdot (-y)$
(c) $(x - y) + (y - z) = x - z$
(d) $x \cdot (y - z) = x \cdot y - x \cdot z$
(e) $(x - y) + (z - w) = (x + z) - (y + w)$

(f) $(x - y) - (z - w) = (x + w) - (y + z)$
(g) $(x - y)\cdot(z - w) = (x\cdot z + y\cdot w) - (x\cdot w + y\cdot z)$

3. Using the definition:

$$x \leq y \quad \textit{if and only if} \quad x = y \quad \text{or} \quad x < y,$$

prove:

$$\textit{if} \quad x + y \leq x + z \quad \textit{then} \quad y \leq z.$$

4. Prove the following:

(a) If $x + x = 0$ then $x = 0$.
(b) If $x\cdot x = 0$ then $x = 0$.
(c) If $u \neq 0$ and $v \neq 0$ then $u\cdot v \neq 0$.

5. Prove the following:

(a) If $0 < x$ and $x\cdot y < x\cdot z$ then $y < z$.
(b) If $x < 0$ and $x\cdot y < x\cdot z$ then $z < y$.
(c) If $0 < x$, $y < 0$, and $z < 0$ then $x\cdot y < y\cdot z$.
(d) If $x < 0$, $0 < y$, and $0 < z$ then $x\cdot y < y\cdot z$.
(e) $x - y < x - z$ if and only if $z < y$.

6. Define the operation of taking the absolute value of a number, and prove:

(a) $|x\cdot y| = |x|\cdot|y|$
(b) $|x + y| \leq |x| + |y|$
(c) $|x| - |y| \leq |x - y|$
(d) $x\cdot|y| \leq |x\cdot y|$
(e) $|x - y| \leq |x - z| + |y - z|$

CHAPTER 8

THEORY OF DEFINITION

§ 8.1 **Traditional Ideas.** In the older logic a definition is the delimitation of a species by stating the genus which includes it and the specific difference or distinguishing characteristic of the species. A typical example is the definition of man as a rational animal. The genus is the animal genus and the distinguishing characteristic is rationality. (What has been stated in capsule form is the Aristotelian theory of definition.) Many textbooks which are not so traditional as to demand strict adherence to the Aristotelian analysis, do seriously promulgate the four traditional "rules" of definition:

1. A definition must give the essence of that which is to be defined.
2. A definition must not be circular.
3. A definition must not be in the negative when it can be in the positive.
4. A definition must not be expressed in figurative or obscure language.

Certainly these rules have serious use as practical precepts. They rule out as definitions statements like:

> Beauty is eternity gazing at itself in a mirror—KHALIL GIBRAN, *The Prophet*,

which violates Rule 4, or:

> Force is not a kinematical notion,

which violates Rule 3. On the other hand, these rules are of little help in clarifying the formal notion of a proper definition within a precisely stated theory, such as the theory of the real numbers partially developed in the last chapter.

For example, we may define in arithmetic the pseudo-operation $*$ as follows:

(1) $\quad x * y = z$ if and only if $x < z$ and $y < z$.

From our intuitive knowledge of arithmetic we may easily use this pseudo-operation to derive a contradiction.

$$1 \star 2 = 3$$

since

$$1 < 3 \quad \text{and} \quad 2 < 3;$$

but also,

$$1 \star 2 = 4,$$

since

$$1 < 4 \quad \text{and} \quad 2 < 4.$$

Hence, we infer:

$$3 = 4,$$

which contradicts the familiar fact that $3 \neq 4$. But all four of the traditional rules seem to be satisfied by (1). Whatever sort of entities essences are, the essence of the $\star$ operation is surely stated by (1). Furthermore, the definition is neither circular, negative, obscure, nor in figurative language. It is transparently clear from this simple example that in order to develop an adequate formal theory of definition we must penetrate beyond the semi-platitudinous level of the four traditional rules.

§ **8.2 Criteria for Proper Definitions.** A traditional definition *per genus et differentiam* is often called a *real* definition because it is said to characterize the essence of a species. The kind of definition common in mathematics, that is, definitions which introduce a new symbol, are often called *verbal* or *nominal* definitions. However, it is not clear how a sharp distinction between the two kinds of definitions can be made. For our purposes, it is sufficient to understand that a definition is a statement which establishes the meaning of an expression. The definition accomplishes this by relating the expression it defines (the *definiendum*) to other expressions (the *definiens*) already available.

At least two questions immediately arise from this vague statement about what definitions are. What is meant by 'other expressions already available'? What restrictions if any are there on the logical form of sentences which may serve as definitions? The answer to the first question is that we have in mind the introduction of a definition within a specified theory, like the elementary theory of arithmetic considered in the previous chapter. As understood here, a *theory* is characterized in terms of its primitive, non-logical symbols and its axioms. In the case of the theory of the last chapter the primitive symbols are the relation symbol '$<$', the operation symbols '$+$' and '$\cdot$', and the individual constants '0' and '1'. The axioms are just the fifteen axioms given at the beginning of the chapter. The theory of groups introduced in Chapter 5 had three primitive symbols and three axioms. In the present chapter, first-order predicate logic with identity as developed in preceding chapters is assumed, and we only consider theories which can be formalized within the framework of this logic.

The first definition in a theory is, then, a sentence of a certain form which establishes the meaning of a new symbol of the theory in terms of the primitive symbols of the theory. The second definition in a theory is a sentence of a certain form which establishes the meaning of a second new symbol of the theory in terms of the primitive symbols and the first defined symbol of the theory. And similarly for subsequent definitions. The point to be noted is that the definitions in a theory are introduced one at a time in some fixed sequence. Because of this fixed sequence we may always speak meaningfully of *preceding* definitions in the theory. Often it is convenient to adopt the viewpoint that any defined symbol must be defined in terms only of the primitive symbols of the theory. In this case there is no need to introduce definitions in some fixed sequence. However, the common mathematical practice is to use previously defined symbols in defining new symbols; and to give an exact account of this practice, a fixed sequence of definitions is needed.

It was remarked in the last chapter that from the standpoint of the logic of inference a definition in a theory is simply regarded as a new axiom or premise. But it is not intended that a definition shall strengthen the theory in any substantive way. The point of introducing a new symbol is to facilitate deductive investigation of the structure of the theory, but not to add to that structure. Two criteria which make more specific these intuitive ideas about the character of definitions are that (i) a defined symbol should always be eliminable from any formula of the theory, and (ii) a new definition does not permit the proof of relationships among the old symbols which were previously unprovable; that is, it does not function as a creative axiom.* In the previous chapter, for instance, we introduced the symbol for subtraction by the equivalence:

$$(1) \qquad x - y = z \quad \text{if and only if} \quad x = y + z.$$

We may use (1) to eliminate any occurrence of the subtraction symbol. Thus by virtue of (1) we eliminate '$-$' from:

$$\text{If} \quad y \neq 0 \quad \text{then} \quad x - y \neq x,$$

and obtain the arithmetically equivalent statement:

$$\text{If} \quad y \neq 0 \quad \text{then} \quad x \neq y + x.$$

It seems reasonable to require that any definition introducing a new symbol may be used to eliminate all subsequent meaningful occurrences of the new symbols. To be eliminable is a characteristic property of a defined

* These two criteria were first formulated by the Polish logician S. Leśniewski (1886–1939); he was also the first person to give rules of definition satisfying the criteria.

symbol, as opposed to a primitive symbol. We may now formalize the concept of eliminability.

Criterion of Eliminability. *A formula* S *introducing a new symbol of a theory satisfies the criterion of eliminability if and only if: whenever* S_1 *is a formula in which the new symbol occurs, then there is a formula* S_2 *in which the new symbol does not occur such that* $\mathsf{S} \rightarrow (\mathsf{S}_1 \leftrightarrow \mathsf{S}_2)$ *is derivable from the axioms and preceding definitions of the theory.*

As the wording of this criterion suggests, hereafter we do not call definitions new axioms although they function as additional axioms in logical inference. The reason for this terminological restriction is obvious. In laying out a given theory for investigation we want to state the creative axioms at the beginning and always refer to them as "the axioms". Since the definitions are theoretically dispensable, we do not want to give them the same status as the basic axioms of the theory.

The notion of a definition not being creative is formalized in the following statement.

Criterion of Non-creativity. *A formula* S *introducing a new symbol of a theory satisfies the criterion of non-creativity if and only if: there is no formula* T *in which the new symbol does not occur such that* $\mathsf{S} \rightarrow \mathsf{T}$ *is derivable from the axioms and preceding definitions of the theory but* T *is not so derivable.*

In other words, we cannot permit a formula S introducing a new symbol to make possible the derivation of some previously unprovable theorem stated wholly in terms of primitive and previously defined symbols. An example of a formula which does not satisfy this criterion of non-creativity is the second axiom for groups if we consider a more limited theory than that of groups. The single primitive symbol of our theory is the binary symbol '$\circ$' and the single axiom the associative axiom:

$$x \circ (y \circ z) = (x \circ y) \circ z. \tag{1}$$

As the first definition of this theory we now propose the following formula introducing the new individual constant 'e':

$$x \circ e = x. \tag{2}$$

However, applying the criterion of non-creativity we reject (2) as a proposed definition in our theory, for from (2) we may derive at once:

$$(\exists y)(x)(x \circ y = x). \tag{3}$$

We note that (3) is a formula whose only non-logical symbol is the primitive symbol of the theory, but it is trivial to find an interpretation showing that (3) cannot be derived from (1). Thus (2) is creative and must be rejected as a proper definition.

It should be noticed that a special consequence of the criterion of non-creativity is the criterion of relative consistency. If the axioms and preceding definitions are consistent and if a formula introducing a new symbol may be used to derive a contradiction, then the new formula does not satisfy the criterion of non-creativity. For from a contradiction any formula may be derived, since $(P \& -P) \rightarrow Q$ is a tautology. Thus we do not need as an independent third criterion that of relative consistency.

In the next section we turn to the task of stating rules of definition which will guarantee satisfaction of the two criteria of eliminability and non-creativity.

EXERCISES

1. Use the definition of subtraction (p. 147) to eliminate the subtraction symbol from the following statements.

   (a) $x - 0 = x$.
   (b) $x - 1 \neq 1$.
   (c) If $x \neq y$ then $x - y \neq y - x$.
   (d) If $0 < x$ and $0 < y$ then $x - y \neq x + y$.

2. Given the theory consisting of the single axiom on the binary operation symbol '$\circ$':

$$x \circ y = y \circ x,$$

prove that the formula:

$$x \circ y = e$$

introducing the individual constant 'e' is creative in this theory.

§ 8.3 Rules for Proper Definitions. In theories stated in precise language (whether the subject matter is pure mathematics, physics, or psychology) we ordinarily introduce three kinds of defined symbols: relation symbols, operation symbols, and individual constants. Thus in the theory of the arithmetic of the real numbers begun in the last chapter, '$\leq$' and '$\geq$' are defined relation symbols, the symbols for subtraction and division are defined operation symbols, and names of any numbers except 0 and 1 are defined individual constants (for instance, '2', '3', and '4').

For simplicity of statement we shall introduce separate rules for each of these three kinds of symbols. We first state rules requiring that a proper definition be an equivalence. Subsequently we discuss the use of identities to define operation symbols and individual constants. In dealing with definitions which are equivalences it is customary to introduce the new symbol on the left side of the equivalence and to call this side the *definiendum* ("thing to be defined"). The right side is called the *definiens* ("thing defining"). Thus in the definition of subtraction, '$x - y = z$' is the definiendum and '$x = y + z$' is the definiens. We use this language of 'definiendum' and 'definiens' throughout this chapter.

RULE FOR DEFINING RELATION SYMBOLS. *An equivalence* D *introducing a new n-place relation symbol* P *is a proper definition in a theory if and only if* D *is of the form* $\mathsf{P}(\mathsf{v}_1, \ldots, \mathsf{v}_n) \leftrightarrow \mathsf{S}$, *and the following restrictions are satisfied:* (i) $\mathsf{v}_1, \ldots, \mathsf{v}_n$ *are distinct variables;* (ii) S *has no free variables other than* $\mathsf{v}_1, \ldots, \mathsf{v}_n$; *and* (iii) S *is a formula in which the only non-logical constants are primitive symbols and previously defined symbols of the theory.*

Note that the definiendum $\mathsf{P}(\mathsf{v}_1, \ldots, \mathsf{v}_n)$ is an atomic formula, which form is needed to guarantee elimination of the defined relation symbol from every possible context. Corresponding to the convention set up in the previous chapter, the variables $\mathsf{v}_1, \ldots, \mathsf{v}_n$ are free in the equivalence D. Strict conformity to the formal rules of inference could be obtained by adding universal quantifiers in front. Some examples coupled with discussion will help clarify the three restrictions on the rule. The requirement that the variables $\mathsf{v}_1, \ldots, \mathsf{v}_n$ be distinct prevents definitions like:

$$(1) \qquad x \leq x \quad \text{if and only if} \quad x = x \quad \text{or} \quad x < x.$$

Formula (1) does not really define the binary relation symbol '$\leq$', since only one variable occurs in the definiendum. With (1) at hand, we would not know how to eliminate '$\leq$' from the formula $x \leq y$. The definiens of (1) must be regarded as defining a unary relation symbol, say, 'U' (unary relations are just properties, i.e., a property is a one-place relation):

$$U(x) \quad \text{if and only if} \quad x = x \quad \text{or} \quad x < x.$$

Of course, the property U is a trivial universal property possessed by every number. As a second example, consider the definition of the quaternary relation which holds between four numbers if the difference between the first and the second is less than that between the third and fourth. We use the letter 'A' as the relation symbol.

$$(2) \qquad A(x, y, u, v) \quad \text{if and only if} \quad x - y < u - v.$$

The generality of (2) and thus the general eliminability of the relation symbol 'A' would be ruined if (2) were replaced by:

$$(3) \qquad A(x, y, u, x) \quad \text{if and only if} \quad x - y < u - x.$$

The definiens of (3) really defines the ternary relation T:

$$(4) \qquad T(x, y, z) \quad \text{if and only if} \quad x - y < z - x.$$

(Notice the intuitive meaning of the relation T: $T(x, y, z)$ just when x is less than the mean of y and z (i.e., $x < (y + z)/2$).)

The second restriction prevents definitions like:

(5) $R(x)$ if and only if $x + y = 0$.

When (5) is added to the axioms of arithmetic we may derive a contradiction. The source of the trouble is the appearance of the variable 'y' in the definiens but not in the definiendum. Now (5) is logically equivalent to the pair of statements:

(6) If $x + y = 0$ then $R(x)$,

(7) If $R(x)$ then $x + y = 0$.

But from the logic of quantifiers we know that (6) is equivalent to:

(8) If there is a y such that $x + y = 0$ then $R(x)$,

and (7) is equivalent to:

(9) If $R(x)$ then for every y, $x + y = 0$.

From (8) and (9) we immediately infer the patent falsehood:

(10) If there is a y such that $x + y = 0$ then for every y, $x + y = 0$.

(Note that the variable 'x' is left free in this discussion, since it appears in a proper manner in both the definiendum and definiens of (5).)

On the other hand, the second restriction does not prevent variables from being free in the definiendum but not in the definiens. Thus we admit as a proper definition:

(11) $Q(x, y)$ if and only if $x > 0$.

There is no definite formal reason for prohibiting variables from being free in the definiendum but allowing them to be free in the definiens. On the other hand, such variables are not used to express anything intuitively meaningful; the triviality of their role is underscored by the fact we can always find an equivalence which has the same logical content and which has the same variables free in the definiendum and the definiens. Thus we may convert (11) into a formula having the same variables free in definiens and definiendum by conjoining to the definiens the logical truth: $y = y$. The new formula:

(12) $Q(x, y)$ if and only if $x > 0 \;\&\; y = y$

is logically equivalent to (11). A similar conjunction of logical identities can be used to convert any equivalence having more free variables in the definiendum than the definiens into one having the same number in both.

The third restriction simply prohibits two kinds of circularity of definition. We could not admit as a proper definition:

(13) $R(x)$ if and only if $R(x)$;

a logical truth such as (13) would not be creative. Its defect is that it does not satisfy the criterion of eliminability. Formula (13) does not yield a procedure for eliminating the relation symbol 'R'. Of a similar sort is the pair of equivalences:

(14) $R(x)$ if and only if it is not the case $P(x)$,

(15) $P(x)$ if and only if it is not the case $R(x)$.

If we define the relation symbol 'R' in terms of the new relation symbol 'P', and vice versa, then we are not able to eliminate either in favor of the primitive notation. Thus we have the requirement that no other new symbol appear in the definition and that the definitions be given in a fixed sequence.

We now turn to the rule for defining operation symbols. One essentially new restriction has to be added to the three needed for relation symbols. In stating the rule we use the standard notation

$$(\mathrm{E}!\mathsf{w})\mathsf{S}$$

for

There is exactly one w *such that* S.

RULE FOR DEFINING OPERATION SYMBOLS. *An equivalence* D *introducing a new n-place operation symbol* O *is a proper definition in a theory if, and only if,* D *is of the form*

$$\mathsf{O}(\mathsf{v}_1, \ldots, \mathsf{v}_n) = \mathsf{w} \leftrightarrow \mathsf{S},$$

and the following restrictions are satisfied: (i) $\mathsf{v}_1, \ldots, \mathsf{v}_n$, w *are distinct variables,* (ii) S *has no free variables other than* $\mathsf{v}_1, \ldots, \mathsf{v}_n$, w, (iii) S *is a formula in which the only non-logical constants are primitive symbols and previously defined symbols of the theory, and* (iv) *the formula* $(\mathrm{E}!\mathsf{w})\mathsf{S}$ *is derivable from the axioms and preceding definitions of the theory.*

Consideration of the pseudo-operation $\star$ introduced in § 8.1 is sufficient to justify the fourth restriction. By use of the pseudo-operation $\star$ we were able to derive a contradiction in § 8.1 just because '$x \star y$' does not designate a unique entity. In the case of the definition of the operation $\star$:

(16) $x \star y = z$ if and only if $x < z$ and $y < z$

we cannot prove that there is exactly one z such that $x < z$ and $y < z$, and thus (16) is not a proper definition. The function of the fourth restriction is to require that the definition of any new operation be preceded by a theorem which guarantees that the operation is uniquely defined. In

Chapter 7 the definitions of the negative operation and subtraction were both preceded by such justifying theorems. The generality of the defining rule, which applies to operations of whatever complexity, should not obscure the fact that we are usually concerned with binary operations, whose definitions are of the form:

$$x \circ y = z \quad \text{if and only if} \quad \mathsf{S}(x, y, z),$$

and we need a preceding theorem to the effect that for every x and y there is exactly one z such that $\mathsf{S}(x, y, z)$.

We can regard individual constants as operation symbols of rank zero. However, because of the somewhat startling form of definitions of individual constants when the definitions are equivalences rather than identities, it seems advisable to state the rule explicitly. The reason for insisting on the consideration of definitions of individual constants in the form of equivalences is that it is not always possible to introduce them by means of identities, as we shall shortly see. In order to make the general rule intuitively clearer, we may first indicate how we could define the constants '0' and '1' in arithmetic if we had formulated our axioms without using them as primitive symbols. We would introduce the following two equivalences:

$$(17) \qquad 0 = y \quad \text{if and only if for every} \quad x, x + y = x,$$

$$(18) \qquad 1 = y \quad \text{if and only if for every} \quad x, x \cdot y = x.$$

Note that the variable 'x' is bound in the definiens of both (17) and (18). The restriction on the uniqueness of y must apply to individual constants. If the restriction were dropped, we could introduce a constant 'b' by a definition such as the following:

$$(19) \qquad b = y \quad \text{if and only if} \quad y > 0$$

and derive a contradiction. For, it follows from (19) that

$$b = 1$$

and

$$b = 2.$$

Hence,

$$1 = 2,$$

which is absurd.

RULE FOR DEFINING INDIVIDUAL CONSTANTS. *An equivalence* D *introducing a new individual constant* c *is a proper definition in a theory if and only if* D *is of the form*

$$\mathsf{c} = \mathsf{w} \leftrightarrow \mathsf{S},$$

and the following restrictions are satisfied: (i) S *has no free variable other than* w, (ii) S *is a formula in which the only non-logical constants are primitive symbols and previously defined symbols of the theory, and* (iii) *the formula* (E!w)S *is derivable from the axioms and preceding definitions of the theory.*

In the next section we introduce rules for defining operation symbols and individual constants by means of identities rather than equivalences. Several applications of the rules stated in this section are to be found in the exercises.

EXERCISES

1. Which of the following definitions of relation symbols are improper? If improper, state what restriction is violated.

(a) $R(x, y)$ if and only if there is a z such that $x + y = z$.
(b) $R(x, y)$ if and only if $x + y > z$.
(c) $R(x, y)$ if and only if $y > 2$.
(d) $R(x, y)$ and $x \neq y$ if and only if $x + y > 0$.
(e) $R(x, x)$ if and only if $x + 1 > 0$.

2. Which of the following definitions of operation symbols are improper? If a definition is improper, use it to derive intuitively a contradiction.

(a) $x \circ y = z$ if and only if $x < 2$ and $y < 3$ and $z < 4$.
(b) $x \circ y = z$ if and only if $x = 1$ and $y = 2$ and $z = 7$.
(c) $x \circ y = z$ if and only if $x < 1$ and $y < 2$ and $z = 7$.
(d) $x \circ y = z$ if and only if $x = 1$ and $y = 2$ and $z < 7$.
(e) $x \circ y = z$ if and only if $x + y = x + z$.
(f) $x \circ y = z$ if and only if $x \cdot z = y$.

3. Which of the following definitions of individual constants are improper? If improper, state what restriction is violated.

(a) $c = y$ if and only if there is an x such that $x + y = 1$.
(b) $c = y$ if and only if for every x if $x > 0$, then $x + y > 0$.
(c) $c = y$ if and only if $y = 1 + 1$.
(d) $c = y$ if and only if $y = y$.
(e) $c = y$ if and only if $y \neq y$.

4. Give an example of an improper definition whose definiens is:

(a) a logical truth.
(b) the negation of a logical truth.
(c) an axiom of the theory in which the definition is proposed (construct a simple theory for this exercise).

5. Give an example of a proper definition whose definiens is:

(a) a logical truth.
(b) the negation of a logical truth.
(c) the negation of an axiom of the theory in which the definition occurs (construct a simple theory for this exercise).

§ 8.4 Definitions Which Are Identities. In the last section we remarked that operation symbols and individual constants are often introduced by identities rather than equivalences. We want now to consider appropriate rules for such identities.

Since the definition:

$$(1) \qquad 2 = 1 + 1$$

seems more natural and elegant than the definition:

$$(2) \qquad 2 = y \quad \text{if and only if} \quad y = 1 + 1,$$

it is reasonable to ask why equivalences are ever used to define individual constants. The answer is very simple: identities alone are not adequate for the job. If '0' and '1' were eliminated as primitive symbols in arithmetic, they could be introduced by equivalences (17) and (18) of the last section, but these equivalences could not be eliminated in favor of identities.*

The same remarks apply to the use of equivalences rather than identities to define operation symbols. For instance, if no other definitions have been given in the theory of arithmetic of Chapter 7, then neither the negative operation nor the subtraction operation symbol can be defined by an identity; but given one, the other can be so defined:

$$x - y = x + (-y),$$

$$-x = 0 - x.$$

Or if the individual constant '-1' is first defined by the equivalence:

$$-1 = x \quad \text{if and only if} \quad x + 1 = 0,$$

then the negative operation symbol may be defined by the identity:

$$-x = -1 \cdot x.$$

As these examples illustrate, the possibility of defining an operation symbol by an identity is relative to the exact character of the preceding definitions of the theory.

* This statement about the inadequacy of identities would be false if our basic logic had been extended to include a description operator 'the object x such that . . .'. This operation is usually symbolized by '(ιx)', a notation first introduced by Peano. With this operator available, we could introduce as a definition of '0':

$$0 = (\iota y)[(x)(x + y = x)].$$

The description operator was not introduced because it is not really needed in the deductive investigation of theories formalizable within first-order predicate logic and because its introduction would further complicate our basic rules of inference.

We now turn to the formal rules. In the case of identities which are used as definitions we call the left side of the identity the definiendum, and the right side the definiens, as would be expected from our previous usage; here the definiendum and the definiens are both *terms*, in the sense of Chapter 3. Corresponding to the notion of atomic formula introduced in Chapter 3, it is convenient to introduce the notion of an *atomic term*. An atomic term is a term which is either an individual constant or which has exactly one occurrence of one operation symbol. Thus '$x \cdot y$', '$x + y$', '$x - y$' and '1' are atomic, while '$x + (x + z)$' and '$x \cdot (y + z)$' are not. When a definition is an identity it is required that the definiendum be an atomic term.

RULE FOR DEFINING OPERATION SYMBOLS. *An identity* D *introducing a new n-place operation symbol* O *is a proper definition in the theory if and only if* D *is of the form*

$$\mathsf{O}(\mathsf{v}_1, \ldots, \mathsf{v}_n) = \mathsf{t},$$

and the following restrictions are satisfied: (i) $\mathsf{v}_1, \ldots, \mathsf{v}_n$ *are distinct variables,* (ii) *the term* t *has no free variables other than* $\mathsf{v}_1, \ldots, \mathsf{v}_n$, (iii) *the only non-logical constants in the term* t *are primitive symbols and previously defined symbols of the theory.*

It is worth remarking that when identities are used to define operation symbols, no justifying theorem is needed to guarantee that the operation symbol is well-defined, for the formula (E!w) (t = w) is a truth of logic.

Since the rule for defining individual constants by use of an identity is very similar to the one just given for operation symbols, its formal statement is left as an exercise. The standard definitions of names of numbers exemplify introduction of individual constants by means of identities:

$$2 = 1 + 1$$
$$3 = 2 + 1$$
$$4 = 3 + 1$$
$$\cdot \quad \cdot \quad \cdot \quad \cdot \quad \cdot$$

EXERCISES

1. Assuming no preceding definitions in the theory of arithmetic of Chapter 7, which of the following arithmetical operations can be defined by identities? Where the answer is positive, state the definition.

(a) Squaring a number.
(b) Cubing a number.
(c) Absolute value of a number.
(d) Operation of adding five to a number.
(e) Operation of multiplying a number by two.

2. Give examples which justify the three restrictions on the rule for defining operation symbols by identities.

3. Formally state the rule for introducing individual constants by definitions which are identities.

4. Given as axioms all true statements of arithmetic not involving the symbol '0', can you define '0' by means of an identity?

5. Given as axioms all true statements of arithmetic involving the primitive symbols '+', '·', '<' and '0', can you define '1' by an identity?

6. Given as axioms all true statements of arithmetic involving the primitive symbols '+', '<', '0' and '1', can you define by an identity the multiplication symbol?

§ 8.5 The Problem of Division by Zero. That everything is not for the best in this best of all possible worlds, even in mathematics, is well illustrated by the vexing problem of defining the operation of division in the elementary theory of arithmetic. If we introduce the definition:

$$(1) \qquad x/y = z \quad \text{if and only if} \quad x = y \cdot z,$$

we realize immediately that (1) does not satisfy the fourth restriction for equivalences defining operation symbols. For we cannot prove that given any two numbers x and y there is a unique z such that $x = y \cdot z$. For instance, there is no z such that $1 = 0 \cdot z$; and any number z has the property that $0 = 0 \cdot z$.

An obvious modification of (1) is:

$$(2) \qquad x/y = z \quad \text{if and only if} \quad y \neq 0 \quad \text{and} \quad x = y \cdot z.$$

But we cannot prove that given any two x and y there is a unique z such that $y \neq 0$ and $x = y \cdot z$. To see this, we need merely consider the case of $y = 0$.

The apparent naturalness of (2) suggests a weakening of the fourth restriction to the requirement that there is *at most* one z such that $y \neq 0$ and $x = y \cdot z$. It is easy to show that (2) satisfies this weakened restriction, but unfortunately we can derive a contradiction from (2). The inference runs as follows. It is a logical truth that

$$(3) \qquad \frac{1}{0} = \frac{1}{0}.$$

Hence, there is an x such that

$$(4) \qquad \frac{1}{0} = x,$$

and thus by (2)

$$0 \neq 0 \quad \text{and} \quad 1 = 0 \cdot x,$$

which is absurd. The status of (3) needs to be noted. It is a fundamental assumption of the logic developed in this book that all terms designate objects. Given any operation symbol we can always write down the analogue of (3) as a logical truth, and then infer the kind of existential statement represented by (4). Such inferences were discussed in Chapter 5, and their logical validity need not be re-examined here. The awkward possibility of changing the basic logic to solve the problem of division by zero is mentioned along with other possible solutions in § 8.7.

In spite of the difficulties besetting us there is a formally satisfactory way of defining division by zero. It is a so-called axiom-free definition since it requires no previous theorem to justify it. It does require that when $y = 0$ then $x/y = 0$. Because of the several quantifiers occurring in the definiens we use logical notation to state the definition.

$$(5) \qquad x/y = z \leftrightarrow (z')[z' = z \leftrightarrow x = y \cdot z']$$
$$\vee\ [-(\exists w)(z')(z' = w \leftrightarrow x = y \cdot z')\ \&\ z = 0].$$

The complicated character of (5) argues strongly for some other solution, which we turn to in the next section. Although (5) is unwieldy, it represents an approach that is theoretically important in providing a method for giving definitions of operation symbols which do not depend on previously established theorems. The general formulation, of which (5) is a specific instance, is the following.

$$\mathsf{O}(x_1, \ldots, x_n) = y \leftrightarrow (z)[z = y \leftrightarrow \mathsf{S}(x_1, \ldots, x_n, z)]$$
$$\vee\ [-(\exists w)(z)(z = w \leftrightarrow \mathsf{S}(x_1, \ldots, x_n, z)\ \&\ y = 0].$$

Sometimes the setting of $y = 0$ when the operation is not "defined" in the usual sense seems a little strange. Thus many mathematicians would be uneasy at seeing:

$$\frac{2}{0} = 0,$$

and thus, say:

$$1 + \frac{2}{0} = 1.$$

These matters are discussed further in §8.7. Here it will suffice to say that there seems to be no method of handling division by zero which is uniformly satisfactory.

EXERCISES

1. Give an axiom-free definition of the square root operation. Here the problem is the square root of any negative number.
2. Prove that there is exactly one z satisfying the definiens of (5).
3. Give an axiom-free definition of subtraction.

§ 8.6 Conditional Definitions. The customary practice in mathematics is to use conditional definitions rather than the awkward axiom-free definitions introduced at the end of the previous section. The technique of conditional definition is to preface an ordinary proper definition by a hypothesis. Thus a possible conditional definition of division is:

(1) If $y \neq 0$ then $x/y = z$ if and only if $x = y \cdot z$.

The main disadvantage of conditional definitions is that they do not fully satisfy the criterion of eliminability. For example, if we use (1) to define division, then we cannot eliminate the symbol for division from the sentence:

$$\frac{1}{0} = \frac{1}{0}.$$

On the other hand, it is apparent we can eliminate it in all "interesting" cases, namely, all cases which satisfy the hypothesis of (1).

Although conditional definitions are used as much in defining relation symbols as operation symbols, we shall only state the rules for operation symbols, and leave the case of relation symbols as an exercise.

RULES FOR CONDITIONAL DEFINITIONS OF OPERATION SYMBOLS. *An implication* C *introducing a new operation symbol* O *is a conditional definition in a theory if and only if* C *is of the form*

$$\mathsf{H} \rightarrow [\mathsf{O}(\mathsf{v}_1, \ldots, \mathsf{v}_n) = \mathsf{w} \leftrightarrow \mathsf{S}]$$

and the following restrictions are satisfied: (i) *the variable* w *is not free in* H, (ii) *the variables* $\mathsf{v}_1, \ldots, \mathsf{v}_n, \mathsf{w}$ *are distinct,* (iii) S *has no free variables other than* $\mathsf{v}_1, \ldots, \mathsf{v}_n, \mathsf{w}$, (iv) S *and* H *are formulas in which the only non-logical constants are primitive symbols and previously defined symbols of the theory, and* (v) *the formula* $\mathsf{H} \rightarrow (\mathrm{E!}\mathsf{w})\mathsf{S}$ *is derivable from the axioms and preceding definitions of the theory.*

In view of the several detailed explanations of similar rules in preceding sections, specific comments are not needed. Several applications are given in the exercises.*

* A philosophical concept related to conditional definitions is Rudolf Carnap's notion of a *reduction sentence,* which provides a method of relating dispositional predicates like 'being soluble in water' to directly observable predicates. A possible reduction sentence for the solubility predicate is:

If x is placed in water, then x is soluble in water if and only if x dissolves.

For further details see Carnap's article, "Testability and Meaning," *Philosophy of Science,* Vol. 3 (1936) pp. 419–471 and Vol. 4 (1937) pp. 1–40.

EXERCISES

1. State rules for conditional definitions of relation symbols.
2. State rules for conditional definitions of individual constants.
3. Give a conditional definition of the property of a number being odd. (HINT: The hypothesis will be: x is an integer.)
4. Give a conditional definition of the property of a number being prime.
5. Give a conditional definition of the square root operation.
6. Assuming exponential operations are already defined give a conditional definition of the logarithmic operation to the base 10.

§ 8.7 Five Approaches to Division by Zero. We have already mentioned several aspects of the problem of division by zero, and have remarked that there is no uniformly satisfactory solution. In this section we want to examine five approaches to the problem. The next to last of the five yields the solution which is probably most consonant with ordinary mathematical practice.

The first approach differs from the others in that it recommends a change in the basic logic to deny meaning to expressions like:

$$\frac{1}{0} = \frac{1}{0}.$$

Without attempting to characterize the basic changes necessary, we may still offer some general objections to this approach. The first objection is that it is undesirable to complicate the basic rules of logic unless it is absolutely necessary. In other words, change the foundations of inference only if all other approaches have failed. Second, if such a change were adopted, the very meaningfulness of expressions would sometimes be difficult if not impossible to decide. For example, assume that we have added to our axioms of Chapter 7 sufficient axioms to obtain the expected theorems on the natural numbers (i.e., the positive integers). Consider now the expression:

(1) *For every natural number* n, $\dfrac{1}{n^*} = \dfrac{1}{n^*}$,

where n^* is the unary operation defined as follows: $n^* = 1$ if n is an odd integer or n is an even integer which is the sum of two prime numbers; $n^* = 0$ if n is an even integer which is not the sum of two primes. The problem of the existence of even integers which are not the sum of two primes is a famous problem of mathematics which is still unsolved (Goldbach's hypothesis). Thus on the basis of the first approach the *meaningfulness* (not the truth or falsity) of (1) is an open question.

The second approach is to let $x/0$ be a real number, but to define division by the conditional definition stated in the last section:

$$\text{If} \quad y \neq 0 \quad \text{then} \quad x/y = z \quad \text{if and only if} \quad x = y \cdot z.$$

In this case for every number x, $x/0$ is a real number, but we are not able to prove what number it is. In fact, we cannot even decide on the truth or falsity of the simple assertion:

$$\frac{1}{0} = \frac{2}{0}. \tag{2}$$

The inability to prove or disprove (2) is an argument against the second approach, since we want our axioms to be as complete as possible.

The third approach agrees with the second in making $x/0$ a real number, but it differs in making $x/0 = 0$ for all x. This eliminates the undecidability of statements like (2). For the third approach the appropriate definition of division is a proper definition similar to the axiom-free definition discussed at the end of § 8.5:

$$x/y = z \leftrightarrow [(y \neq 0 \rightarrow x = y \cdot z) \ \& \ (y = 0 \rightarrow z = 0)].$$

An advantage of the third approach is that it permits the definition of zero by a straightforward proper definition fully satisfying the criteria of eliminability and non-creativity. The main disadvantage of this approach is the one mentioned in § 8.5: many mathematicians feel uneasy with the identity:

$$\frac{x}{0} = 0.$$

The fourth approach agrees with the second and third in requiring no basic change of logic; it differs in placing the object $x/0$ outside the domain of real numbers; more precisely it differs in not making it possible to prove that $x/0$ is a real number. The basic idea is to introduce a predicate 'R' which means 'is a real number'. Hypotheses using this predicate must be added to all the axioms stated in Chapter 7. Furthermore, to guarantee that addition and multiplication of numbers yield numbers, we must add the two closure axioms:

$$R(x) \ \& \ R(y) \rightarrow R(x + y),$$

$$R(x) \ \& \ R(y) \rightarrow R(x \cdot y).$$

The introduction of the predicate 'R' widens our domain of individuals, for now it does not follow that everything in the domain is a real number (in particular that $1/0$ is a real number). The introduction of 'R' has the further consequence of making it natural to make all our definitions of arithmetical relations and operations conditional. There is no point in defining the relation of equal to or less than, for instance, for things which are not numbers. Thus we would have:

$$[R(x) \ \& \ R(y)] \rightarrow [x \leq y \leftrightarrow (x = y \lor x < y)].$$

The definition of division would be:

$$(3) \qquad [R(x) \;\&\; R(y) \;\&\; R(z) \;\&\; y \neq 0] \rightarrow [x/y = z \leftrightarrow x = y \cdot z].$$

If x is a real number, with (3) at hand we cannot prove:

$$\frac{x}{0} \text{ is a real number}$$

and we cannot prove:

$$\frac{x}{0} \text{ is not a real number,}$$

but we are not faced with the counterintuitive situation of being forced to call $x/0$ a real number. The situation can be improved by introducing into our system a primitive symbol for some object which is not a real number. Without saying what the object is, let us designate it by 'ν'. Then we have the axiom:

$$-R(\nu),$$

that is, the assertion that ν is not a real number. We may now define division by:

$$(4) \qquad [R(x) \;\&\; R(y)] \rightarrow [x/y = z$$
$$\leftrightarrow [(y \neq 0 \rightarrow (R(z) \;\&\; x = y \cdot z)) \;\&\; (y = 0 \rightarrow z = \nu)]].$$

The virtue of (4) is that it definitely places $x/0$ outside the domain of real numbers for any number x. Such a consequence would seem to be in closest accord with ordinary mathematical usage. Definition (4) also has the virtue of making

$$\frac{x}{0} = \frac{y}{0},$$

where x and y are real numbers, thus eliminating a vast proliferation of odd mathematical entities. On the other hand, to see once for all that all is not for the best in this best of all possible worlds, notice that if we adopt (4) we cannot decide whether or not ν/ν is a real number. If we use conditional definitions and insist on not tampering with the basic law of identity for terms, then we must be prepared for the undecidability of the status of entities like ν/ν. We may take the attitude that everything is in good order in the domain where we intend to use the division operation, and we really do not care what is going on elsewhere if no inconsistencies can creep in.

A fifth approach should be mentioned which is of considerable theoretical importance but does not correspond at all to ordinary mathematical practice. The idea is simple: banish operation symbols and individual constants, and use only relation symbols. Thus '0' is replaced by the primitive one-place predicate 'Z', where it is intended that $Z(x)$ means that x is an identity element with respect to addition. The ternary relation symbol 'A' is used to replace the addition operation symbol:

$$A(x, y, z) \leftrightarrow x + y = z.$$

Similarly the ternary relation symbol 'M' is used for multiplication:

$$M(x, y, z) \leftrightarrow x \cdot y = z.$$

With this apparatus we may easily give a proper definition of the division relation symbol:

$$D(x, y, z) \leftrightarrow -Z(y) \ \& \ M(y, z, x).$$

In this approach there is no need for unusual mathematical entities, but it is extraordinarily awkward to work continually with relation symbols rather than operation symbols. For example, the associativity of addition has to be expressed in some manner like the following:

$$A(x, y, w) \ \& \ A(w, z, s_1) \ \& \ A(y, z, v) \ \& \ A(x, v, s_2) \rightarrow s_1 = s_2.$$

§ 8.8 Padoa's Principle and Independence of Primitive Symbols. When the primitive symbols of a theory are given, it is natural to ask if it would be possible to define one of them in terms of the others. The Italian logician Alessandro Padoa formulated in 1900 a principle applying the method of interpretation which may be used to show that the primitive symbols are independent, that is, that one may not be defined in terms of the other. The principle is simple: to prove that a given primitive symbol is independent of the remaining primitives, find two interpretations of the axioms of the theory such that the given primitive has two different interpretations and the remaining primitive symbols have the same interpretation. For instance, consider the theory of preference based on the primitive relation symbols 'P' (for strict preference) and 'I' (for indifference). The axioms of the theory are:

A1. *If xPy & yPz, then xPz.*
A2. *If xIy & yIz, then xIz.*
A3. *Exactly one of the following:*

$$xPy, yPx, xIy.$$

We want to show that 'P' is independent of 'I', that is, cannot be defined in terms of 'I'. Let the domain of interpretation for both interpretations be the set $\{1, 2\}$. Let 'I' be interpreted as identity in both cases. In one case let 'P' be interpreted as '$<$' and in the other case as '$>$'. In the first interpretation, we have:

$$1\,P\,2$$

since

$$1 < 2,$$

and consequently by Axiom A3

$$\text{not}\quad 2\,P\,1.$$

But in the second interpretation, we have:

$$2\,P\,1$$

since

$$2 > 1.$$

Now if 'P' were definable in terms of 'I' then 'P' would have to be the same in both interpretations, since 'I' is. However, 'P' is not the same, and we conclude that 'P' cannot be defined in terms of 'I'.

To make clear the procedure for applying Padoa's principle to any theory formalized in first-order predicate logic with identity, we want now to make more precise the general definition of independence of a primitive symbol and also to characterize more sharply the notion of two interpretations of a theory being *different* for a given primitive symbol of the theory.

Let R be an n-place primitive relation symbol of a theory. Then we say that R is *dependent* on the other primitive symbols of the theory if a formula of the form

$$\mathsf{R}(\mathsf{v}_1, \ldots, \mathsf{v}_n) \leftrightarrow \mathsf{S}$$

may be derived from the axioms, where (i) $\mathsf{v}_1, \ldots, \mathsf{v}_n$ are distinct variables, (ii) the only free variables in S are $\mathsf{v}_1, \ldots, \mathsf{v}_n$, and (iii) the only non-logical constants occurring in S are the other primitive symbols of the theory.

The close relation between this definition of dependence of a primitive relation symbol and the rule for defining new relation symbols in a theory is obvious and expected. The definitions of dependence of operation symbols and individual constants are similar and will be left as exercises.

We now want to use the definition of dependence for relation symbols to sharpen the description of Padoa's principle for proving independence.* To prove an n-place primitive relation symbol 'R' independent of the other primitive symbols of a theory, we need to find two interpretations of the theory, that is, two interpretations of the axioms of the theory such that:

* Padoa's original discussion of these matters is not entirely adequate. The standard references for a complete discussion are J. C. C. McKinsey "On the Independence of Undefined Ideas," *Bulletin of the American Mathematical Society*, Vol. 41 (1935) pp. 291–297, and Alfred Tarski, "Einige methodologische Untersuchungen über die Definierbarkeit der Begriffe," *Erkenntnis*, Vol. 5 (1935–1936) pp. 80–100. An English translation of Tarski's article is to be found in Alfred Tarski, *Logic, Semantics, Metamathematics*, Oxford, 1956.

(i) The domain of both interpretations is the same.

(ii) The two interpretations are the same for all other primitive symbols of the theory.

(iii) Let 'R_1' be the first interpretation of 'R' and 'R_2' the second, then 'R_1' and 'R_2' must be different in the following respect: there are elements $x_1, \ldots, x_n$ in the domain of interpretation such that

(a) $\quad$ '$R_1(x_1, \ldots, x_n)$' is true,

and

(b) $\quad$ '$R_2(x_1, \ldots, x_n)$' is false.

To see that two such interpretations establish the independence of 'R', suppose that 'R' is dependent on the other primitive symbols of the theory; that is, suppose that there is a formula:

$$R(x_1, \ldots, x_n) \leftrightarrow \mathsf{S} \tag{1}$$

of the kind demanded by the definition of dependence such that (1) is derivable from the axioms of the theory. As before, let the subscript '1' refer to the first interpretation and '2' to the second interpretation. In both interpretations we must have:

$$R_1(x_1, \ldots, x_n) \leftrightarrow \mathsf{S}_1, \tag{2}$$

and

$$R_2(x_1, \ldots, x_n) \leftrightarrow \mathsf{S}_2, \tag{3}$$

since (1) is a logical consequence of the axioms of the theory. Moreover, since all primitive symbols except 'R' are the same in both interpretations, we also have:

$$\mathsf{S}_1 \leftrightarrow \mathsf{S}_2. \tag{4}$$

From (2), (3), and (4) we infer that

$$R_1(x_1, \ldots, x_n) \leftrightarrow R_2(x_1, \ldots, x_n), \tag{5}$$

which contradicts (a) and (b) of (iii) and proves that our supposition of dependence is absurd.

The definition of *differentness* of interpretation is similar for operation symbols and individual constants; * precisely the same kind of argument as that just given shows that finding the two appropriately different interpretations is adequate to prove the independence of operation symbols or individual constants.

A simple example will illustrate how Padoa's principle is used to prove the independence of an operation symbol. We consider the theory whose

* A classical way of describing this difference is that the primitive symbol being proved independent must have a different *extension* in the two interpretations.

primitive symbols are a one-place relation symbol 'P' (that is, 'P' denotes a property) and a binary operation symbol '$\circ$'. The axioms of the theory are:

A1. $P(x)$ & $P(y) \rightarrow P(x \circ y)$.
A2. $P(x)$ & $-P(y) \rightarrow -P(x \circ y)$.
A3. $x \circ y = y \circ x$.

We indicate the first and second interpretations by subscripts as previously. The common domain of interpretation is the set of positive integers, and

$$P_1(x) \leftrightarrow P_2(x) \leftrightarrow x \quad \text{is an even integer,}$$
$$x \circ_1 y = x + y,$$
$$x \circ_2 y = x + y + 2.$$

We easily verify that both interpretations satisfy the axioms. The independence of '$\circ$' follows from the fact that we have:

$$1 \circ_1 2 = 3$$
$$1 \circ_2 2 = 5 \neq 3.$$

Without going into details it may be mentioned that Padoa's principle may be easily extended to theories which assume in their formalization not only first-order predicate logic with identity but also a good deal of classical mathematics. Probably some of the most interesting applications of Padoa's principle to such "advanced" theories are in the domain of empirical science, for there is a fair amount of confused discussion regarding the interdefinability of various empirical concepts. In discussions of the foundations of mechanics, for instance, it is often claimed, following Ernst Mach, that the concept of mass can be defined in terms of the concept of acceleration, or that the concept of force may be defined in terms of the concepts of mass and acceleration. However, it can be shown by application of Padoa's principle that under several plausible axiomatizations the concepts of mass and force are each independent of the other primitive concepts of mechanics.*

EXERCISES

Be careful not to violate Rule VII, §4.2 in doing these exercises.

1. Consider the weak theory of preference discussed in this section. Prove that the primitive relation symbol 'I' is dependent on 'P'.
2. Define the notion of dependence for operation symbols.
3. Characterize the two different interpretations needed to prove that an operation symbol is an independent primitive symbol of a theory.
4. Consider the theory of groups as given in Chapter 5 and based on three primitive symbols. Is the operation symbol '$^{-1}$' for the inverse of an element of a group dependent or independent?

* See §12.5.

5. Consider the axioms for the measurement of mass given in Exercise 9, § 4.5. Prove that the primitive symbols 'Q' and '$\star$' are both independent.

6. Define the notion of dependence for individual constants.

7. Characterize the two different interpretations needed to prove that an individual constant is an independent primitive symbol of a theory.

8. Consider the theory of groups discussed in Chapter 5. Is the individual constant 'e' dependent or independent in this theory?

9. Consider the axiomatization of Huntington's for the "informal part" of *Principia Mathematica*, already discussed as Exercise 12, § 4.1. The three primitive symbols are a one-place relation symbol 'C', a binary operation symbol '$+$', and a unary operation symbol '$^{\prime}$'; and the five axioms are:

A1. *If* $C(x + y)$ *then* $C(y + x)$.
A2. *If* $C(x)$ *then* $C(x + y)$.
A3. *If* $C(x^{\prime})$ *then* $-C(x)$.
A4. *If* $-C(x^{\prime})$ *then* $C(x)$.
A5. *If* $C(x + y)$ & $C(x^{\prime})$ *then* $C(y)$.

To show that the primitive symbol 'C' is independent, consider the domain of interpretation consisting of the numbers 1, 2, 3, 4. In the first interpretation let 1 and 2 have the property C, i.e., $C_1(1)$ & $C_1(2)$. The binary operation symbol '$+$' is given by the following table for both interpretations (the use of such finite tables is common in proving independence of axioms or primitive symbols).

+	1	2	3	4
1	1	1	1	1
2	1	2	1	2
3	1	1	3	3
4	1	2	3	4

The table is used in the following manner. To find what element $3 + 2$ is, we look at the entry occurring in the third row and second column and find:

$$3 + 2 = 1.$$

For the joint interpretation of the unary operation $^{\prime}$ we have the following table:

x	$x^{\prime}$
1	4
2	3
3	2
4	1

In the second interpretation the only change is in 'C'. We now have: $C_2(1)$ & $C_2(3)$. It is easily verified that both interpretations satisfy the axioms. Since

$$C_1(2) \ \& \ -C_2(2),$$

we conclude that 'C' is an independent primitive symbol. Prove that the two operation symbols are also independent.

10. Using tables similar to those given in the previous exercise, prove that the five axioms of Exercise 9 are independent.

PART II

ELEMENTARY INTUITIVE SET THEORY

CHAPTER 9

SETS

§ 9.1 **Introduction.** In this and the next two chapters we develop the elementary theory of sets in an intuitive manner. The present chapter is concerned with arbitrary sets; Chapter 10 with those sets which are relations; and Chapter 11 with those sets which are functions.

Since in Chapter 7 a certain portion of arithmetic was developed in a logical fashion from a small list of axioms, some readers may feel that it would be more appropriate to develop the theory of sets axiomatically rather than intuitively. However, there are good grounds for introducing the concepts of set theory informally. The concepts of arithmetic are familiar to everyone; an axiomatic presentation of arithmetic may continually call upon familiar facts to guide and motivate its lines of development. Although the concepts of set theory are logically simpler in several respects than those of arithmetic, they are not generally familiar. The purpose of Chapters 9–12 is to provide such familiarization.

§ 9.2 **Membership.** By a *set* we mean any kind of a collection of entities of any sort.* Thus we can speak of the set of all Americans, or the set of all integers, or the set of all Americans and integers, or the set of all straight lines, or the set of all circles which pass through a given point. Many other words are used synonymously with 'set': for instance, 'class', 'collection', and 'aggregate'. We shall sometimes use these other words for the sake of literary variety.

We say of the members of a set that they *belong to* the set; it is customary to use the symbol '$\in$' resembling the Greek letter epsilon, as an abbreviation for 'belongs to'. Thus we write:

Elizabeth II belongs to the class of women,

or simply:

(1) Elizabeth II $\in$ the class of women.

* Although the notion of a set was introduced in Chapter 2 in connection with the theory of inference, the discussion here is self-contained.

In ordinary language (1) would be expressed by:

Elizabeth II is a woman.

Thus the verb 'to be' often has the meaning of set membership.

We use the word 'set' in such a way that a set is completely determined when its members are given; i.e., if A and B are sets which have exactly the same members, then $A = B$. Thus we write:

The set of equilateral triangles = the set of equiangular triangles,

for something belongs to the first set if and only if it belongs to the second, since a triangle is equilateral if and only if it is equiangular. This general principle of identity for sets is usually called the *principle of extensionality for sets;* it may be formulated symbolically thus:

$$A = B \leftrightarrow (x)(x \in A \leftrightarrow x \in B). \tag{2}$$

Sometimes one finds it convenient to speak of a set even when it is not known that this set has any members. A geneticist may wish to talk about the set of women whose fathers, brothers, and husbands are all hemophiliacs, even though he does not know of an example of such a woman. And a mathematician may wish to talk about maps which cannot be colored in fewer than five colors, even though he cannot prove that such maps exist (the question whether there are such maps, as a matter of fact, is a famous unsolved mathematical problem). Thus it is convenient to make our usage of the term 'set' wide enough to include *empty* sets, i.e., sets which have no members.

From our analysis of implications in Chapter 1, it is clear that if A is a set which has no members, then the following statement is true, since the antecedent is always false:

$$(x)(x \in A \rightarrow x \in B). \tag{3}$$

And, correspondingly, if B is empty, i.e., has no members, then it is true that:

$$(x)(x \in B \rightarrow x \in A). \tag{4}$$

From (2), (3), and (4) we conclude that if two sets A and B are empty, then:

$$A = B;$$

that is to say, there is just one empty set; for given two empty sets, it follows from the principle of extensionality for sets that the two sets are identical. Hence we shall speak of *the* empty set, which we denote by a capital Greek lambda:

$$\Lambda.$$

Λ is the set such that for every x, x does not belong to Λ; that is, symbolically:

$$(x)-(x \in \Lambda),$$

and we abbreviate '$-(x \in \Lambda)$' to '$x \notin \Lambda$', and write:

$$(x)(x \notin \Lambda).$$

We shall find it convenient in general to use the notation '$\notin$' to indicate that something does *not belong to* a set.

Often we shall describe a set by writing down names of its members, separated by commas, and enclosing the whole in braces. For instance, by:

$$\{\text{Roosevelt, Parker}\}$$

we mean the set consisting of the two major candidates in the 1904 American Presidential election. By:

$$\{1, 3, 5\}$$

we mean the set consisting of the first three odd positive integers. It is clear that

$$\{1, 3, 5\} = \{1, 5, 3\}$$

(for both sets have the same members: the order in which we write down the members of a set is of no importance). Moreover,

$$\{1, 1, 3, 5\} = \{1, 3, 5\}$$

(for we do not count an element of a set twice).

The members of a set can themselves be sets. Thus a political party can be conceived as a certain set of people, and it may be convenient to speak of the set of political parties in a given country. Similarly we can have sets whose members are sets of integers; for instance, by:

$$\{\{1, 2\}, \{3, 4\}, \{5, 6\}\}$$

we mean the set which has just three members, namely, $\{1, 2\}$, $\{3, 4\}$, and $\{5, 6\}$. By:

$$\{\{1, 2\}, \{2, 3\}\}$$

we mean the set whose two members are $\{1, 2\}$ and $\{2, 3\}$. By:

$$\{\{1, 2\}, \{1\}\}$$

we mean the set whose two members are the sets $\{1, 2\}$ and $\{1\}$.

A set having just one member is not to be considered identical with that member. Thus the set $\{\{1, 2\}\}$ is not identical with the set $\{1, 2\}$; this is clear from the fact that $\{1, 2\}$ has two members, whereas $\{\{1, 2\}\}$ has just one member (namely, $\{1, 2\}$). Similarly,

$$\{\text{Elizabeth II}\} \neq \text{Elizabeth II},$$

for Elizabeth II is a woman, while {Elizabeth II} is a set.

Ordinarily it is not true that a set is a member of itself. Thus the set of chairs is not a member of the set of chairs: i.e., the set of chairs is not itself a chair. This remark illustrates the very great difference between identity and membership; for the assertion that

$$A = A$$

is always true, whereas that

$$A \in A$$

is usually false.*

The relation of membership also differs from the relation of identity in that it is not symmetric: from $A \in B$ it does not follow that $B \in A$. For instance, we have:

$$2 \in \{1, 2\},$$

but:

$$\{1, 2\} \notin 2.$$

Moreover, the relation of membership is not transitive: from $A \in B$ and $B \in C$ it does not follow that $A \in C$. Thus, for example, we have:

$$2 \in \{1, 2\}$$

and:

$$\{1, 2\} \in \{\{1, 2\}, \{3, 4\}\}$$

but:

$$2 \notin \{\{1, 2\}, \{3, 4\}\},$$

for the only members of $\{\{1, 2\}, \{3, 4\}\}$ are $\{1, 2\}$ and $\{3, 4\}$, and neither of these sets is identical with 2.

It should be noticed that if, for instance, $\{a, b\}$ is any set with two members, then, for every x, $x \in \{a, b\}$ if and only if either $x = a$ or $x = b$, that is, symbolically:

$$(x)(x \in \{a, b\} \leftrightarrow (x = a \lor x = b)).$$

Similarly, if $\{a, b, c\}$ is a set with three members, then $x \in \{a, b, c\}$ if and only if either $x = a$ or $x = b$ or $x = c$. It is for this reason that we just said that $2 \notin \{\{1, 2\}, \{3, 4\}\}$; for if $x \in \{\{1, 2\}, \{3, 4\}\}$, then either $x = \{1, 2\}$ or $x = \{3, 4\}$; and since $2 \neq \{1, 2\}$ and $2 \neq \{3, 4\}$, it follows that $2 \notin \{\{1, 2\}, \{3, 4\}\}$.

It should also be noticed that there is a close relationship between saying that something has a property and saying that it belongs to a set: a thing

* In most standard systems of axiomatic set theory no set may be a member of itself.

has a given property if and only if it belongs to the set of things having the property.* Thus to say that 6 has the property of being an even number amounts to saying that 6 belongs to the set of even numbers. Since we can always in this way express things in terms of membership in sets instead of in terms of the possession of properties, we do not find it necessary to give any more detailed discussion of properties.

In § 5.1 we expressed the principle of the identity of indiscernibles in terms of properties. Expressed in terms of membership the principle becomes: If y belongs to every set to which x belongs, then $y = x$. Put in this form, the principle has perhaps a more obvious character than it has when put in terms of properties. For $x \in \{x\}$ (i.e., x belongs to the set whose only member is x), and hence, if y belongs to every set to which x belongs, we conclude that $y \in \{x\}$, so that $y = x$.

§ 9.3 Inclusion. If A and B are sets such that every member of A is also a member of B, then we call A a *subset* of B, or say that A *is included in* B. We often use the sign '$\subseteq$' as an abbreviation for 'is included in'. Thus we can write, for instance:

The set of Americans is a subset of the set of men,

or:

The set of Americans is included in the set of men,

or simply:

The set of Americans $\subseteq$ the set of men.

Symbolically we have:

$$(1) \qquad A \subseteq B \leftrightarrow (x)(x \in A \rightarrow x \in B).$$

It is clear that every set is a subset of itself; i.e., for every set A we have: $A \subseteq A$. Moreover, the relation of inclusion is transitive; i.e., if $A \subseteq B$ and $B \subseteq C$, then $A \subseteq C$ (for if every member of A is a member of B, and every member of B is a member of C, then every member of A is a member of C). The relation of inclusion is not symmetric, however; thus $\{1, 2\} \subseteq \{1, 2, 3\}$, but it is not the case that $\{1, 2, 3\} \subseteq \{1, 2\}$.

It is intuitively obvious that identity, membership, and inclusion are distinct and different notions, but it is still somewhat interesting to observe that their distinction may be inferred simply from considering the questions of symmetry and transitivity. Thus inclusion is not the same as identity, since identity is symmetric while inclusion is not. And inclusion is not the same as membership, since inclusion is transitive while membership is not. And we have seen earlier that identity is not the same as

* This principle is sometimes called *Cantor's axiom for sets*, after the founder of set theory, G. Cantor (1845–1918). This principle must be suitably restricted to avoid contradiction.

membership, since identity is both symmetric and transitive, while membership is neither. In everyday language all three notions are expressed by the one overburdened verb 'to be'. Thus in everyday language we write:

Elizabeth II is the present Queen of England,
Elizabeth II is a woman,
Women are human beings.

But in the more exact language being developed here:

Elizabeth II = the present Queen of England,
Elizabeth II $\in$ the class of women,
The class of women $\subseteq$ the class of human beings.

When $A \subseteq B$, the possibility is not excluded that $A = B$; it may happen also that $B \subseteq A$, so that A and B have exactly the same members, and hence are identical.

When $A \subseteq B$ but $A \neq B$, we call A a *proper subset* of B. We use '$\subset$' as an abbreviation for 'is a proper subset of'. Thus:

$$\{1, 2\} \subset \{1, 2, 3\}$$

is true, as is also:

$$\{1, 2\} \subseteq \{1, 2, 3\};$$

but:

$$\{1, 2, 3\} \subset \{1, 2, 3\}$$

is false, although:

$$\{1, 2, 3\} \subseteq \{1, 2, 3\}$$

is of course true.

Symbolically we have:

$$A \subset B \leftrightarrow A \subseteq B \;\&\; A \neq B.$$

EXERCISES

1. Which of the following statements are true (for all sets A, B, and C)?

(a) If $A = B$ and $B = C$, then $A = C$.
(b) If $A \in B$ and $B \in C$, then $A \in C$.
(c) If $A \subseteq B$ and $B \subseteq C$, then $A \subseteq C$.
(d) If $A = B$ and $B \in C$, then $A \in C$.
(e) If $A \in B$ and $B = C$, then $A \in C$.
(f) If $A \in B$ and $B \subseteq C$, then $A \in C$.
(g) If $A \subseteq B$ and $B \in C$, then $A \in C$.
(h) If $A \subset B$ and $B \in C$, then $B \subset C$.
(i) If $A \subset B$ and $B \subseteq C$, then $A \subset C$.
(j) If $A \in B$ and $B \subset C$, then $A \in C$.
(k) If $A \in B$ and $B \subset C$, then $A \subset C$.

2. For each of the statements in Exercise 1 which is false give an example of particular sets A, B, and C which show that the statement is not in general true.

3. Give an example of sets A, B, C, D satisfying the conditions:

$$A \subset B$$
$$B \in C$$
$$C \subset D$$
$$D = E$$

4. What is wrong with the following argument?

Socrates is a man. Men are numerous. Therefore, Socrates is numerous.

5. What is wrong with the following argument?

Tomcats are cats. Cats are a species. Therefore, tomcats are a species.

6. In each of the following examples decide which of the following statements are true: $A \in B$, $A \subseteq B$, $A \subset B$, $A = B$.

EXAMPLE (A).

$A = \{1, \{1\}, \text{Roosevelt}, 4\}$,
$B = \{1, \{1\}, \text{Roosevelt}, \text{Churchill}\}$.

EXAMPLE (B).

$A =$ the set of positive integers,
$B =$ the set of positive and negative integers.

EXAMPLE (C).

A is the set consisting of the following: the number 5, the set consisting of Roosevelt, the set consisting of the set consisting of the number 1.

B is the set consisting of the following: the number 5, Roosevelt, the set consisting of the set consisting of Roosevelt, the number 1.

EXAMPLE (D).

$A = \{1, 3, 4, 2, 9\}$
$B = \{1, 2 + 1, 1 + 8, 10, 2 + 0, 1008, 4\}$

EXAMPLE (E).

$A =$ the set whose members are the following: the set of all Presidents of the United States in the nineteenth century, Truman, McKinley, the smallest positive integer divisible by 5.

$B =$ the set whose members are the following: the set of all Presidents of the United States up to 1953, the number 10.

7. Which of the following statements are true for all sets A, B, and C?

(a) $A \not\subset B \,\&\, B \not\subset C \rightarrow A \not\subset C$
(b) $A \neq B \,\&\, B \neq C \rightarrow A \neq C$
(c) $A \in B \,\&\, \neg(B \subseteq C) \rightarrow A \notin C$
(d) $A \subset B \,\&\, B \subseteq C \rightarrow \neg(C \subset A)$
(e) $A \subseteq B \,\&\, B \in C \rightarrow A \notin C$

§ 9.4 The Empty Set. As mentioned earlier, the empty set, Λ, is characterized by the property that, for every x, $x \notin \Lambda$.

It should be noticed that, although nothing belongs to the empty set, the empty set can itself very well be a member of another set. Thus if we speak of the set of all subsets of the set $\{1, 2\}$, we are speaking of the set $\{\{1, 2\}, \{1\}, \{2\}, \Lambda\}$ which has four members; the three-member set $\{\{1, 2\}, \{1\}, \{2\}\}$ on the other hand, is the set of all non-empty subsets of $\{1, 2\}$.

We recall the fact that a set A is a subset of a set B if and only if every member of A is also a member of B, i.e., if and only if: for every x, if $x \in A$, then $x \in B$. In particular, the empty set Λ is a subset of a set B if and only if: for every x, if $x \in \Lambda$, then $x \in B$. Since always $x \notin \Lambda$, however, it is always true that if $x \in \Lambda$, then $x \in B$. Thus, for every set B, we have:

$$\Lambda \subseteq B.$$

That is, the empty set is a subset of every set. In addition, the empty set is the only set which is a subset of the empty set; for if $B \subseteq \Lambda$, then, since we also have: $\Lambda \subseteq B$, we can conclude that $B = \Lambda$.

We can summarize these facts about the empty set as follows:

$$\text{(i)} \qquad (x)(x \notin \Lambda),$$

$$\text{(ii)} \qquad (\exists A)(\Lambda \in A) \mathbin{\&} (\exists A)(\Lambda \notin A),$$

$$\text{(iii)} \qquad (A)(\Lambda \subseteq A),$$

$$\text{(iv)} \qquad (A)(A \subseteq \Lambda \leftrightarrow A = \Lambda).$$

§ 9.5 Operations on Sets. If A and B are sets, then by the *intersection* of A and B (in symbols: $A \cap B$) we mean the set of all things which belong both to A and to B. Thus, for every x, $x \in (A \cap B)$ if and only if $x \in A$ and $x \in B$; that is, symbolically:

$$\text{(1)} \qquad (x)(x \in A \cap B \leftrightarrow x \in A \mathbin{\&} x \in B).$$

If A is the set of all Americans, and B is the set of all blue-eyed people, then $A \cap B$ is the set of all blue-eyed Americans.

If A is the set of all men, and B is the set of all animals which weigh over ten tons, then $A \cap B$ is the set of all men who weigh over ten tons. In this case we notice that $A \cap B$ is the empty set (despite the fact that $A \neq \Lambda$, and $B \neq \Lambda$, since some whales weigh more than ten tons). When $A \cap B = \Lambda$, we say that A and B are *mutually exclusive.*

Our use of the term 'intersection' is similar to its use in elementary geometry, where by the intersection of two circles, for instance, we mean the points which lie on both circles. Some authors use, instead of '$\cap$', the dot '$\cdot$' which is used in algebra for multiplication; such authors often speak of the "product" of two sets, instead of their intersection.

If A and B are sets, then by the *union* of A and B (in symbols: $A \cup B$) we mean the set of all things which belong to at least one of the sets A and B. Thus, for every x, $x \in (A \cup B)$ if and only if either $x \in A$ or $x \in B$. (Notice that, as explained in Chapter 1, we use the connective 'or' in its *non-exclusive* sense: '$x \in A$ or $x \in B$' is false only in case both $x \notin A$ and $x \notin B$.) Symbolically:

$$(2) \qquad (x)(x \in A \cup B \leftrightarrow x \in A \vee x \in B).$$

If A is the set of all animals, and B is the set of all plants, then $A \cup B$ is the set of all living organisms. One often wishes to consider the union of two sets, however, even when they are not mutually exclusive. For instance, if A is the set of all human adults, and B is the set of all people less than 40 years old, then $A \cup B$ is the set of all human beings.

Some authors use the addition sign '+' instead of '$\cup$', and call the union of two sets their "sum".

If A and B are two sets, then by the *difference* of A and B (in symbols: $A \sim B$) we mean the set of all things which belong to A but not to B. Thus, for every x, $x \in A \sim B$ if and only if $x \in A$ and $x \notin B$; that is, symbolically:

$$(3) \qquad (x)(x \in A \sim B \leftrightarrow x \in A \;\&\; x \notin B).$$

If A is the set of all human beings, and B is the set of all human females, then $A \sim B$ is the set of all human males. One often wishes to consider the difference of two sets A and B, however, even when B is not a subset of A. For instance, if A is the set of human beings, and B is the set of all female animals, then $A \sim B$ is still the set of all human males, and $B \sim A$ is the set of all female animals which belong to a non-human species.

These operations on sets (intersection, union, and difference) can of course be iterated. Thus, suppose, for instance, that

$$\begin{aligned} A &= \{1, 2\}, \\ B &= \{1, 3, 5\}, \\ C &= \{2, 3, 5, 7\}, \\ D &= \{4, 5, 6, 7\}; \end{aligned}$$

then

$$A \cup B = \{1, 2\} \cup \{1, 3, 5\} = \{1, 2, 3, 5\}$$

and hence

$$C \cap (A \cup B) = \{2, 3, 5, 7\} \cap \{1, 2, 3, 5\} = \{2, 3, 5\}$$

and hence

$$D \sim [C \cap (A \cup B)] = \{4, 5, 6, 7\} \sim \{2, 3, 5\} = \{4, 6, 7\}.$$

Similarly, since

$$C \cup D = \{2, 3, 5, 7\} \cup \{4, 5, 6, 7\} = \{2, 3, 4, 5, 6, 7\},$$

we have:

$$(A \cup B) \cap (C \cup D) = \{1, 2, 3, 5\} \cap \{2, 3, 4, 5, 6, 7\} = \{2, 3, 5\}.$$

EXERCISES

1. If A is the set of all even positive integers, and B is the set of all integers which are greater than 10, what are the following?

$$A \cup B \qquad A \sim B$$
$$A \cap B \qquad B \sim A$$

2. If A is any set, what are the following?

$$A \cap \Lambda \qquad A \sim \Lambda$$
$$A \cup \Lambda \qquad \Lambda \sim A$$

3. Letting:

A = the set of all positive integers $\qquad C = \{2, 4\}$
$B = \{3, 5\} \qquad D = \{1, 2\}$

find the following:

$$A \sim B \qquad (B \cup C) \cap (B \cup D)$$
$$A \sim C \qquad A \sim (C \cap D)$$
$$A \sim D \qquad (A \sim C) \cup (A \sim D)$$

4. Letting

$A = \{1, 2\} \qquad F = \{\{1, 3\}, 1, 2\}$
$B = \{\{3, 4\}, 1, 7\} \qquad G = \{\{1, 2\}, 1, 7\}$
$C = \{\{3, 4\}, 1, 2\} \qquad H = \{\{1, 2\}, 1, 2\}$
$D = \{\{1, 3\}, 4, 7\} \qquad I = \{\{1, 2\}, \{1\}, \{2\}\}$
$E = \{\{1, 3\}, 1, 7\} \qquad J = \{\{1\}, \{2\}\}$

find the following:

$$A \cap B \qquad A \cap G$$
$$A \cap C \qquad A \cap H$$
$$A \cap D \qquad A \cap I$$
$$A \cap E \qquad A \cap J$$
$$A \cap F$$

5. Find the following:

$$\Lambda \cap \{\Lambda\}$$
$$\{\Lambda\} \cap \{\Lambda\}$$
$$\{\Lambda, \{\Lambda\}\} \sim \Lambda$$
$$\{\Lambda, \{\Lambda\}\} \sim \{\Lambda\}$$
$$\{\Lambda, \{\Lambda\}\} \sim \{\{\Lambda\}\}$$

6. Letting

$A = \{1\} \qquad C = \{1, 2\} \qquad E = \{1, \{1, \{1\}\}\}$
$B = \{1, \{1\}\} \qquad D = \{1, 2, \{1\}\}$

find the following:

$$A \cap B \qquad (C \cup D) \sim B$$
$$A \cup B \qquad (A \cap D) \sim E$$
$$(A \cup B) \cap C \qquad \{B\} \cap E$$
$$\{A\} \cap B \qquad (\{A\} \cup D) \cap (E \sim C).$$

7. Using the sets of Exercise 6, which of the following statements are true?

(a) $A \in B$
(b) $A \subseteq B$
(c) $B \in E$
(d) $B \subseteq E$
(e) $C \in D$
(f) $C \subseteq D$
(g) $B \subset D$
(h) $B \sim A \in D$
(i) $E \sim B \subseteq A$

8. Taking the binary relation symbol '$\in$' as the single primitive symbol of set theory, define the operation symbols of intersection, union, and difference in a manner which satisfies the rules of definition for operation symbols given in Chapter 8.

§ 9.6 Domains of Individuals. Often one is interested, not in all possible sets, but merely in all the subsets of some fixed set. Thus in sociology, for instance, it might be natural to be talking mostly about sets of human beings; and to speak with the understanding that when a set was mentioned it was to be taken to be a set of people, unless an explicit statement to the contrary was made. In such discourse one might say, for example, 'the set of albinos', and it would be understood that one was referring only to the set of albino people, and not also to albino monkeys, albino mice, and other albino animals.

Similarly, in some geometrical discourse it is natural to use the word 'set' to mean 'set of points'. (Sometimes in mathematics people press into service in some specialized sense some of the various words mentioned above as being here taken to be synonymous with 'set': a geometrician might, for example, adopt the convention of speaking of *sets* of points, *classes* of sets of points, and *aggregates*—or perhaps *families*—of geometrical curves.)

When a fixed set D is taken as given in this way, and one confines himself to the discussion of subsets of D, we shall call D the *domain of individuals*, or sometimes the *domain of discourse.* Thus the domain of individuals of the sociological discussion mentioned above is the set of all human beings.

We shall denote the domain of individuals by 'V'. It is important to remember that though 'Λ', as seen earlier, stands for a uniquely determined entity (the empty set), the symbol 'V' is interpreted differently in different discussions. In one context 'V' may stand for the set of all human beings, in another for the set of points of space, and in another for the set of positive integers.

When dealing with a fixed domain of individuals V, it is convenient to introduce a special symbol for the difference of V and a set A:

$$\sim A = V \sim A.$$

We call $\sim A$ the *complement of* A. More generally, the difference $B \sim A$ of B and A is called the *complement of A relative to B*; so *the* complement of a set is simply its complement relative to the given domain of individuals.

As should be expected, we may find the complement of a set which itself results from operations on other sets. For example, let

$$\begin{aligned} V &= \{1, 2, 3\} \\ A &= \{1, 2\} \\ B &= \{2, 3\}. \end{aligned}$$

Then

$$\sim A = V \sim A = \{1, 2, 3\} \sim \{1, 2\} = \{3\},$$

and correspondingly,

$$\begin{aligned} \sim(A \cup B) = V \sim (A \cup B) &= \{1, 2, 3\} \sim (\{1, 2\} \cup \{2, 3\}) \\ &= \{1, 2, 3\} \sim \{1, 2, 3\} = \Lambda. \end{aligned}$$

Some further facts about the operation of complementation are mentioned in § 9.9.

EXERCISES

1. If

$$V = \{1, 2, 3, 4, 5\} \qquad \begin{aligned} A &= \{1, 2\} \\ B &= \{2, 3\} \end{aligned}$$

what are the following:

$$\begin{array}{lll} \sim A & \sim A \cap \sim B & A \sim (\sim B) \\ \sim B & \sim(A \cap B) & \sim A \sim B \\ & & \sim A \sim (\sim B) \end{array}$$

2. Let V be the set of all positive integers, and let

A = set of all even positive integers,
B = set of all odd positive integers,
C = set of all positive integers greater than 10,
D = set of all positive integers less than 15.

Find:

(a) $\sim A$
(b) $\sim(A \cup B)$
(c) $\sim(A \cap B)$
(d) $\sim C$
(e) $D \sim C$
(f) $\sim(D \sim C)$
(g) $C \cup \sim D$
(h) $\sim C \cap D$
(i) $A \sim (\sim C)$
(j) $(A \cap D) \sim (\sim B)$
(k) $A \sim (\sim C \cup D)$

3. Consider the axioms for groups given in Chapter 5. Let the domain of interpretation be the set of all subsets of some given domain V of individuals. Inter-

pret the inverse operation symbol as complementation. Let the empty set be the identity element. If the binary operation symbol is interpreted as:

(a) intersection of sets
(b) union of sets
(c) difference of sets

are the three group axioms satisfied? If not, state which axioms are not satisfied and give explicit counterexamples.

4. Same as Exercise 3 except that the identity element is interpreted as the domain V of individuals.

§ 9.7 Translating Everyday Language.* This section is devoted to the problem of translating sentences of everyday language into the symbolism we have been developing in this chapter. It should be realized, as was pointed out in Chapter 3, that the usage of everyday language is not so uniform that one can give unambiguous and categorical rules of translation. In everyday language we often use the same word for essentially different notions ('is', for example, for both '$\in$' and '$\subseteq$'); and, sometimes for literary elegance, we often use different words for the same notion ('is', 'is a subset of', and 'is included in', for example, for '$\subseteq$').

We consider here only those sentences which can be translated into a symbolism consisting just of letters standing for sets, parentheses, and the following symbols:

$$\cap,\ \cup,\ \sim,\ \Lambda,\ =,\ \neq,\ \subseteq.$$

Such a symbolism can handle statements involving one-place predicates very well, but it is not adequate to the complexities of thought that can be expressed by means of many-place predicates. This symbolism is essentially equivalent to the language of the classical theory of the syllogism. It is important to note that we are not using here the notion of membership; we restrict ourselves to sets all of which are on the same level—subsets of some fixed domain of individuals.

An English statement of the form 'All ... are ...', where the two blanks are filled with common nouns such as 'men' or 'Americans' or 'philosophers', means, of course, that the set of things described by the first noun is a subset of the set of things described by the second noun. Thus, for example:

(1) All Americans are philosophers

means:

The set of Americans $\subseteq$ the set of philosophers,

or, using 'A' as an abbreviation for 'the set of Americans', and 'P' as an

* This section may be omitted without loss of continuity.

abbreviation for 'the set of philosophers':

$$A \subseteq P.$$

We can also express the meaning of this statement in other, equivalent, ways:

$$A \cup P = P,$$

or:

$$A \cap \sim P = \Lambda,$$

etc., and these other modes of expression often turn out to be useful. In such discussions a domain of individuals is easily fixed and thus the size of complement sets such as $\sim P$. Here, for example, V can be taken as the set of all human beings.

We use the same mode of translation of statements of the form 'All ... are ...' also when the second blank is filled with an adjective. For example, we take:

(2) All Americans are mortal

to mean:

The class of Americans $\subseteq$ the class of mortal beings,

or, using obvious abbreviations:

$$A \subseteq M.$$

Sometimes, however, in contexts of this sort people suppress the word "all"—writing, for instance:

Tyrants are mortal

instead of:

(3) All tyrants are mortal,

or:

Women are fickle

instead of:

(4) All women are fickle,

which we should translate, respectively, by:

$$T \subseteq M$$

and:

$$W \subseteq F.$$

One must be on guard when translating statements of this kind, however; for ordinary language uses the same form also to express essentially different ideas. Thus, as we have seen before:

(5) Men are numerous

does not mean:

The set of men $\subseteq$ the set of numerous things

(i.e., that every man is numerous) but rather, letting M be the set of men and N be the set of sets which have numerous members:

$$M \in N.$$

Similarly:

(6) The apostles are twelve

means that the set of apostles belongs to the set of sets having just twelve members.

Corresponding to the distinction which we have made between membership and inclusion, the older logic made a distinction between the "distributive" and "collective" application of the predicate to the subject. Using this terminology, one says that in (1), (2), (3), and (4) the predicate is applied to the subject distributively, and that in (5) and (6) it is applied to the subject collectively.

An English statement of the form 'Some ... are ...', where the blanks are filled by common nouns (or perhaps the second blank is filled by an adjective) means that there exists something which is described by both terms: i.e., that the intersection of the two corresponding sets is not empty. Thus, for instance:

(7) Some Americans are philosophers

means that there exists at least one person who is both an American and a philosopher, and is accordingly translated:

$$A \cap P \neq \Lambda.$$

Although a statement of the form of (7) implies that the sets corresponding to subject and predicate are not empty, no such inference is to be drawn from a statement of the form of (1). Thus, for example, it is true that

All three-headed, six-eyed men are three-headed men,

but it is not true that

Some three-headed, six-eyed men are three-headed men.

(Modern logic differs in this point from the older logic, which allowed the inference of 'Some S are P' from 'All S are P'.)

An English statement of the form 'No ... are ...' (where, as before, the blanks are filled by common nouns) means that nothing belongs both to the set corresponding to the first noun, and to the set corresponding to the second noun: i.e., that the intersection of these two sets is empty. For instance, the sentence:

(8) No Americans are philosophers

is translated:

$$A \cap P = \Lambda.$$

Thus (2) has the same meaning as:

No Americans are immortal

since both can be translated:

$$A \cap \sim M = \Lambda.$$

An English statement of the form 'Some ... are not ...' (where the blanks are filled by common nouns) means that there exists something which belongs to the set corresponding to the first noun, and does not belong to the set corresponding to the second noun: i.e., that the intersection of the first set with the complement of the second is not empty. The sentence:

Some Americans are not philosophers

is translated:

$$A \cap \sim P \neq \Lambda.$$

We turn now to the problem of translating some statements of a more complicated sort.

The word 'and' often corresponds to the intersection of sets. Thus:

All Americans are clean and strong

is translated (using obvious abbreviations):

$$A \subseteq C \cap S.$$

The same applies to the word 'but'; thus:

Freshmen are ignorant but enthusiastic

is translated:

$$F \subseteq I \cap E.$$

The situation is quite different, however, when the 'and' occurs in the subject rather than in the predicate. Thus:

(9) Fools and drunk men are truth tellers

is translated, not by:

(10) $$(F \cap D) \subseteq T$$

but rather by:

(11) $$(F \cup D) \subseteq T.$$

For (9) means that both the following statements are true:

(12) All fools are truth tellers

and:

(13) All drunk men are truth tellers;

and (12) and (13) are translated, respectively, by:

(14) $$F \subseteq T$$

and:

(15) $$D \subseteq T;$$

and (14) and (15) are together equivalent to (11). (It should be noticed that (10) says less than (11); for:

$$F \cap D \subseteq F \cup D$$

is true for every F and D—and hence (10) is true whenever (11) is true—while a statement of the form (10) can be true even when the corresponding statement of the form (11) is false.)

Often the statement to be translated does not contain any form of the verb 'to be' at all. Thus the statement:

Some Frenchmen drink wine

can be translated:

$$F \cap W \neq \Lambda,$$

if we think of 'F' as standing for the set of Frenchmen and 'W' as standing for the set of wine drinkers.

The statement:

Some Americans drink both coffee and milk

can be translated:

$$A \cap C \cap M \neq \Lambda,$$

where 'A' stands for the set of Americans, 'C' for the set of people who drink coffee, and 'M' for the set of people who drink milk. (Here we have adopted the practice, which is frequently employed, of suppressing parentheses in representing the intersection of three or more sets, writing simply: $A \cap C \cap M$ instead of: $A \cap (C \cap M)$; we shall sometimes adopt a similar practice in connection with the representation of the union of three or more sets. We shall discuss this matter further in § 9.9.)

Still more complicated examples are possible. Consider:

(16) Some Americans who drink tea do not drink either coffee or milk.

The general form of this statement is: Some S are not P. The subject is translated:

$$A \cap T,$$

where T = the set of tea drinkers, and the predicate is translated:

$$C \cup M.$$

The whole sentence (16) is then translated:

$$(A \cap T) \cap \sim(C \cup M) \neq \Lambda,$$

which is also equivalent to:

$$A \cap T \cap \sim C \cap \sim M \neq \Lambda,$$

since, corresponding to De Morgan's laws for the sentential connectives, we have:

$$(17) \qquad \sim(C \cup M) = \sim C \cap \sim M.$$

Identities like (17) are discussed in more detail in § 9.9.

EXERCISES

1. Suppose a psychologist performs a learning experiment with a group of rats in a T-maze. On the basis of a theory which he has developed he expects certain responses with certain characteristics. For simplicity let us think of these characteristics as left or right (corresponding to a left or right turn in the maze), and reinforced or unreinforced. For accurate comparison of what is predicted by the theory and what is observed experimentally, he distinguishes between the set of predicted responses and the set of observed responses. Letting

V = the set of all possible responses,
P = the set of all predicted responses,
O = the set of all observed responses,
L = the set of all left responses,

R = the set of all right responses,
E = the set of all reinforced responses,
U = the set of all unreinforced responses,

translate into symbolic form the following statements which the psychologist might make about his experiment.

(a) All predicted responses were observed.
(b) Some left responses were predicted.
(c) All predicted left responses were reinforced.
(d) No observed responses were not predicted.
(e) No observed right responses were not predicted.
(f) Some left responses which were observed were not predicted.
(g) No right responses were either observed or predicted.
(h) Some right responses which were reinforced were observed but not predicted.
(i) Some left responses which were not predicted were unreinforced.
(j) All left responses which were predicted were observed and they were found to be reinforced.

2. Letting

V = the set of all people,
A = the set of all Americans,
C = the set of all people who drink coffee,
F = the set of all Frenchmen,
M = the set of all murderers,
P = the set of all philosophers,
T = the set of all people who drink tea,
W = the set of all people who drink wine,

translate the following statements into symbolic form:

(a) Some American wine-drinkers are philosophers;
(b) No Frenchman is an American;
(c) People who drink wine and coffee also drink tea;
(d) All French murderers drink coffee, tea, and wine;
(e) Some American murderers drink coffee and tea, but not wine;
(f) Some French murderers who drink wine do not drink either coffee or tea;
(g) A philosopher drinks neither tea nor coffee;
(h) Some Frenchmen are either philosophers or murderers;
(i) All coffee drinkers drink either tea or wine.

§ **9.8 Venn Diagrams.*** In studying sets and relations between them, it is sometimes helpful to represent the sets diagrammatically; one draws a rectangle to represent the domain of individuals, and then draws circles, or other figures, inside the rectangle—thinking of the points inside the various figures as corresponding to the members of the sets being represented by the figures.† Thus if we know of two sets A and B, for instance,

* This section may be omitted without loss of continuity.

† The basic idea of using circles in this way was due to the eighteenth-century Swiss mathematician Euler. Some of the refinements to be explained below are due to the nineteenth-century British logician Venn. The diagrams are sometimes called ***Euler diagrams***, or *Venn diagrams*.

that they are mutually exclusive, we can represent this situation by the following diagram:

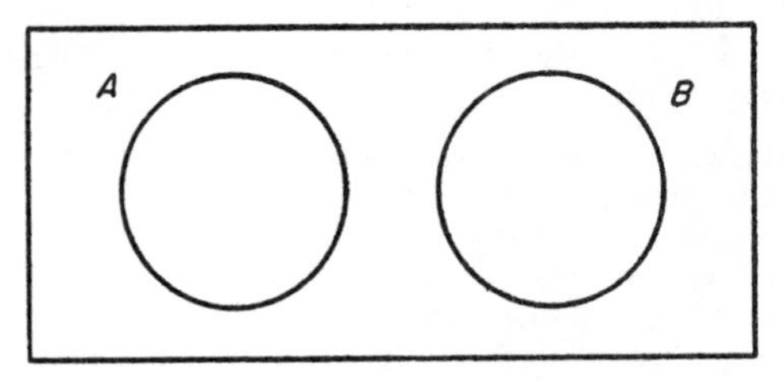

FIGURE 1

If we know that $A \subseteq B$, we can represent the situation by Figure 2.

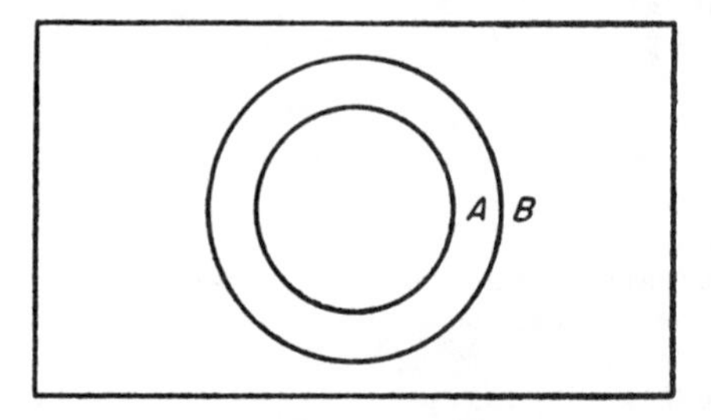

FIGURE 2

A more traditional way of describing Figure 2 is to say that all A are B, i.e., all members of A are members of B. To reinforce the discussion in the last section, we shall, when it is convenient, place such traditional statements in parentheses.

If we know of three sets A, B, and C, that $A \subseteq B$ (all A are B) and $B \cap C = \Lambda$, (i.e., no B are C), we can represent the situation by Figure 3.

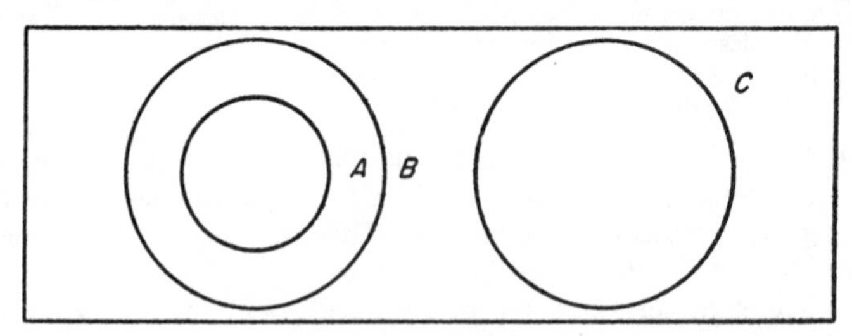

FIGURE 3

From the figure it is clear that in this case we must also have: $A \cap C = \Lambda$ (no A are C).

Sometimes, instead of trying to incorporate the given information into the diagram simply by drawing the circles in an appropriate manner, it is convenient to draw the figures in a rather arbitrary way (so that they will divide the interior of the rectangle into a maximum number of parts) and then get the information into the figure by other methods, such as the shading of areas. Having decided, let us say, to indicate by *horizontal shading* that an area corresponds to the empty set, we indicate that $A \cap B = \Lambda$ by Figure 4; and that $A \subseteq B$ by Figure 5; for to say that A is a subset of B means that no part of A lies outside B.

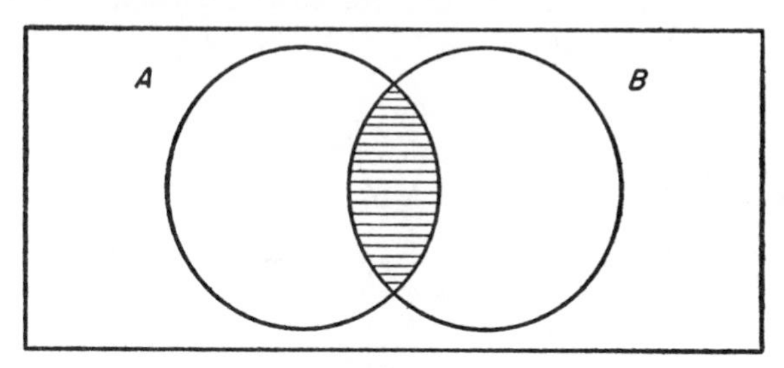

FIGURE 4

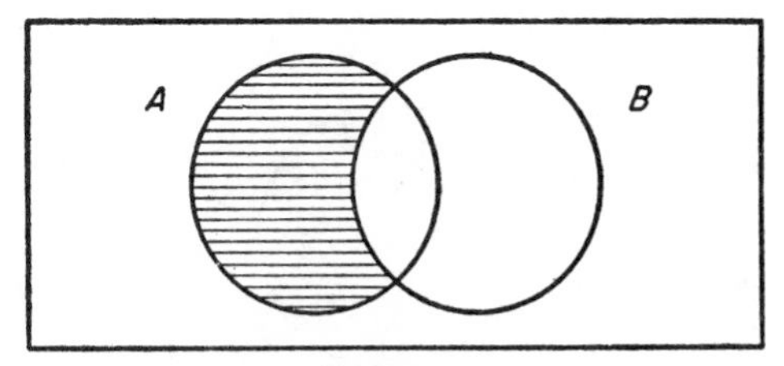

FIGURE 5

Using such diagrams, it is often easy to see what conclusions can be drawn from given information about two or more sets. Thus suppose, for example, that it is given, of two sets A and B both that $A \cap B = \Lambda$, and that $A \subseteq B$. The first statement (as indicated in Figure 4) means that the common part of A and B is to be shaded; and the second statement (as indicated in Figure 5) means that the part of A which is outside of B is to be shaded. Thus we obtain Figure 6, where we notice that all of A is shaded. Thus we see that the two given statements jointly imply that A is the empty set.

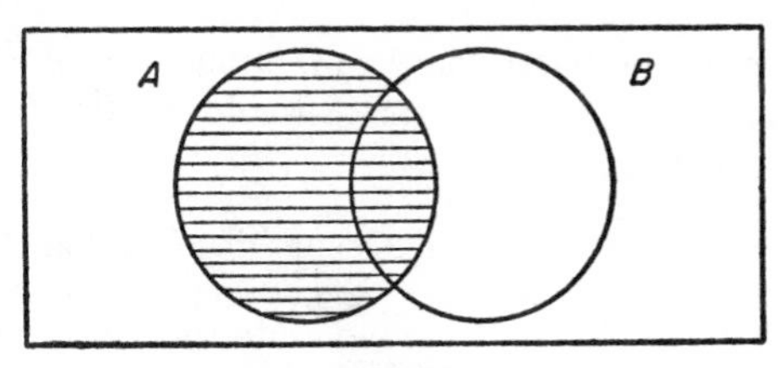

FIGURE 6

Horizontal shading is used to indicate emptiness of a region. Another kind of symbol is needed for *non-emptiness.* We shall use a device of *linked crosses.** Thus if $A \cap B \neq \Lambda$ (some A are B) we represent this situation by Figure 7; the cross indicates that the region common to A and B is not empty. We represent the more complicated situation: $A \cap (B \cup C) \neq \Lambda$, (some A are either B or C) by Figure 8.

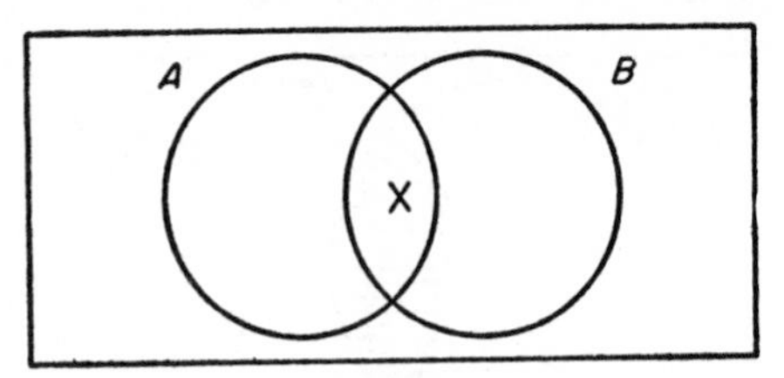

FIGURE 7

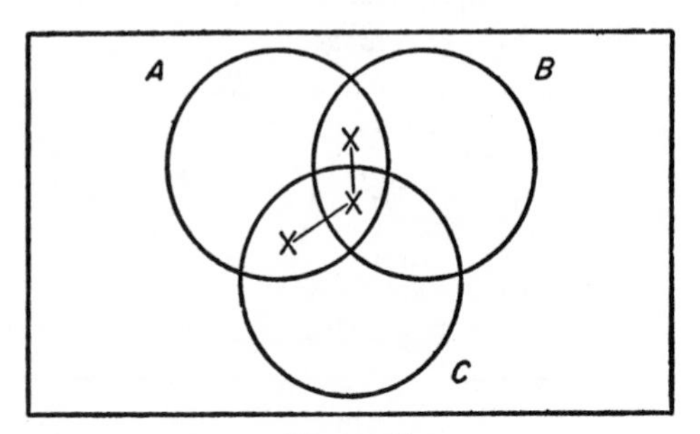

FIGURE 8

The three crosses in Figure 8 are *linked* to show that at least one of the three small regions is non-empty. If the linkage had been omitted in Figure 8, the figure would represent much more than that $A \cap (B \cup C) \neq \Lambda$. If the linkage were omitted, we could infer:

(1)	$(A \cap B) \sim C \neq \Lambda$	(the top cross)
(2)	$A \cap (B \cap C) \neq \Lambda$	(the middle cross)
(3)	$(A \cap C) \sim B \neq \Lambda$	(the bottom cross)

Obviously, any one of the assertions (1)–(3) implies that $A \cap (B \cup C) \neq \Lambda$ and more. Without the linkage Figure 8 would say far too much.

The situation described by

$A \cup B \neq \Lambda$ (Something is either A or B)
$A \cup \sim C \neq \Lambda$ (Something is either A or not C)

* This device is due to Professor Robert McNaughton.

is represented by Figure 9. Note the two separate linkages, one for each of the two existential statements.

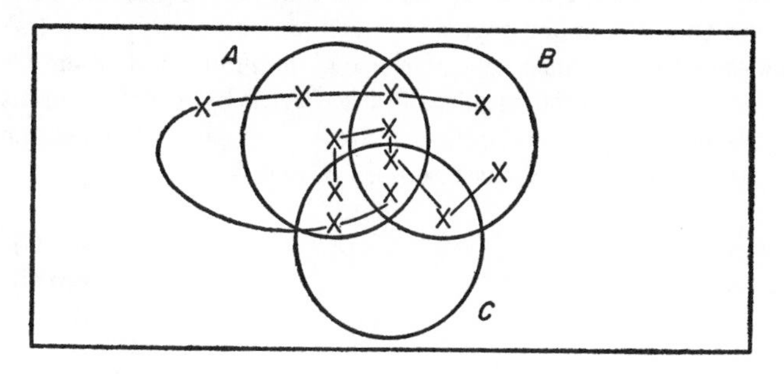

FIGURE 9

What interpretation should be given to a diagram in which a cross and shading occur in the same region? Suppose, for example, that we have:

(4) $A \cap C \neq \Lambda$ (Some A are C)
(5) $C \subseteq B$ (All C are B)

We obtain Figure 10, in which the part of C which is outside of B has been shaded horizontally, to show that it is empty, and linked crosses have been placed in the two parts of the common region of A and C, to show that it is not empty. The problem of interpretation centers around Region ①. Consideration of (4) and (5) clearly urges the stipulation that shading

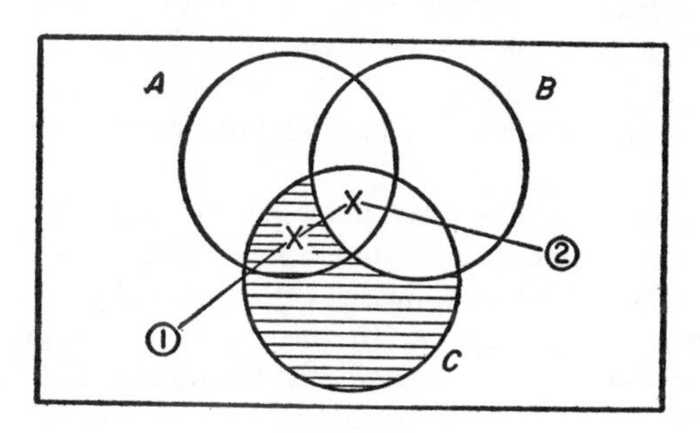

FIGURE 10

dominates a cross, and hence Region ① is empty. We are thus able to conclude that Region ② is not empty, that is, (4) and (5) imply that $A \cap (B \cap C) \neq \Lambda$ (some A are B and C).

There is one set of circumstances in which we do not want to say simply that shading dominates a cross. When *every* cross in a linkage of crosses is "covered" by shading we must conclude that the diagram is inconsistent rather than that the linked regions are completely empty, for a linkage of crosses *means* that at least one of the regions linked is non-empty. We may in fact use these circumstances to investigate by use of Venn diagrams the consistency of a set of conditions imposed on sets. Thus suppose, for example, that it is given of three sets A, B, and C that:

(6) $A \subseteq C$ (All A are C)
(7) $A \cap C = \Lambda$ (No A are C)
(8) $A \cap B \neq \Lambda$ (Some A are B)

This situation is represented by a Venn diagram in Figure 11.

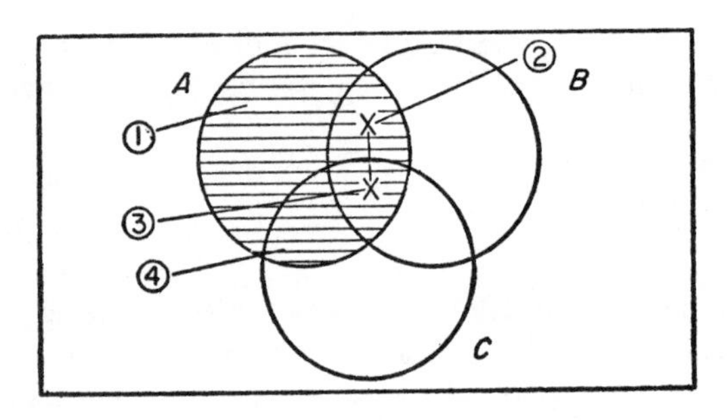

FIGURE 11

Assumption (6) leads us to shade Regions ① and ②; assumption (7) leads us to shade Regions ③ and ④; and assumption (8) leads us to place two linked crosses in Regions ② and ③. Thus the given assumptions imply that Regions ② and ③ are both empty and non-empty, which is a contradiction.

There is, of course, a very great difference between saying that certain conditions on sets are inconsistent and saying merely that they imply that some set is empty. Thus assumptions (6) and (7) above imply that A is empty, but these two assumptions by themselves are not inconsistent.

With the notation for Venn diagrams now complete, it is of some interest to show how the apparatus may be used to establish the validity of classical syllogisms. As an example, consider the syllogism:

(9) No B are C
(10) All A are B
(11) Therefore no A are C

Premises (9) and (10) are represented by Figure 12.

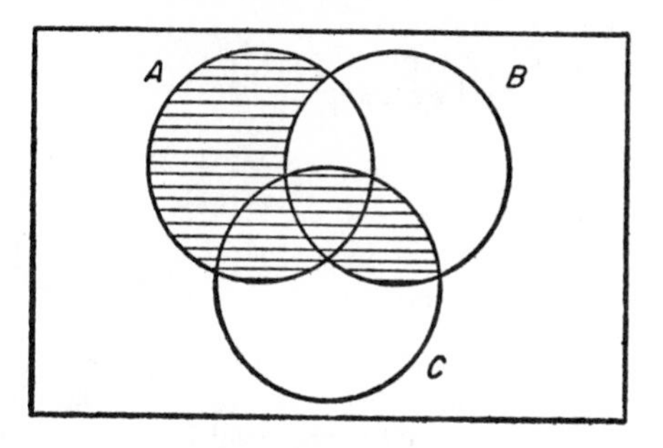

FIGURE 12

We now examine the diagram to see if it implies that no A are C. We see at once that the region common to A and C is horizontally shaded, and we conclude that the conclusion of the syllogism is valid. All other valid syllogisms may be tested in the same way, but there is no need to restrict the use of Venn diagrams to testing the validity of those arguments which have the classical syllogistic form. Venn diagrams may be used to represent any argument which does not involve more than three sets. Moreover, by a careful use of ellipses in place of circles relations among four sets can be represented diagrammatically, but relations among five or more sets can often not be represented by any simple diagrammatic device.

EXERCISES

1. Let A and B be sets such that $A \cap \sim B = \Lambda$. Represent this situation by a Venn diagram.
2. Let A and B be sets such that $A \cap \sim B = \Lambda$ and $B \cap \sim A = \Lambda$. Represent this situation by a Venn diagram. Express the relation between A and B in a simpler manner.
3. If it is given that $A \subseteq B$ and that $C \cap \sim B \neq \Lambda$, what relation can be concluded to hold between A and C?
4. Draw a Venn diagram representing that $A \subseteq C$, and $B \cap \sim C \neq \Lambda$.
5. Are the following assumptions mutually consistent?

$$B \cap C = \Lambda$$
$$(A \cap C) \sim B = \Lambda$$
$$(A \cap B) \sim C = \Lambda$$
$$(A \cap B) \cup (A \cap C) \cup (B \cap C) \neq \Lambda$$

6. Are the following assumptions mutually consistent?

$$C \neq \Lambda$$
$$A \cap B \neq \Lambda$$
$$A \cap C = \Lambda$$
$$(A \cap B) \sim C = \Lambda$$

7. Are the following assumptions mutually consistent?

Some Americans are virtuous.
No virtuous people steal from the poor.
Some Americans steal from the poor.

8. Given:

All unicorns are dead
No unicorns are dead

can you infer that there are no unicorns?

9. Are the assumptions of Exercise 8 mutually consistent?

10. Test the validity of the following arguments by Venn diagrams. State in terms of regions of the diagram why the argument is valid or invalid.

(a) All witnesses are prejudiced.
Some witnesses are not liars.
∴ Some liars are not prejudiced.

(b) All witnesses are prejudiced.
Some liars are not prejudiced.
∴ Some liars are not witnesses.

(c) All liars are prejudiced.
Some witnesses are not liars.
∴ Some witnesses are not prejudiced.

(d) $A \cap B \subseteq \sim C$
$A \cup C \subseteq B$
$\therefore A \cap C = \Lambda$

(e) $A \subseteq \sim(B \cup C)$
$B \subseteq \sim(A \cup C)$
$\therefore B = \Lambda$

(f) $\sim A \subseteq \sim(B \cap C)$
$C \subseteq \sim A$
$B \subseteq A \cup \sim C$
$\therefore \sim(B \cup C) \subseteq \sim A$

§ 9.9 Elementary Principles About Operations on Sets. We have seen how it is possible to use diagrams in order to determine what conclusions can be drawn from given assumptions. A somewhat similar application of diagrams could be made to the problem of determining whether certain equations, or principles of other forms, are always true (for all sets). However, a method which has more general applicability and more interest from a mathematical standpoint is to reduce the test of the truth of an equation for all sets to the problem of deciding if a corresponding formula is a tautological equivalence, which can always be done by sentential methods. For instance, suppose we want to show that for all sets A and B

$$A \cup B = B \cup A. \tag{1}$$

From the definition of the union operation we know that for an arbitrary individual x

$$(2) \qquad x \in A \cup B \leftrightarrow x \in A \vee x \in B.$$

Moreover, it is a tautological equivalence that

$$(3) \qquad x \in A \vee x \in B \leftrightarrow x \in B \vee x \in A.$$

And corresponding to (2), we have:

$$(4) \qquad x \in B \cup A \leftrightarrow x \in B \vee x \in A.$$

Combining (2)–(4) from the transitivity and symmetry of equivalence we conclude:

$$(5) \qquad x \in A \cup B \leftrightarrow x \in B \cup A,$$

and we conclude from (5) by the principle of extensionality for sets that

$$(6) \qquad A \cup B = B \cup A.$$

The inference from (5) to (6) is routine and usually omitted, and similarly for the inference from (1) to (2). Moreover, these equivalences may be presented in compact form like the identities of Chapter 7. The informal proof of (1) may then be written:

$$\begin{aligned} x \in A \cup B &\leftrightarrow x \in A \vee x \in B \\ &\leftrightarrow x \in B \vee x \in A \\ &\leftrightarrow x \in B \cup A. \end{aligned}$$

Since the steps of the inference either involve tautological equivalences or one of the four characterizing equivalences:

$$\begin{aligned} x \in A \cup B &\leftrightarrow x \in A \vee x \in B \\ x \in A \cap B &\leftrightarrow x \in A \;\&\; x \in B \\ x \in A \sim B &\leftrightarrow x \in A \;\&\; x \notin B \\ x \in {\sim}A &\leftrightarrow x \notin A, \end{aligned}$$

each line need not be tagged by a justifying reason. To illustrate these methods on a more complicated example we may prove that for all sets A and B

$$A \sim (A \cap B) = A \sim B.$$

PROOF.

$$\begin{aligned} &(1) \quad x \in A \sim (A \cap B) &&\leftrightarrow x \in A \;\&\; x \notin A \cap B \\ &(2) &&\leftrightarrow x \in A \;\&\; -(x \in A \;\&\; x \in B) \\ &(3) &&\leftrightarrow x \in A \;\&\; (x \notin A \vee x \notin B) \\ &(4) &&\leftrightarrow (x \in A \;\&\; x \notin A) \vee (x \in A \;\&\; x \notin B) \\ &(5) &&\leftrightarrow x \in A \;\&\; x \notin B \\ &(6) &&\leftrightarrow x \in A \sim B. \end{aligned}$$

Q.E.D.

Note that in line (1) we eliminated only one set operation. The intersection was then eliminated in line (2). Replacement of one set operation at a time by the appropriate sentential connective avoids mistakes. In going from (4) to (5) we used the tautological equivalence $(P \mathbin{\&} -P) \vee Q \leftrightarrow Q$. For handy reference we list the more useful identities and some other closely related principles. For each of the identities there is a corresponding tautological equivalence if we let $P \mathbin{\&} -P$ correspond to Λ and $P \vee -P$, say, correspond to V. Proofs of the identities by the sentential methods just discussed are left as exercises. The proofs of those principles which are not identities is only slightly more complicated.

LIST OF SET IDENTITIES AND OTHER PRINCIPLES

(1) $A \cup \Lambda = A$
(2) $A \cap V = A$
(3) $A \cup B = B \cup A$
(4) $A \cap B = B \cap A$
(5) $A \cup (B \cap C) = (A \cup B) \cap (A \cup C)$
(6) $A \cap (B \cup C) = (A \cap B) \cup (A \cap C)$
(7) $A \cup \sim A = V$
(8) $A \cap \sim A = \Lambda$
(9) $A \cup A = A$
(10) $A \cap A = A$
(11) $A \cup V = V$
(12) $A \cap \Lambda = \Lambda$
(13) $\Lambda \neq V$
(14) $\sim\sim A = A$
(15) $A = \sim B \rightarrow B = \sim A$
(16) $A \cup B \neq \Lambda \rightarrow A \neq \Lambda \vee B \neq \Lambda$
(17) $A \cap B \neq \Lambda \rightarrow A \neq \Lambda$
(18) $A \cup (B \cup C) = (A \cup B) \cup C$
(19) $A \cap (B \cap C) = (A \cap B) \cap C$
(20) $A \cup (A \cap B) = A$
(21) $A \cap (A \cup B) = A$
(22) $\sim A \neq A$
(23) $\sim(A \cup B) = \sim A \cap \sim B$
(24) $\sim(A \cap B) = \sim A \cup \sim B$
(25) $A \sim A = \Lambda$
(26) $A \sim (A \cap B) = A \sim B$
(27) $A \cap (A \sim B) = A \sim B$
(28) $(A \sim B) \sim B = A \sim B$
(29) $(A \sim B) \sim A = \Lambda$
(30) $(A \sim B) \cup B = A \cup B$
(31) $(A \cup B) \sim B = A \sim B$

These thirty-one principles are deliberately arranged in a certain order, for the first eight may be taken as axioms of a theory formalized in first-order predicate logic with identity; with the addition of the axiom: $(\exists A)(\exists B)(A \neq B)$ and the appropriate definition of set difference the remaining twenty-three principles may be derived as theorems.* Any model of these axioms is called a *Boolean algebra,* and the axioms are sometimes called the axioms for the algebra of sets. A much more powerful

* This set of axioms is essentially due to E. V. Huntington, "Sets of Independent Postulates for the Algebra of Logic," *Transactions of the American Mathematical Society,* Vol. 5 (1904) pp. 288–309.

set of axioms can be based on the notion of set membership, which is not used in any of the above principles.

Most of the equations (1)–(31) are so important, and so frequently referred to, that they have been given special names, some of which are identical with the names of the corresponding tautologies. Others have names derived from the fact that they express important structural properties of the operations on sets like associativity or commutativity. Identity (1) says that the empty set is a right-hand *identity element* with respect to the union operation, and (2) says that the domain V of individuals is a right-hand *identity element* with respect to intersection. (3) and (4) are the *commutative* laws for union and intersection. Identities (5) and (6) are dual *distributive* laws. Notice that if we compare union to arithmetical addition and intersection to arithmetical multiplication, the analogue of (5) does not hold in arithmetic. Identity (7) is the set formulation of the law of *excluded middle;* everything in a given domain is either in a set or its complement. In (8) we have the law of *contradiction;* nothing can be a member of both a set and its complement. Equations (9) and (10) express what is usually called the *idempotency* of union and intersection. The problem of finding an arithmetical operation which is idempotent is left as an exercise. Principle (14) is the law of *double negation*, and (15) expresses a principle of *contraposition*. In (18) and (19) we have *associative* laws, which justify the omission of parentheses, and in (20) and (21) laws of *absorption*. Equations (23) and (24) are *De Morgan's laws.*

It should be remarked that for proof of those principles involving the empty set Λ or the domain V of individuals, we need to use that for every x

$$x \notin \Lambda,$$
$$x \in \mathrm{V}.$$

Consider Principle (13): $\Lambda \neq \mathrm{V}$

PROOF. Suppose that $\Lambda = \mathrm{V}$. Then by the principle of extensionality

$$x \in \Lambda \leftrightarrow x \in \mathrm{V};$$

but

$$x \notin \Lambda,$$

whence

$$x \notin \mathrm{V},$$

which is absurd. Q.E.D.

By considering an arbitrary member of a set and using tautological implications rather than tautological equivalences, useful principles of inclusion for sets are easily established. As an example we may consider:

$$A \cap B \subseteq A.$$

PROOF.

$$(1) \quad x \in A \cap B \leftrightarrow x \in A \ \& \ x \in B$$

$$(2) \quad \rightarrow x \in A. \qquad \text{Q.E.D.}$$

Line (1) is an equivalence; the right-hand side of (1) tautologically implies the right-hand side of (2) but is not tautologically equivalent to it. The characterizing equivalence used for inclusion is, of course:

$$A \subseteq B \leftrightarrow (x)(x \in A \rightarrow x \in B).$$

In writing down proofs of inclusion relations between sets, one point needs to be remarked. Consider:

$$(A \cap B) \sim (B \cap C) \subseteq A.$$

PROOF.

$$\begin{aligned} &(1) \quad x \in (A \cap B) \sim (B \cap C) \leftrightarrow x \in A \cap B \ \& \ x \notin B \cap C \\ &(2) \quad \rightarrow x \in A \cap B \\ &(3) \quad \rightarrow x \in A \ \& \ x \in B \\ &(4) \quad \rightarrow x \in A \qquad \text{Q.E.D.} \end{aligned}$$

The point is that although line (3) is tautologically equivalent to line (2), we use an implication sign in line (3), because this implication sign indicates the relation of (3) to the left side of (1), not to the right side of (2).

Inclusion relations corresponding to all the useful tautological implications given in Chapter 2 are easily verified by the method just indicated.

EXERCISES

1. Prove by the methods of this section the validity of each of principles (1)–(31).

2. If union is interpreted as addition, intersection as multiplication, difference of sets as subtraction, the empty set as zero, and the domain V of individuals as one, which of the principles (1)–(31) express truths of arithmetic?

3. Can you give an example of an arithmetical operation which is idempotent?

4. Give a counterexample to show that the operation of difference (of sets) is not in general commutative.

5. Give a counterexample to show that the operation of difference is not in general associative.

6. Give a counterexample to show that the operation of difference is not in general distributive with respect to union, that is, it is not true for all sets A, B, C, that

$$A \sim (B \cup C) = (A \sim B) \cup (A \sim C).$$

7. Is the operation of difference distributive with respect to intersection?

8. Prove by the methods of this section the validity of the following inclusion relations for any sets A, B, C:

(a) $A \subseteq A \cup B$
(b) $A \sim B \subseteq A$
(c) $A \sim (B \cup C) \subseteq A \sim (B \cap C)$
(d) $(A \cap B) \sim C \subseteq (A \cup B) \sim C$

9. Using Principles (1)–(8) as axioms of a formalized theory, which has primitive operation symbols '$\cup$', '$\cap$' and '$\sim$', and which has individual constants 'Λ' and 'V', derive Principles (9)–(24) as theorems, using as an additional axiom

$$(\exists A)(\exists B)(A \neq B). \tag{8.1}$$

10. Let us add to our axioms for Boolean algebras of the preceding exercise:

DEFINITION 1. $A \sim B = A \cap \sim B.$

Prove Principles (25)–(31).

11. Continuing the development of the theory of Boolean algebras, we add:

DEFINITION 2. $A \subseteq B \leftrightarrow A \cup B = B.$

Prove the following theorems (numbered consecutively with the thirty-one principles):

(32) $A \subseteq A$
(33) $A \subseteq B \;\&\; B \subseteq A \rightarrow A = B$
(34) $A \subseteq B \;\&\; B \subseteq C \rightarrow A \subseteq C$
(35) $\Lambda \subseteq A$
(36) $A \subseteq \Lambda \rightarrow A = \Lambda$

12. Still continuing the theory of Boolean algebras, we add:

DEFINITION 3. $A \subset B \leftrightarrow A \subseteq B \;\&\; A \neq B.$

Prove:

(37) $-(A \subset A)$
(38) $A \subset B \rightarrow -(B \subset A)$
(39) $A \subset B \;\&\; B \subset C \rightarrow A \subset C$
(40) $A \subset B \rightarrow A \subseteq B$
(41) $A \cap B \subset A \rightarrow A \neq B$
(42) $A \cap B \neq \Lambda \rightarrow A \sim B \subset A$

13. Referring to the last section of the previous chapter, prove that none of the five primitive symbols of the theory of Boolean algebras presented in the above exercises is independent of the others.

CHAPTER 10

RELATIONS

§ 10.1 Ordered Couples. In Chapter 9 we were concerned almost entirely with sets and the notion of membership in sets; the other notions (such as the notion of the empty set, of the intersection of two sets, and so on) were *defined* in terms of these two (thus, the empty set was defined to be the set Λ such that, for all $x, x \notin \Lambda$; the intersection of two sets was defined to be the set of all things which belong to both of the given sets; and so on). In this chapter, however, we introduce a new notion, the notion of an *ordered couple*, which we do not here define in terms of set and membership.* Intuitively, an ordered couple is simply two objects given in a fixed order. We use pointed brackets to denote ordered couples. Thus $\langle x, y\rangle$ is the ordered couple whose first member is x and whose second member is y.† In § 9.2 we defined two sets as identical when they have the same members. The requirement of identity for ordered couples is stricter. Two ordered couples are identical just when the first member of one is identical with the first member of the other, and the second member of one is identical with the second member of the other. In symbols:

(1) $$\langle x, y\rangle = \langle u, v\rangle \leftrightarrow (x = u \;\&\; y = v).$$

We have, for example:

$$\{1, 2\} = \{2, 1\},$$

but:

$$\langle 1, 2\rangle \neq \langle 2, 1\rangle.$$

We may define ordered triples, and in general ordered n-tuples, in terms of ordered couples. An ordered triple, for instance, is an ordered couple whose first member is an ordered couple, that is,

(2) $$\langle x, y, z\rangle = \langle\langle x, y\rangle, z\rangle.$$

* Such a definition is possible.

† The notation: (x, y) is also widely used.

Proceeding in this way, we define ordered quadruples:

$$\langle x, y, z, w\rangle = \langle\langle x, y, z\rangle, w\rangle,$$

and in general ordered n-tuples:

$$\langle x_1, x_2, \ldots, x_n\rangle = \langle\langle x_1, x_2, \ldots, x_{n-1}\rangle, x_n\rangle.$$

In this book we shall do very little work with ordered triples, ordered quadruples, etc., but it is perhaps useful to indicate how the above definitions are used to prove facts about ordered n-tuples. We select the simple task of showing that two ordered triples are identical just when their corresponding members are identical. Let $\langle x, y, z\rangle$ and $\langle u, v, w\rangle$ be two arbitrary ordered triples. It is clear from the logic of identity that if $x = u$, $y = v$ and $z = w$ then $\langle x, y, z\rangle = \langle u, v, w\rangle$. On the other hand, let us begin with: $\langle x, y, z\rangle = \langle u, v, w\rangle$. Then by (2) we have:

$$\langle\langle x, y\rangle, z\rangle = \langle\langle u, v\rangle, w\rangle,$$

and using (1) we obtain:

$$\langle x, y\rangle = \langle u, v\rangle \tag{3}$$

and:

$$z = w.$$

But (3) and (1) together yield:

$$x = u \quad \text{and} \quad y = v.$$

Thus we have shown:

$$\langle x, y, z\rangle = \langle u, v, w\rangle \leftrightarrow (x = u \;\&\; y = v \;\&\; z = w).$$

It is important to notice that the repetition of the same element adds nothing in describing sets but it does in the case of ordered triples. For example,

$$\{1, 2, 2\} = \{1, 2\},$$

but

$$\langle 1, 2, 2\rangle \neq \langle 1, 2\rangle,$$

since $\langle 1, 2, 2\rangle = \langle\langle 1, 2\rangle, 2\rangle$ and $\langle 1, 2\rangle \neq 1$.

The notion of a *finite sequence* may be defined in terms of ordered n-tuples. S is a *finite sequence* if and only if there is a positive integer n such that S is an ordered n-tuple. Thus, for example, ⟨Socrates, Plato, Democritus, Aristotle⟩ is a finite sequence of Greek philosophers. In particular, it is an ordered quadruple.

It is often useful to consider the set of all ordered couples which can be formed from two sets in a fixed order. The *Cartesian* (or *cross*) *product* of two sets A and B (in symbols: $A \times B$) is the set of all ordered couples $\langle x, y \rangle$ such that $x \in A$ and $y \in B$. For example, if

$$A = \{1, 2\}$$
$$B = \{\text{Gandhi, Nehru}\}$$

then

$$A \times B = \{\langle 1, \text{Gandhi} \rangle, \langle 1, \text{Nehru} \rangle, \langle 2, \text{Gandhi} \rangle, \langle 2, \text{Nehru} \rangle\}.$$

§ 10.2 Definition of Relations. In ordinary discourse we often speak of *relations*, which we think of as holding *between* two things, or among several things. Thus we say that Elizabeth II stands in the relation of mother to Prince Charles—by which we mean that Elizabeth II is the mother of Prince Charles. Or we may say that the *relation of coincidence* holds among three lines—by which we mean that they all intersect in one point.

For our discussions, it is convenient to introduce letters to designate relations. Thus we may introduce the letter 'M', for the relation of mother, and then write:

$$(\text{Elizabeth II})\ M\ (\text{Prince Charles})$$

to indicate that Elizabeth II is the mother of Prince Charles. Similarly we may wish to speak of the relation A such that, for every x and y, $x\ A\ y$ if and only if x is an ancestor of y.*

In dealing with a relation that holds among three or more things, it is convenient to put the letter standing for the relation, and then the names, in proper order, of the things among which it holds. Thus we may speak of the relation P such that, for every x, y, and z, $P(x, y, z)$ if and only if x and y are the parents of z. Similarly, we may speak of the relation B such that, for every x, y, z, and w, $B(x, y, z, w)$ if and only if x owes y dollars to z for w; so that, for example:

$$B(\text{John, 5, Henry, shoes})$$

means that John owes Henry five dollars for shoes.

When we speak of relations in everyday contexts, we insist that there be some intuitive way of describing what sort of connection there is between things which we say stand in some given relation to each other. Unfortunately, this intuitive idea of connectedness is often vague, even though it may have a very precise meaning in particular cases. This same situation arose, if you remember, in our discussion of the sentential connectives. Our decision in Chapter 1 was to permit a conjunction of any two sentences, regardless of how unconnected they might seem intuitively. Similarly, we now find it expedient, again because of the difficulties of

* In the language of Chapter 4, the letters 'M' and 'A' are predicates.

formulating a precise general notion of connection or dependence, to insist that *any* set of ordered couples is a binary relation. Thus the set

$$\{\langle \text{Aristotle}, \Lambda\rangle, \langle 7, \text{Julius Caesar}\rangle\}$$

is a relation, although no one would claim it has any intuitive significance. Our general definition is then:

(I) *A binary relation is a set of ordered couples.*

According to this definition the relation of *loving* is the set of ordered couples $\langle x, y\rangle$ such that x loves y. The relation of *being less than* is the set of all ordered couples $\langle x, y\rangle$ of numbers such that, for some positive number z,

$$x + z = y.$$

The obvious extension of (I) is that a relation which holds among three things is a set of ordered triples, and a relation which holds among n things is a set of ordered n-tuples.

A relation is called 'n-ary' if its members are n-tuples. For the special cases $n = 2$ and $n = 3$ we use special names, speaking of '*binary*' and '*ternary*' relations.

Since a relation is a *set* of ordered n-tuples, we can also use the "$\in$" notation to indicate that certain things stand in a given relation. Thus we can write:

$$\langle \text{John}, \text{Mary}\rangle \in L,$$

instead of:

$$\text{John } L \text{ Mary}$$

to indicate that John loves Mary. Similarly we can write:

$$\langle \text{George}, \text{Mary}, \text{Elizabeth}\rangle \in P,$$

instead of:

$$P(\text{George}, \text{Mary}, \text{Elizabeth})$$

to indicate, let us say, that George and Mary are the parents of Elizabeth.

It is necessary to remember that an ordered couple is not a relation, but the set consisting of the ordered couple is. For instance,

$$\langle \text{Thomas Aquinas}, 4\rangle \text{ is not a relation;}$$
$$\{\langle \text{Thomas Aquinas}, 4\rangle\} \text{ is a relation;}$$
$$\{\{\langle \text{Thomas Aquinas}, 4\rangle\}\} \text{ is not a relation.}$$

The last example of the three is not a relation because the only member of the set is itself a set, which is not an ordered couple.

We now introduce some useful special terminology for binary relations. If R is a binary relation, then the *domain* of R—in symbols: $D(R)$—is the set of all things x such that, for some y, $\langle x, y\rangle \in R$. Thus if M is the relation which consists of all couples $\langle x, y\rangle$ such that x is the mother of y, then the domain of M is the set of all women who are not childless. If

$$R_1 = \{\langle \Lambda, \text{Plato}\rangle, \langle \text{Jane Austen}, 101\rangle$$
$$\langle \text{the youngest bride in Tibet}, \text{Richelieu}\rangle\},$$

then

$$D(R_1) = \{\Lambda, \text{Jane Austen}, \text{the youngest bride in Tibet}\}.$$

The *counterdomain* (or *converse domain*) of a binary relation R (in symbols: $C(R)$) is the set of all things y such that, for some x, $\langle x, y\rangle \in R$. The counterdomain of the relation M considered just above is the set of all people—since everyone has a mother. If B is the relation which consists of all couples $\langle x, y\rangle$ such that x is the brother of y, then the domain of B is the set of all men who have at least one brother or sister, and the counterdomain is the set of all people who have at least one brother. We have for the relation R_1 defined above:

$$C(R_1) = \{\text{Plato}, 101, \text{Richelieu}\}.$$

The *field* of a binary relation R (in symbols: $\mathscr{F}(R)$) is the union of its domain and its counterdomain. Thus z belongs to the field of a binary relation R if and only if either $\langle x, z\rangle \in R$ for some x or $\langle z, y\rangle \in R$ for some y. The field of the relation B considered just above is the set of all people who belong to families containing at least two children, at least one of which is male. As another example,

$$\mathscr{F}(R_1) = \{\Lambda, \text{Jane Austen}, \text{the youngest bride in Tibet}, \text{Plato}, 101, \text{Richelieu}\}.$$

EXERCISES

1. Let:

$$A_1 = \{1, 2\}$$
$$A_2 = \{\Lambda\}$$
$$R = \{\langle 1, 2\rangle, \langle 2, \Lambda\rangle\}.$$

(a) Is R a subset of the Cartesian product $A_1 \times A_2$?
(b) Is $D(R)$ a subset of A_1?
(c) Is $C(R)$ a subset of A_2?
(d) Is $\mathscr{F}(R)$ a subset of $A_1 \cup A_2$?

2. What are the domain, counterdomain, and field of the relation of being a father?

3. What are the domain, counterdomain, and field of the relation of being a grandfather?

4. Is the domain of the relation of being a grandfather a proper subset of the domain of the relation of being a father?

§ 10.3 Properties of Binary Relations. We now turn to some important properties of binary relations. As certain of our examples and exercises will show, these properties play a useful role in a wide variety of scientific contexts.

A (binary) relation R is *reflexive in the set A* if for every x in A, $x\,R\,x$ (i.e., $\langle x, x\rangle \in R$). In symbols:

$$R \text{ reflexive in } A \leftrightarrow (x)(x \in A \rightarrow x\,R\,x).$$

The relation $\leq$, for instance, is reflexive in the set of all real numbers, since for every number x, $x \leq x$. If

$$A_1 = \{\text{Descartes, Mersenne}\},$$
$$A_2 = \{\text{Descartes}, 5\},$$

and

$$R_2 = \{\langle\text{Descartes, Descartes}\rangle, \langle\text{Mersenne, Mersenne}\rangle, \langle 5, \Lambda\rangle\},$$

then R_2 is reflexive in A_1 but is not reflexive in A_2, since the ordered couple $\langle 5, 5\rangle$ is not a member of R_2. The relation of loving is probably reflexive in the set of all people; indeed, some moralists maintain that we all love ourselves somewhat too well.

A relation R is *irreflexive in the set A* if, for every x in A it is not the case that $x\,R\,x$. In symbols:

$$R \text{ irreflexive in } A \leftrightarrow (x)(x \in A \rightarrow -(x\,R\,x)).$$

The relation of being a mother is irreflexive in the set of people, since no one is his own mother. The relation $<$ is irreflexive in the set of real numbers, since no number is less than itself. We have already seen that R_2 is not reflexive in A_2, but it is also not irreflexive in A_2, because $\langle\text{Descartes, Descartes}\rangle \in R_2$. Consider:

$$A_3 = \{5, \text{Elizabeth I}\}.$$

It is clear that R_2 is irreflexive in A_3. From the example of R_2 and A_2, it should be obvious that for any relation R and set A there are three possibilities:

(1) R is reflexive in A.
(2) R is irreflexive in A.
(3) Neither (1) nor (2).

These three possibilities are mutually exclusive, with one exception: every relation R is both reflexive and irreflexive in Λ.

A relation R is *symmetric in the set A* if for every x and y in A, whenever $x\,R\,y$, then $y\,R\,x$. In symbols:

$$R \text{ symmetric in } A \leftrightarrow (x)(y)[x \in A \,\&\, y \in A \,\&\, x\,R\,y \rightarrow y\,R\,x].$$

The relation of being cousins is symmetric, but the relation of loving is not, an unfortunate fact which has been remarked upon by many novelists. The relation of being a brother is not symmetric, since any woman who has a brother affords a counterexample. (Notice that in these last three examples we have omitted explicit reference to a set in which the relations are or are not symmetric. We shall often do this when the set we have in mind is obvious, in this case the set of all people.) R_2 is symmetric in all three sets, A_1, A_2, and A_3.

A relation R is *asymmetric in the set* A if, for every x and y in A, whenever $x \, R \, y$, then it is not the case $y \, R \, x$. In symbols:

$$R \text{ asymmetric in } A \leftrightarrow (x)(y)[x \in A \,\&\, y \in A \,\&\, x \, R \, y \rightarrow -(y \, R \, x)].$$

The relation of being a mother is asymmetric, for obvious biological reasons. On the other hand, the relation of loving is neither symmetric nor asymmetric, which fact partly accounts for the dramatic interest of the subject. If we want to show that a particular relation is neither symmetric nor asymmetric, we need to give a definite counterexample to show that it is not symmetric, and a different one to show that it is not asymmetric. (Similarly we need to give two distinct counterexamples to show that a relation is neither reflexive nor irreflexive.) For example, the relation $\leq$ in the set of numbers is not symmetric, since $1 \leq 2$ but not $2 \leq 1$. On the other hand, it is not asymmetric, since from $3 \leq 3$ it clearly does not follow that not $3 \leq 3$. In the first counterexample we substituted '1' for 'x' and '2' for 'y'. In the second, we substituted '3' for both 'x' and 'y'. In trying to grasp the exact sense of these definitions of properties of relations it is important to remember that the *same* term can be substituted for different variables such as 'x' and 'y'.

The relation $\leq$, which is neither symmetric nor asymmetric, has a closely related property which we now define. A relation R is *antisymmetric in the set* A if for every x and y in A, whenever $x \, R \, y$ and $y \, R \, x$, then $x = y$. In symbols:

$$R \text{ antisymmetric in } A \leftrightarrow (x)(y)[x \in A \,\&\, y \in A \,\&\, x \, R \, y \,\&\, y \, R \, x \rightarrow x = y].$$

As already remarked, $\leq$ is an example of an antisymmetric relation. The relation $\subseteq$ of inclusion is a second example. The relation R_2 is antisymmetric in A_1, A_2, and A_3. On the other hand, only in a world of completely egocentric, narcissistic people would loving be antisymmetric. Notice that vacuously every asymmetric relation is also antisymmetric. I say 'vacuously' because if a relation is asymmetric in a set A then there are no two objects x and y in A such that $x \, R \, y$ and $y \, R \, x$; that is, it is never the case that $x \in A$ and $y \in A$ and $x \, R \, y$ and $y \, R \, x$. Hence by a simple application of truth tables (an implication is true when the antecedent is false) it is always the case that if $x \in A$ and $y \in A$ and $x \, R \, y$ and $y \, R \, x$, then $x = y$.

It is also possible for a relation to be neither symmetric, asymmetric, nor antisymmetric in a set. An example is the relation of loving already mentioned several times. Provided a relation R holds between at least two (not necessarily distinct) elements of A, the mutually exclusive and exhaustive possibilities in the case of symmetry conditions are the following:

(1) R is symmetric in A.
(2) R is asymmetric in A.
(3) R is antisymmetric but neither symmetric nor asymmetric in A.
(4) Neither (1), (2), nor (3).

In order to make the conditions mutually exclusive, we had to require that R be antisymmetric but not symmetric as well as not asymmetric. We leave as an exercise the construction of a relation which is both symmetric and antisymmetric in a set and holds between two not necessarily distinct elements of the set.

A relation R is *transitive in the set* A if, for every x, y, and z in A, whenever $x \, R \, y$ and $y \, R \, z$, then $x \, R \, z$. In symbols:

R transitive in

$$A \leftrightarrow (x)(y)(z)[x \in A \; \& \; y \in A \; \& \; z \in A \; \& \; x \, R \, y \; \& \; y \, R \, z \rightarrow x \, R \, z].$$

The relations $\leq$ and $\subseteq$ are obviously transitive. The relation of identity is also transitive. On the other hand, the relation of being a mother is not, since if x is the mother of y and y is the mother of z it cannot be the case that x is also the mother of z. The relation R_2 is transitive in A_1, A_2, and A_3 in what may be called a vacuous sense, for there are no two ordered couples in R_2 which afford a test case, so to speak, of transitivity by having the second member of one ordered couple (the y of $x \, R \, y$) identical with the first member of another ordered couple (the y of $y \, R \, z$), and thereby permit the test of having the remaining two members (x and z) stand in the given relation ($x \, R \, z$). Let us consider a case in which such a test arises:

$$A_4 = \{2, 7, \text{Goethe}\}$$
$$A_5 = \{2, 7, \text{Edgar Guest}\}$$
$$R_3 = \{\langle 2, \text{Goethe}\rangle, \langle \text{Goethe}, 7\rangle, \langle \text{Edgar Guest}, 2\rangle, \langle 2, 7\rangle\}.$$

It should be obvious that R_3 is transitive in A_4 and not in A_5. For R_3 to be transitive in A_5, we would need to add the couple $\langle$Edgar Guest, 7$\rangle$ to R_3. The test case for the transitivity of R_3 in A_4 is provided by the couples $\langle$2, Goethe$\rangle$ and $\langle$Goethe, 7$\rangle$. For R_3 to be transitive in A_4 the couple $\langle 2, 7\rangle$ must also be in R_3. The example of R_3 can be misleading.

In general we cannot decide if a relation is transitive in a given set by considering a single test case; it is often necessary to consider several cases or even to decide in a systematic way what the situation is in an infinity of cases (as for $\leq$).

Related to transitivity is the less important notion of intransitivity. A relation R is *intransitive in the set* A if for every x, y, and z in A, whenever $x\,R\,y$ and $y\,R\,z$, then it is not the case that $x\,R\,z$. In symbols:

R intransitive in

$$A \leftrightarrow (x)(y)(z)[x \in A \;\&\; y \in A \;\&\; z \in A \;\&\; x\,R\,y \;\&\; y\,R\,z \rightarrow -(x\,R\,z)].$$

The relation of being a mother is a familiar example of an intransitive relation. There is a general tendency, particularly in the literature of the social sciences, to confuse non-transitive and intransitive relations. Clearly a relation may be non-transitive without being intransitive.

A relation R is *connected in the set* A if for every x and y in A, whenever $x \neq y$, then $x\,R\,y$ or $y\,R\,x$. In symbols:

$$R \text{ connected in } A \leftrightarrow (x)(y)(x \in A \;\&\; y \in A \;\&\; x \neq y \rightarrow x\,R\,y \vee y\,R\,x).$$

From the definition it is obvious that a relation is connected in a set when it connects any two distinct members of the set; that is, given any two distinct members, one stands in the relation to the other. The relations $\leq$ and $<$ are both connected in the set of numbers. On the other hand, the relation of being a mother is not connected in the set of people, since given two people chosen at random it is seldom the case that one is the mother of the other. The relation R_2 is not connected in either A_1, A_2, or A_3. For instance, it is not connected in A_1 because neither $\langle$Descartes, Mersenne$\rangle \in R_2$ nor $\langle$Mersenne, Descartes$\rangle \in R_2$. However, the relation R_3 is connected in A_4, but it is not connected in A_5. For R_3 to be connected in A_5 we would need to have either $\langle$Edgar Guest, 7$\rangle \in R_3$ or $\langle$7, Edgar Guest$\rangle \in R_3$.

We now introduce a property very similar to connectedness. A relation R is *strongly connected in the set* A if for every x and y in A, either $x\,R\,y$ or $y\,R\,x$.* In symbols:

$$R \text{ strongly connected in } A \leftrightarrow (x)(y)(x \in A \;\&\; y \in A \rightarrow x\,R\,y \vee y\,R\,x).$$

It should be clear that if R is strongly connected in A then R is also connected in A. The relation $\leq$ is strongly connected in the set of all numbers. On the other hand, the relation $<$ is not strongly connected in the set of numbers, since not $1 < 1$, that is, if $x = y = 1$ then neither $x < y$ nor $y < x$.

* The terminology 'strongly connected' is not widely used. There does not seem to be a generally agreed upon name for this property.

In concluding this section we want to make two general points which may be illustrated by considering the property of connectedness. Let

$$A_6 = \{2, \text{the author of } \textit{Hamlet}, \text{Francis Bacon}\}$$
$$R_4 = \{\langle 2, \text{Francis Bacon}\rangle, \langle 2, 2\rangle\}.$$

The problem: Is R_4 connected in A_6? The answer depends not on logic but on Shakespearian scholarship: Did Francis Bacon actually write the plays attributed to Shakespeare? If so, then he is identical with the author of *Hamlet*, and R_4 is connected in A_6. However, most scholars do not believe that Bacon wrote the plays, and the most acceptable answer is that R_4 is not connected in A_6. From our standpoint the interest of this example is in emphasizing that it often is not a question of logic or mathematics but of empirical fact as to whether a relation has a given property such as asymmetry, transitivity, or connectedness. The point to be noted is that different names or descriptions may be used in referring to a single individual. If Bacon had written the plays, then

$$\text{Francis Bacon} = \text{the author of } \textit{Hamlet},$$

and in describing A_6 we would have been referring to the man Bacon in two different ways. A final example also illustrating this last point is the following. Let

$$A_7 = \{1, 2\}$$
$$R_4 = \{\langle 1, 1\rangle, \langle 1 + 1, 2\rangle, \langle 1, 2\rangle, \langle 1 + 1, 1\rangle\}.$$

Since $2 = 1 + 1$, R_4 is both reflexive and symmetric in A_7.

EXERCISES

1. Classify the following relations according to the properties they do or do not have (e.g., reflexive, symmetric, not antisymmetric, not transitive, etc.)

(a) The relation of being a grandfather in the set of all persons.
(b) Let A = the set of all real numbers, and $x\,R\,y$ if and only if $x < y + 1$.
(c) Let A = the set of all real numbers, and $x\,R\,y$ if and only if $x < y - 1$.
(d) Let $A = \{1, 2, \text{Mark Twain}\}$, and $R = \{\langle 1, 2\rangle, \langle 2, 1\rangle, \langle \text{Mark Twain}, 1\rangle\}$.
(e) Let $A = \{3, 5, 8\}$, and $R = \{\langle 3, 3\rangle, \langle 5, 5\rangle, \langle 8, 8\rangle\}$.
(f) Let $A = \{\text{Madison}, \text{Pinckney}, 2\}$, and $R = \{\langle 1, 1\rangle, \langle \text{Madison}, \text{Pinckney}\rangle, \langle 1, 2\rangle\}$.
(g) The relation of being the same height in the set of all persons.
(h) The relation of being exactly one year younger in the set of all persons.
(i) The relation of exact divisibility in the set of positive integers.

2. State the precise circumstances under which a relation is both symmetric and asymmetric in a set A.

3. Let $A = \{1, 2, \{1\}\}$.

(a) Give an example of a binary relation which is reflexive and transitive, but not symmetric in A.
(b) Give an example of a binary relation which is reflexive and symmetric, but not transitive in A.
(c) Give an example of a binary relation which is reflexive, but neither symmetric nor transitive in A.
(d) Give an example of a binary relation which is neither reflexive, symmetric, nor transitive in A.

4. Let $A = \{\text{Napoleon, Wellington}\}$. Can you give an example of a relation which is irreflexive, symmetric, and transitive in A?

5. Give an example of a relation which is both symmetric and antisymmetric in a set and is such that it holds between at least two elements of the set.

6. Give an example of a family relationship which is both transitive and intransitive.

7. Any equal-arm balance used for measuring mass is not perfectly sensitive. Three objects x, y and z may always be found such that x exactly balances y, y exactly balances z, but x does not exactly balance z. The ordered triple $\langle x, y, z\rangle$ is called an *intransitive triad*. Given a balance which is only sensitive to differences of mass greater than or equal to .001 gram, state the masses in grams of a possible intransitive triad for this balance.

§ 10.4 Equivalence Relations. Binary relations which have certain combinations of the properties defined in the previous section are both significant and useful. We shall define a few of the more common types in this and the next section.

A relation which is reflexive, symmetric, and transitive in the set A is an *equivalence relation* in A. The relation of identity is an equivalence relation. The same is true of the relation of having the same mass in the set of physical bodies, the relation of having the same volume, and the relation which holds between two sets if and only if they have the same number of elements. The relation of parallelism between straight lines is another example. Both in mathematical language and ordinary language equivalence relations are often called relations of *equality*. Thus, as was pointed out in § 5.1, lines which have the same length are called equal. To avoid the common confusion between equivalence in some aspect of two objects and identity of objects, we always use 'equality' as a synonym for 'identity'.

Between the relation of identity and equivalence relations there is an intimate kind of connection which has far-reaching applications. We shall describe this connection without developing its consequences. Let R be an equivalence relation in A, and for $x \in A$, we characterize the set $[x]$ as follows:

$$y \in [x] \leftrightarrow (y \in A \ \&\ x\,R\,y).$$

The special square brackets enclosing 'x' are used to designate the *R-equivalence class of x in A;* in other words, $[x]$ is the R-equivalence class *generated* by x. We may easily show that each element of A belongs to exactly one R-equivalence class in A. Furthermore, we have the important result that

$$(1) \qquad [x] = [y] \quad \text{if and only if} \quad x\, R\, y.$$

The general philosophical significance of (1) is that it formulates a principle of abstraction: objects which are equivalent in some respect generate identical classes.

One of the reasons for the importance of this principle of abstraction is that application of it can often lead to a substantial reduction in the number of entities being considered by passing from objects to equivalence classes of objects. For example, suppose we are comparing the apparent fairness of a large number of coins by flipping each of them fifteen times. The complete result of our test of each coin may be represented by an ordered 15-tuple. 'H' is used to show that the outcome of a flip was heads and 'T' to show that it was tails. Thus, for example,

$$\langle H, T, H, \ldots, H \rangle$$

indicates that the first, third, and fifteenth flips came out heads and the second flip tails. Since the outcome of each flip is either H or T, there are 2^{15} (= 32,768) possible outcomes of the test, i.e., 32,768 ordered 15-tuples of the kind described. Developing a theory for comparing these 32,768 possible outcomes seems rather complicated. But an essential simplification suggests itself. In deciding on the fairness of a coin we are really only interested in the ratio of the number of flips with outcome heads (or tails) to the total number of flips. For example, if a coin came up heads fifteen times in a row we would be strongly inclined to conclude that it was not a fair coin, i.e., a coin with respect to which the chances of heads or tails on a flip were about even. Since we are interested only in the ratio of heads to the total number of flips, we may "throw away" the information concerning in what order H and T occurred. We define two of our ordered tuples as *ratio-equivalent* if H occurs the same number of times in each. Thus,

$$\langle H, H, H, T, T, T, T, H, T, T, T, H, H, T, T \rangle$$

and

$$\langle T, T, H, T, H, H, T, T, T, T, H, H, T, H, T \rangle$$

are ratio-equivalent since six H's occur in both the ordered 15-tuples. It is obvious that there are 16 ratio-equivalence classes corresponding to 0, 1, 2, ..., 15 occurrences of heads, and the small number of such equivalence classes compares impressively with the 32,768 ordered tuples. In particular, it is possible to develop rules for accepting or rejecting a coin as fair simply according to which equivalence class its test belongs to. The actual construction of such rules is beyond the scope of our discussion, since it is essentially a problem of statistics rather than logic.

EXERCISES

1. Which, if any, of the relations in Exercise 1 of the preceding section are equivalence relations?

2. Can you give an example of a family relationship which is an equivalence relation in the set of all persons?

3. Let S be the relation such that $x\, S\, y$ if and only if x and y are born in the same state, and let A be the set of all persons born in some state of the United States.

(a) How many S-equivalence classes of A are there?
(b) What S-equivalence class is [Truman]?
(c) What S-equivalence class is [Eisenhower]?

4. A manufacturer of high-quality dice tests the fairness of each die by rolling it a hundred times and tabulating the results. There are thus 6^{100} possible outcomes of a given test. Describe an intuitively acceptable equivalence relation which will group the possible outcomes into a relatively small number of equivalence classes.

§ 10.5 Ordering Relations. In this section we define relations which *order* sets in various weak and strong senses. Historically the strong orderings, such as $\leq$ and $<$ for the real numbers, were discovered first. The weaker orderings discussed below are generalizations of such strong orderings. The ordering relations defined in this section all have important applications both in mathematics and the empirical sciences, particularly the social sciences. Some of the exercises at the end of the section illustrate possible applications.

No new fundamental properties of relations are introduced in this section. You will quickly be able to understand the notions of ordering defined if you explicitly notice that each ordering simply represents a certain combination of the fundamental properties defined in § 10.3. We begin with the ordering relations which are logically the simplest.

A relation R is a *quasi-ordering of the set A* if and only if R is reflexive and transitive in A. The relation $\subseteq$ of inclusion among sets is an example of a quasi-ordering (of the set of all sets). The relation $\leq$ is a quasi-ordering of the set of all numbers. The relation of being at least as tall as is a quasi-ordering of the set of all persons. Since all three of these relations have other properties than those of being reflexive and transitive, they are, as we shall see, more than just quasi-orderings. For another example, let

$$A_1 = \{1, 2, \text{Robin Hood}\}$$

$$R_1 = \{\langle 1, 1\rangle, \langle 2, 2\rangle, \langle \text{Robin Hood}, \text{Robin Hood}\rangle, \langle 1, 2\rangle, \langle 2, \text{Robin Hood}\rangle, \langle 2, 1\rangle, \langle 1, \text{Robin Hood}\rangle\}.$$

Then R_1 is a quasi-ordering of A_1.

A relation R is a *partial ordering of the set A* if and only if R is reflexive, antisymmetric, and transitive in A. In other words, a partial ordering of A is an antisymmetric quasi-ordering of A. The relations $\subseteq$ and $\leq$ just mentioned are partial orderings. On the other hand, R_1 is not a partial ordering of A_1, since it is not antisymmetric in A_1: $\langle 1, 2 \rangle \in R_1$ and $\langle 2, 1 \rangle \in R_1$, but $1 \neq 2$. Also, the relation of being at least as tall as is not a partial ordering of the set of persons, since two distinct men may have the same height, and thus the relation is not antisymmetric.

A partial ordering R of a finite set A may be represented by a Hasse diagram.* Small circles represent elements of A and if y may be reached from x by a continually ascending, not necessarily straight line,

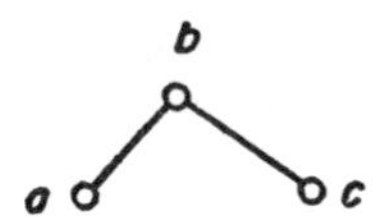

then $x \, R \, y$. For instance, the diagram above represents the partial ordering of a, b, c such that $a \, R \, b$ and $c \, R \, b$. R is the set $\{\langle a, a \rangle, \langle b, b \rangle, \langle c, c \rangle, \langle a, b \rangle, \langle c, b \rangle\}$.

Other examples of partial orderings are given by the other diagrams.

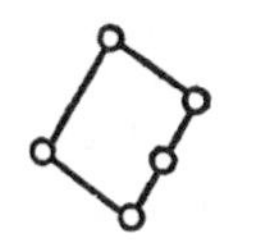

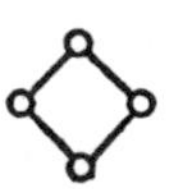
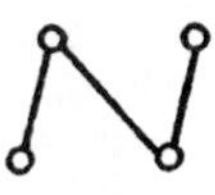

A relation R is a *simple ordering of the set A* if and only if R is reflexive, antisymmetric, transitive, and connected in A. In other words, a simple ordering of A is a connected partial ordering of A. The relation $\leq$ is a simple ordering of the set of all numbers. On the other hand, the relation $\subseteq$ of inclusion is not a simple ordering of the set of all sets, since neither $\{1\} \subseteq \{2\}$ nor $\{2\} \subseteq \{1\}$; i.e., since $\subseteq$ is not connected in the set of all sets. None of the above Hasse diagrams represents a simple ordering, since each diagram has at least two elements not connected by an ascending line. In fact, Hasse diagrams of simple orderings are dull. They always look like:

that is, a Hasse diagram of a simple ordering is just a single vertical line.

* These diagrams are so called after the mathematician H. Hasse.

The relation $\subset$ of proper inclusion and the relation $<$ of less than are closely connected with $\subseteq$ and $\leq$ respectively, but $\subset$ and $<$ are not partial orderings. Since these latter relations are important examples of ordering relations, they suggest the notion of a *strict* partial ordering. We use the word 'strict' because the relation $<$ is sometimes called a strict inequality. Thus a relation R is a *strict partial ordering of the set A* if and only if R is asymmetric and transitive in A. The relation of being taller than is a strict partial ordering in the set of human beings. It is of some interest to note that the relation of being a husband is a strict partial ordering in the set of human beings. The asymmetric character of the relation is obvious, and the relation is transitive in a vacuous sense, since there are not three individuals x, y, and z such that x is the husband of y and y is the husband of z, for this would require y to be both male and female.

It should be noticed that in the phrase 'strict partial ordering' the word 'strict' does not function as an adjective modifying 'partial ordering', for strict partial orderings are not certain special partial orderings. Excluding relations on the empty set, no relation can be both a partial ordering and a strict partial ordering, for the former must be reflexive and the latter irreflexive. On the other hand, the distinction between partial orderings and strict partial orderings is substantively trivial. To each partial ordering there corresponds a unique strict partial ordering, and conversely. Furthermore, the Hasse diagrams of a partial ordering and the corresponding strict partial ordering are identical.

The relation between $\leq$ and $<$ suggested partial orderings and strict partial orderings. It also suggests simple orderings and *strict* simple orderings. A relation R is a *strict simple ordering of the set A* if and only if R is asymmetric, transitive, and connected in A. As is to be expected, a strict simple ordering of A is just a connected strict partial ordering of A. The relation $<$ in the set of numbers is a prime example of a strict simple ordering.

A relation R is a *weak ordering of the set A* if and only if R is transitive and strongly connected in A.* Since if a relation is reflexive and connected, it is strongly connected, any simple ordering is also a weak ordering, just as any simple ordering is also a partial ordering; but there are weak orderings which are neither partial nor simple orderings—for instance, the relation of being at least as tall as, already mentioned. Also, R_1 is a weak ordering of A_1. If $x = y$, then the assumption that $x\,R\,y$ or $y\,R\,x$ implies that $x\,R\,x$, which leads us to the conclusion that any weak ordering of a set A is also a quasi-ordering of A.

The following diagram makes clear the relations between the various kinds of orderings introduced in this section. When we can pass from one ordering to another by a continuously rising path, this means the second

* The term 'weak ordering' is not entirely standard.

is a special case of the first. For example, every partial ordering is also a quasi-ordering, but, of course, not every quasi-ordering is a partial ordering. Since all the ordering relations considered in this section are transitive, we make transitive relations the most general case in the diagram.

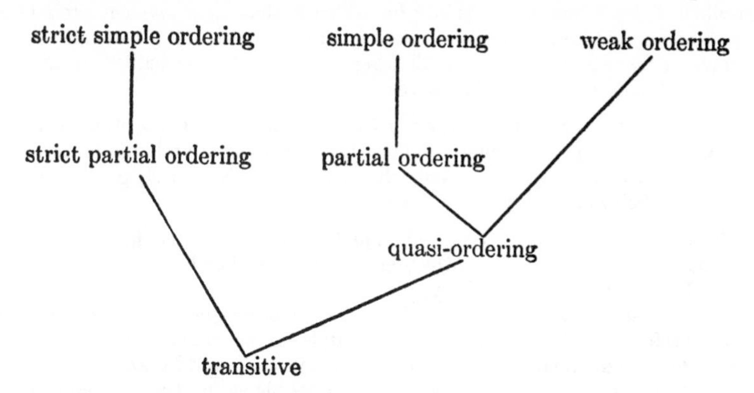

Sometimes, instead of saying, for instance, that a relation R is a partial ordering of A, it is convenient to say that A is *partially ordered* by R. Similar terminological remarks apply to the other kinds of ordering.

EXERCISES

1. In the text no example is given of a quasi-ordering which is not a partial ordering, a weak ordering, or an equivalence relation. Give such an example.

2. Determine what kind of ordering relation, if any, each of the relations is in Exercise 1 of § 10.3. Characterize each relation as fully as possible. (For example, if a relation is a partial ordering, it is not sufficient to say it is a quasi-ordering.)

3. Let A be the set of married couples, and let $x\,P\,y$ if and only if the male of couple x is shorter than the male of couple y, and the female of couple x is shorter than the female of couple y. Is P a strict simple ordering of A? If not, why not?

4. Since equivalence relations provide a kind of ordering, where would you put them on the diagram of this page?

5. Consider the following relations. If they are partial orderings or strict partial orderings of their fields draw their Hasse diagrams.

$R_1 = \{\langle 1, 2\rangle, \langle 1, 3\rangle, \langle 2, 4\rangle, \langle 1, 4,\rangle\ \langle 3, 4\rangle\}$
$R_2 = \{\langle \text{Lincoln, Jefferson}\rangle, \langle \text{Jefferson, Madison}\rangle, \langle \text{Madison, Cleveland}\rangle\}$
$R_3 = \{\langle 1, 1\rangle, \langle 2, 2\rangle, \langle 3, 3\rangle, \langle 1, 2\rangle\}$
$R_4 = \{\langle 1, 2\rangle, \langle 2, 1\rangle\}$
$R_5 = \{\langle 1, 2\rangle, \langle 1, 5\rangle, \langle 5, 3\rangle, \langle 1, 3\rangle, \langle 5, 4\rangle, \langle 1, 4\rangle\}$

6. Economists, psychologists, sociologists, and philosophers have been interested in the kinds of choices people make in varying situations. They have attempted to lay down certain necessary conditions of rationality on a man's choice. (See Exercise 11 of § 4.1.) If these conditions are not satisfied, it is maintained

that he is not acting or choosing in an intuitively reasonable way. For example, it seems reasonable to expect a man's preference relation to be transitive and asymmetric. Correspondingly, it is reasonable, at least as a first approximation, to expect a man's "indifference" relation to be an equivalence relation, that is, reflexive, symmetric, and transitive. By "indifference relation" we mean the relation which holds between two things for which he has an equivalent preference. Thus, Smith might be indifferent as between a green Buick and a blue one.

Summing up, we lay down the following conditions of rationality for an individual faced with a set A of alternatives:

(1) His relation P of preference must be asymmetric and transitive in A.
(2) His relation I of indifference must be an equivalence relation in A.
(3) For any two, not necessarily distinct, alternatives in A, exactly one of the following: $x\ P\ y$, $y\ P\ x$, $x\ I\ y$.

Condition (3) simply says that the individual has to know what he wants in the sense that given any two alternatives he can say whether he prefers one to the other or is indifferent between them.

Now our precise definition of a relation as a set of ordered couples can be empirically useful in studying to what extent individuals are rational in their choices. We present an individual with a list of choices, and we record his answers in terms of ordered couples. Even better, we may observe his actual behavior and record his choices.

The problem to solve is the following: Given the data on a certain individual, i.e., the record of his choices, to decide if he is rational in the sense defined above, and if he is not, what conditions he has violated (e.g., perhaps his relation I of indifference is not symmetric). Since it is senseless to present a subject with a choice between one and the same object, in describing the I relation of an individual we omit couples of the form $\langle x, x\rangle$, and we shall simply assume in the following problems that the I relation of every subject is reflexive. Notice that it is up to the examiner or experimenter to see that at least one ordered couple occurs in either the relation P or I comparing every object in A with every other.

(a) We ask Jones to choose pairwise between four automobiles: Buick, Chrysler, Mercury, Studebaker. These automobiles constitute the set A. From his responses we obtain the following relations P and I. Are his responses rational? If not, what conditions are violated?

$$P = \{\langle \text{Chrysler, Buick}\rangle, \langle \text{Buick, Mercury}\rangle, \langle \text{Mercury, Studebaker}\rangle, \langle \text{Buick, Studebaker}\rangle, \langle \text{Chrysler, Studebaker}\rangle\}$$
$$I = \{\langle \text{Mercury, Chrysler}\rangle, \langle \text{Chrysler, Mercury}\rangle, \langle \text{Studebaker, Chrysler}\rangle\}.$$

(b) We ask factory workers to choose between paired items in terms of their relative importance to them. The items are: ventilation and lighting (V & L), interesting work (W), complete health insurance plan (H), pension fund (F), opportunities for promotion (O).

Mr. Smith responds as follows:

$$P = \{\langle F, O\rangle, \langle H, O\rangle, \langle H, F\rangle, \langle H, V \,\&\, L\rangle, \langle H, W\rangle, \langle F, V \,\&\, L\rangle, \langle O, W\rangle, \langle F, W\rangle, \langle F, O\rangle\}.$$
$$I = \{\langle V \,\&\, L, W\rangle, \langle W, V \,\&\, L\rangle, \langle V \,\&\, L, O\rangle, \langle O, V \,\&\, L\rangle\}.$$

Is he rational? If not, what conditions are violated?

(c) Using the set of automobiles given in (a), construct rational preference and indifference relations for yourself.

7. Economists often use the notion of a *commodity space*. Very simply, a commodity space is a set of ordered n-tuples of goods. For brevity of notation, we may represent an ordered n-tuple of goods by an ordered n-tuple of real numbers. Each number represents the amount of some definite good. For example, $\langle 1, 3, 2\rangle$ might represent 1 TV set, 3 loaves of bread, and 2 bottles of beer. Such ordered n-tuples of goods are often called *commodity bundles*. Now it is sometimes difficult to choose between two commodity bundles; one bundle might have, for instance, more oranges and fewer apples than another. But on the assumption that a larger amount of any good is preferred to a smaller amount, the following weak principle of preference may be introduced:

(I) *Commodity bundle a is to be preferred to commodity bundle b if a has as much of every good as b, and more of at least one good.*

On this principle, the bundle $a_1 = \langle 1, 3, 2\rangle$ is to be preferred to bundle $a_2 = \langle 0, 3, 2\rangle$, since a_1 has as much bread and beer as a_2, and it has in addition a television set.

(a) In general what kind of ordering relation on a commodity space may be derived from Principle (I)?

(b) Let R be the ordering relation on a commodity space derived from (I). A bundle x of a commodity space S is said to be a *Pareto optimal point* if there is no y in S such that $y\,R\,x$, that is, such that y is preferred to x according to Principle (I).

Let

$$S = \{\langle 1, 3, 2\rangle, \langle 0, 3, 2\rangle, \langle 1, 2, 2\rangle, \langle 1, 2, 5\rangle, \langle 2, 0, 4\rangle, \langle 0, 5, 1\rangle, \langle 1, 2, 0\rangle, \langle 0, 1, 0\rangle\}.$$

Which elements of S are Pareto optimal points?

(c) Let $A = \{0, 1, 2, 3, 4, 5, 6\}$, and let $S = A \times A \times A$. Which elements of S are Pareto optimal points?

§ **10.6 Operations on Relations.** If V is any domain of individuals, then by the *universal relation* over V we mean the set of all ordered couples $\langle x, y\rangle$ where $x \in V$ and $y \in V$, that is, the Cartesian product $V \times V$.

The empty set Λ is the relation which never holds. If R, for instance, is the relation which holds between x and y if and only if x is the mother of y, and y is the mother of x, then $R = \Lambda$, for no one is his own grandmother.

Since relations are a special kind of sets, we can consider, as applied to relations, the usual operations on sets. Thus, for example, if R and S are relations, then $R \cap S$ is the relation which consists of the intersection of R and S: i.e., $x(R \cap S)y$ if and only if both $x\,R\,y$ and $x\,S\,y$. Similarly, $R \cup S$ is the union of R and S: $x(R \cup S)y$ if and only if either $x\,R\,y$ or $x\,S\,y$. And $R \sim S$ is the relation such that $x(R \sim S)y$ if and only if $x\,R\,y$ and not $x\,S\,y$.

If $x\,B\,y$ when x is a brother of y, and $x\,S\,y$ when x is a sister of y, then $x(B \cup S)y$ when x is a sibling of y; but $B \cap S$ is the empty relation. We notice that $B \sim S = B$, since the sets of ordered couples B and S are mutually exclusive.

If $R \subseteq S$ we call R a *subrelation* of S. Thus brotherhood is a subrelation of siblinghood; for whenever x is a brother of y, then x is a sibling of y. Every relation is a subrelation of the universal relation over its own field.

If we have chosen a domain of individuals V, and if U is the universal relation over V (i.e., $U = V \times V$), then by $\sim R$, when R is a relation whose field is a subset of V, we mean simply $U \sim R$.

Besides these operations, which are all simply special cases of the operations on sets, it is also possible to define some special operations on binary relations which depend on the fact that they are sets of *ordered couples*. The *converse* of a relation R (in symbols: $\breve{R}$) is the relation such that, for all x and y, $x\,\breve{R}\,y$ if and only if $y\,R\,x$. Thus the converse of a relation is obtained simply by reversing the order of all the ordered couples which constitute it. The converse of the relation H which holds between x and y if and only if x is the husband of y, is the relation W which holds between y and x if and only if y is the wife of x. Thus,

$$W = \breve{H}.$$

If

$$R = \{\langle 1, 2\rangle, \langle 1, 3\rangle, \langle 2, 4\rangle\}$$

then

$$\breve{R} = \{\langle 2, 1\rangle, \langle 3, 1\rangle, \langle 4, 2\rangle\}.$$

If R and S are binary relations, then by the *relative product* of R and S (in symbols: R/S) we mean the relation which holds between x and y if and only if there exists a z such that R holds between x and z, and S holds between z and y. Symbolically,

$$x\,R/S\,y \leftrightarrow (\exists z)(x\,R\,z\,\&\,z\,S\,y).$$

If $x\,P\,y$ means that x is a parent of y, and $x\,S\,y$ means that x is a sister of y, then $x(S/P)y$ means that there is a z such that x is a sister of z and z is a parent of y, and hence such that x is an aunt of y. The formation of relative products is not a commutative operation. For instance, with P and S as above, P/S is not the same as S/P: a man's aunt is seldom a parent of his sister.

If $x\,P\,y$ when x is a parent of y, then $x(P/P)y$ if and only if x is a grandparent of y; $x[(P/P)/P]y$ if and only if x is a great-grandparent of y; and so on. On the supposition that creation had no beginning (i.e., every person has parents), the relations P, P/P, $(P/P)/P$, etc., are all different.

As a final example, if

$$R = \{\langle 1, 3\rangle, \langle 2, 3\rangle\},$$
$$S = \{\langle 3, 1\rangle\},$$

then

$$R/S = \{\langle 1, 1\rangle, \langle 2, 1\rangle\},$$
$$S/R = \{\langle 3, 3\rangle\},$$
$$(S/R)/R = \Lambda.$$

EXERCISES

1. If

$x\,F\,y$	means	x is father of y
$x\,M\,y$	means	x is mother of y
$x\,B\,y$	means	x is a brother of y
$x\,S\,y$	means	x is a sister of y

what does it mean to say that:

(a) $x(F/M)y$
(b) $x(M/S)y$
(c) $x(M/F)y$
(d) $x(S \cup B)y$
(e) $x[S/(M \cup F)]y$
(f) $x[(B \cup S)/(M \cup F)]y$
(g) $x(\breve{M \cup F})y$
(h) $x(\breve{M} \cup \breve{F})y$
(i) $x[(\breve{M} \cup \breve{F})/(B \cup S)]y$
(j) $x((\breve{M} \cup \breve{F})/[(B \cup S)/(M \cup F)])y$

2. Using the letters introduced in Exercise 1, express the following by means of formulas:

(a) x is a grandparent of y
(b) x is a grandchild of y
(c) x is a great-uncle of y
(d) x is a half-sibling of y

3. Consider the following relations:

$$A = \{\langle 1, 2\rangle, \langle 5, 10\rangle, \langle 2, 3\rangle, \langle 1, 3\rangle, \langle 4, 4\rangle\}$$
$$B = \{\langle 1, 1\rangle, \langle 2, 2\rangle, \langle 3, 3\rangle\}$$
$$C = \{\langle 1, 2\rangle, \langle 2, 1\rangle\}$$
$$D = \{\langle 1, 2\rangle, \langle 2, 1\rangle, \langle 1, 1\rangle, \langle 2, 2\rangle\}$$

(a) What properties of reflexivity, symmetry, and transitivity does each of the relations have in its own field?
(b) Is the field of the relation C a subset of the domain of the relation A?
(c) Is the converse domain of D a subset of the field of C?
(d) Is any one of the four relations an equivalence relation in its own field?
(e) Is B a subrelation of A?
(f) Is C a subrelation of D?
(g) What is the universal relation over the field of C?
(h) Which (if any) of the relations is identical with its converse?
(i) What relation is the relative product B/C? List its members.
(j) List the members of C/B. Does $B/C = C/B$?
(k) What relation holds between D/C and C/D?

4. What is the converse of the relation $\leq$ for the real numbers?
5. Is the relation $<$ a subrelation of $\leq$?
6. What relation on the real numbers is the relative product $</<$? (That is, the relative product of less-than with itself.)
7. Let the field of $<$ be the set of all real numbers. What is the field of the relation $</\breve{<}$?
8. Let

$$A = \{\langle 1, 2\rangle, \langle 3, 4\rangle, \langle 2, 2\rangle\}$$

$$B = \{\langle 4, 2\rangle, \langle 2, 5\rangle, \langle 3, 1\rangle, \langle 1, 3\rangle\}.$$

List the members of the following sets:

(a) A/B
(b) $\breve{A}/\breve{B}$
(c) $A/(B \cup A)$
(d) $\breve{B}/A$
(e) $A/(B/A)$
(f) $A \cap (A/B)$

9. Let the relations A and B be defined as in Exercise 8.

(a) Is A/B a subset of the Cartesian product $A \times B$?
(b) Is B/A a subset of the Cartesian product $D(B) \times C(B)$?

10. Let A be the set of all real numbers. If we fix an origin and unit the points of a plane may be considered as the Cartesian product $A \times A$. A straight line in the plane is a certain subset of $A \times A$, whence it is a relation, and we may use the language of this chapter to talk about operations on lines.

(a) What sort of geometrical figures are the following:
  (i) the union of two straight lines?
  (ii) the intersection of two straight lines?
  (iii) the difference of two straight lines?
(b) If L is a straight line, what is its geometric relation to its converse $\breve{L}$?
(c) Given any two straight lines, is their relative product a straight line?

11. One objection to the theory of preference discussed in previous chapters in various exercises is that the relation I of indifference is not transitive for many individuals and situations. R. Duncan Luce has given a simple set of axioms on the relations P and I which do not require transitivity of I, and on the basis of which two relations may be defined which do satisfy the axioms given in Exercise 6, § 10.5.* His axioms are (for every x, y, z, w in the set A of alternatives):

S1. $x\, I\, x$.
S2. *If* $x\, P\, y$ & $y\, I\, z$ & $z\, P\, w$ *then* $x\, P\, w$.
S3. *If* $x\, P\, y$ & $y\, P\, z$ & $y\, I\, w$ *then not both* $x\, I\, w$ & $z\, I\, w$.
S4. *Exactly one of the following:* $x\, P\, y$, $y\, P\, x$, $x\, I\, y$.

We now define:

(i) $x\, \bar{P}\, y \leftrightarrow x\, P\, y \vee [x\, I\, y \,\&\, (\exists z)(z \in A \,\&\, x\, I\, z \,\&\, z\, P\, y)] \vee [x\, I\, y \,\&\, (\exists z)(z \in A \,\&\, x\, P\, z \,\&\, z\, I\, y)]$
(ii) $x\, \bar{I}\, y \leftrightarrow -x\, \bar{P}\, y \,\&\, -y\, \bar{P}\, x$.

Prove:

(a) Exactly one of the following: $x\, \bar{P}\, y$, $y\, \bar{P}\, x$, $x\, \bar{I}\, y$. (Distinguish cases in the proof.)
(b) $\bar{I}$ is reflexive in A.
(c) $\bar{I}$ is symmetric in A.
(d) $\bar{I}$ is transitive in A.
(e) $\bar{P}$ is transitive in A.

* R. Duncan Luce, "Semiorders and a Theory of Utility Discrimination," *Econometrica*, Vol. 24 (1956) pp. 178–191.

CHAPTER 11

FUNCTIONS

§ **11.1 Definition.** Webster's New International Dictionary defines the mathematical notion of function in this manner:

> A magnitude so related to another magnitude that to values of the latter there correspond values of the former.

The two difficult words in this definition are 'magnitude' and 'corresponds'. What are magnitudes and what does it mean for one thing to correspond to another? In the last chapter we added the notion of ordered couple to those of set and membership in order to define relations. Can we avoid adding the seemingly vague notions of magnitude and correspondence in order to define the notion of function? Before deciding this question, let us look at another definition. A well-known textbook * in the differential and integral calculus states:

> A *variable* is a quantity to which an unlimited number of values can be assigned in an investigation .... When two variables are so related that the value of the first variable is determined when the value of the second is given, then the first variable is said to be a function of the second .... The second variable ... is called the *independent variable,* or *argument;* and the first variable ... is called the *dependent variable,* or *function.*

To begin with, in this definition the word 'variable' is not being used as it is in this book, but rather more or less as a synonym for Webster's 'magnitude'. This definition is neither worse nor better than Webster's: we cannot clearly understand either definition without some clarification concerning the elusive notions of magnitude and quantity.

A somewhat more sophisticated definition is the following:

> A function is a rule which assigns to each element of a given set a unique element of some other, not necessarily distinct, set.

Here, at least, there is no loose talk about magnitudes and quantities. In many respects this definition is intuitively satisfactory. On the other

* W. A. Granville, P. F. Smith, and W. R. Longley, *Elements of the Differential and Integral Calculus,* rev. ed., New York, 1941.

hand, the notion of rule it uses is reminiscent of the vague notion of dependence or connection we were careful to avoid in characterizing the sentential connectives and in defining the notion of a relation.

Fortunately there is a way out which bypasses these difficulties. We may define functions as binary relations which relate to each element of their domains a *unique* element of their counterdomains. More formally, a *function* R is a binary relation such that if $x \, R \, y$ and $x \, R \, z$ then $y = z$.

You are probably used to thinking of functions in numerical terms, but there is no necessity for this. For example, in our society the relation of being a husband is a function, since a person has at most one wife (at a given time). On the other hand, the relation of being a mother is not a function since one person can be the mother of several distinct children, but the converse is a function. It will be useful to consider some simple constructed relations. Let

$$R_1 = \{\langle 1, 2\rangle, \langle \text{Madison, Pinckney}\rangle\}$$

$$R_2 = \{\langle 1, 2\rangle, \langle 1, 3\rangle, \langle \text{Plato, Aristotle}\rangle\}.$$

It should be clear at once that R_1 is a function and R_2 is not, but to avoid any possible misunderstandings let us examine the situation in some detail. The domain of R_1 is the set

$$\{1, \text{Madison}\}.$$

To each element in $D(R_1)$ there corresponds a unique element in the counterdomain of R_1 (i.e., $C(R_1)$). Thus,

$$1 \, R_1 \, 2 \quad \text{and} \quad \text{Madison} \, R_1 \, \text{Pinckney}.$$

The domain of R_2 is the set

$$\{1, \text{Plato}\}.$$

To Plato there corresponds a unique element in $C(R_2)$, namely, Aristotle, but to the number one there does *not* correspond a unique element in $C(R_2)$, for we have:

$$1 \, R_2 \, 2 \quad \text{and} \quad 1 \, R_2 \, 3,$$

and

$$2 \neq 3.$$

Thus R_2 is not a function.

We have already used two notations for relations: $x \, R \, y$ and $\langle x, y\rangle \in R$. We now introduce a third and customary notation for the special case of functions. In the case of R_1 above, which is a function, we write:

$$R_1(1) = 2,$$

$$R_1(\text{Madison}) = \text{Pinckney}.$$

Notice that this special functional notation cannot be used for relations which are not functions. For example, if we apply it to R_2 we obtain the following absurdity:

$$R_2(1) = 2,$$

$$R_2(1) = 3,$$

and hence

$$2 = 3.$$

It is often convenient to use lower case letters to denote functions. For example,

$$f_1 = \{\langle \text{Romeo, Juliet} \rangle, \langle \text{Abelard, Heloise} \rangle\},$$

and

$$f_1(\text{Romeo}) = \text{Juliet},$$

$$f_1(\text{Abelard}) = \text{Heloise}.$$

In mathematics functions are frequently defined by means of equations. For example, we might want to consider the function f_2 such that the domain of f_2 is the set of all real numbers and for any real number x,

$$f_2(x) = x^2 + 2x + 5.$$

Thus,

$$f_2(1) = 8 \quad \text{and} \quad f_2(2) = 13.$$

A function need not be defined by a single equation. Consider, for instance, the function g such that the domain of g is the set of all real numbers and for any real number x,

$$(1) \qquad \begin{cases} g(x) = x^2 & \text{for } x \leq 0, \\ g(x) = 5x^3 + 17 & \text{for } x > 0. \end{cases}$$

Thus

$$g(-1) = 1 \quad \text{and} \quad g(1) = 22.$$

(1) is equivalent to the more formal expression:

$$(2) \qquad g(x) = y \leftrightarrow [(x \leq 0 \rightarrow y = x^2) \ \& \ (x > 0 \rightarrow y = 5x^3 + 17)],$$

but for obvious typographical reasons it is customary to use the more informal notation, which is also often written:

$$(3) \qquad g(x) = \begin{cases} x^2 \ \text{if } x \leq 0, \\ 5x^3 + 17 \ \text{if } x > 0. \end{cases}$$

It needs to be emphasized that it is not formally sufficient to give the defining equations of a function. It is also necessary to state explicitly what the domain of the function is. In ordinary mathematical practice this latter statement is often omitted when it is intuitively clear what is the intended domain of the function. For instance, the use of defining equations like (3) usually implicitly entails that the domain of the function is the set of all real numbers. However, we could use (3) to define a function whose domain is the set of all real numbers greater than -10, say. The point is that when the domain of a function is an infinite set it is impossible actually to list the ordered couples which belong to the function. Since most functions of mathematical interest have infinite domains, the definition of a function usually consists of (a) specifying its domain and (b) giving a rule which states the value of the function for each element in its domain.

A second example of a function defined by more than one equation is the function h whose domain is the set of positive real numbers and which is such that for any positive real number x

$$(4) \qquad h(x) = \begin{cases} x \text{ if } x \text{ is an integer,} \\ 1 \text{ if } x \text{ is a rational number but not an integer,} \\ 0 \text{ if } x \text{ is not a rational number.} \end{cases}$$

Thus,

$$h(3) = 3, h(\tfrac{1}{2}) = 1 \quad \text{and} \quad h(\sqrt{2}) = 0.$$

It also needs to be emphasized that a function is a set-theoretical, not a linguistic, entity. Functions exist independent of any language. We may digress for a moment and classify the main entities introduced in this book as set-theoretical or linguistic.

LINGUISTIC	SET-THEORETICAL
Formulas	Individuals
Sentences	Ordered Couples
Sentential Connectives	Sets
Variables	The empty set
Terms	Relations
Quantifiers	Functions

Linguistic entities are, of course, always part of some language, whereas set-theoretical entities in general are not. At first blush it may seem strange to say that an individual like Thomas Aquinas is a set-theoretical entity. The justification is that individuals are the ingredients out of which we build up sets. At level zero we have individuals, at level one sets of individuals, at level two sets of sets of individuals, and so on. Linguistic entities are themselves set-theoretical entities, and it would be more appropriate to entitle the right-hand column above "Set-theoretical entities which are not linguistic entities".*

* The set-theoretical status of linguistic entities is somewhat complex. As an example consider the variable 'x'. On the one hand, we might classify 'x' as an individual, but then what is to be said about occurrences of 'x'? Now if we classify occurrences of 'x'

Since every function is a relation, we may speak of the domain and counterdomain of a function. Certain special terminology is used for functions. The domain is often called the *domain of definition* of the function, and the counterdomain is called the *range of values* of the function, for which we sometimes use the script letter '$\mathscr{R}$.' Thus if f is a function,

$$C(f) = \mathscr{R}(f),$$

that is, the counterdomain of f is the same thing as the range of f. The range of the function h defined by (4) is the set of non-negative integers. If f is the function whose domain of definition is the set of all real numbers and which is such that for every real number x

$$f(x) = x^2,$$

then the range of f is the set of all non-negative real numbers.

A function is sometimes said to *map* its domain onto its range. Thus the function h maps the set of real numbers onto the set of non-negative integers. Using this sort of terminology, we sometimes say that a function f is a *mapping*, and when x is an element of the domain of f, then $f(x)$ is the *image* of x. A function is occasionally called a *correspondence* (which provides, by the way, a sharp definition of 'correspondence'). Also certain special functions are called *transformations* and *operators*, but we shall not go into the basis of this usage.

In logic a function is often called a *many-one* relation. The genesis of this terminology should be clear, from the definition of a function. Analogously, a relation R is *one-many* if whenever $y\, R\, x$ and $z\, R\, x$, then $y = z$. We shall further refer to these notions in the next section.

A particularly important class of functions is the class of *binary operations*. A *binary operation on the set* A is a function whose domain is $A \times A$ and whose range is a subset of A. It is customary to speak of a binary operation *from* $A \times A$ to A, and it is understood when this language is used that the range of the operation need not be the whole of A, but merely a subset of A. For example, if N is the set of positive integers and $+$ is the binary operation of addition of positive integers, then we say $+$ is a binary operation from $N \times N$ to N, although the range of $+$ is $N \sim \{1\}$, for no two positive integers sum to one. Since ordered triples are simply certain special ordered couples, namely those ordered couples

as individuals and the variable 'x' as the set of all such occurrences, a certain ambiguity arises concerning the notion of occurrence. In earlier chapters we have referred to the occurrence of a variable in a sentence, but the sentence itself has many actual occurrences. The safest procedure seems to be to start with the actual physical inscriptions of 'x' and build up from there. Details of this construction are not relevant here.

whose first members are ordered couples, we may also characterize binary operations as certain ternary relations.

A ternary relation T is a binary operation if and only if for every x, y, z, and w if $\langle x, y, z\rangle \in T$ & $\langle x, y, w\rangle \in T$ then $z = w$.

EXERCISES

1. Which of the following relations are functions?

(a) $\{\langle 1, 1\rangle$, $\langle$Judy Canova, 2$\rangle$, $\langle$Madison, 1$\rangle\}$
(b) $\{\langle 1, 1\rangle, \langle 2, 1\rangle, \langle 2, 2\rangle, \langle 2, 3\rangle\}$
(c) $\{\langle 1, 1\rangle, \langle 2, 2\rangle, \langle 3, 2\rangle\}$
(d) the relation of being a spouse
(e) the relation of being a grandfather

2. State what is the domain and range of each of the relations in Exercise 1 which is a function.
3. Which of the relations in Exercise 1 of § 10.3 are functions?
4. Let x and y be real numbers. Which of the following relations are functions?

(a) The relation R such that $x\,R\,y$ if and only if $x > y$.
(b) The relation R such that $x\,R\,y$ if and only if $y \geq x + 2$.
(c) The relation R such that $x\,R\,y$ if and only if $x = y^2$.
(d) The relation R such that $x\,R\,y$ if and only if $y = x^2$.
(e) The relation R such that $x\,R\,y$ if and only if $3x + 2y = 0$.
(f) The relation R such that $x\,R\,y$ if and only if $x^2 + y^2 = 0$.
(g) The relation R such that $x\,R\,y$ if and only if $x^2 = y^2 = 0$.
(h) The relation R such that $x\,R\,y$ if and only if $x = y$.

5. State what the domain is and the range of each of the relations in Exercise 4 which is a function.
6. Let x, y, and z be real numbers. Which of the following ternary relations are binary operations?* The relation T such that $\langle x, y, z\rangle \in T$ if and only if:

(a) $x + 2y = z$.
(b) $x < y$ & $y < z$.
(c) $x < y$ & $z = 2$.
(d) $x \cdot y = 2z$.
(e) $x + y = z - 1$.
(f) $x \cdot y \cdot z = 7$.

7. State the range of each of the ternary relations in Exercise 6 which is a binary operation.

§ 11.2 Operations on Functions. Since a function is simply a special kind of relation, we can speak of the converse of a function. The converse of a function is always a relation, but in general it will not be a function.

* Compare Exercise 2 of § 8.3.

Thus if

$$f = \{\langle 1, 2\rangle, \langle 2, 2\rangle\},$$

then

$$\breve{f} = \{\langle 2, 1\rangle, \langle 2, 2\rangle\};$$

clearly f is a function and $\breve{f}$ is not. As a second example, suppose f is the function such that for every real number x,*

$$f(x) = x^2;$$

if we consider f as a relation, $\langle x, y\rangle \in f$ if and only if

$$x = y^2.$$

Consider, now, $x = 4$. Then $\langle 4, 2\rangle \in \breve{f}$, and also $\langle 4, -2\rangle \in f$. Hence $\breve{f}$ is not a function.

On the other hand, some functions are such that their converses are also functions. For example, if g is the function such that for every real number x,

$$g(x) = 2x + 4.$$

Then $\breve{g}$ is the function such that

$$\breve{g}(x) = \frac{1}{2}x - 2.$$

When a function f is such that its converse relation $\breve{f}$ is also a function, then we say that $\breve{f}$ is *the inverse* of f. To mark this special situation we use '-1' as a superscript in place of the more general cup notation '$\breve{\ }$'. Thus if the converse of f is a function,

$$f^{-1} = \breve{f}.$$

We use the notation '$^{-1}$' only when $\breve{f}$ is a function. It should be noticed that when the inverse of f exists (i.e., the converse of f is a function), we have the following identities: For every x in the domain of f

$$\text{(I)} \qquad f^{-1}(f(x)) = x,$$

and for every x in the range of f

$$\text{(II)} \qquad f(f^{-1}(x)) = x.$$

Principles (I) and (II) are both important in solving for f and f^{-1} as we shall soon see.

At the end of the last section we spoke of many-one and one-many relations. A relation which is both many-one and one-many is *one-one.*

* It is understood that the use of the quantifier 'for every real number x' indicates that the domain of definition of f is the set of all real numbers.

We also speak of a one-one relation as a one-one function. Consider now the following statements about a relation R:

(1) R is a function.
(2) R is a many-one relation.
(3) The converse of R is a function.
(4) R is a one-many relation.
(5) R and its converse are both functions.
(6) R is a function and its inverse exists.
(7) R is a many-one relation and R is a one-many relation.
(8) R is a one-one relation.
(9) R is a one-one function.

The following equivalences hold:

$$(1) \leftrightarrow (2),$$

$$(3) \leftrightarrow (4),$$

$$(5) \leftrightarrow (6) \leftrightarrow (7) \leftrightarrow (8) \leftrightarrow (9).$$

When functions are defined by equations as in some of the examples already discussed, you may have some difficulty in deciding if the inverse of the function exists, and if so, in finding an equation which will define the inverse. Without going into any great detail, one or two practical hints may be of value. To fix our ideas, let f be the function such that for every real number x,

$$(1) \qquad f(x) = 5x - 4.$$

We want to use (1) to find f^{-1}. Since (1) holds for every real number x, we may substitute '$f^{-1}(x)$' for 'x' (application of universal specification) and obtain:

$$(2) \qquad f(f^{-1}(x)) = 5f^{-1}(x) - 4.$$

By virtue of (II) we infer from (2):

$$(3) \qquad x = 5f^{-1}(x) - 4.$$

We now solve equation (3) for '$f^{-1}(x)$', and obtain the equation defining f^{-1}:

$$(4) \qquad f^{-1}(x) = \tfrac{1}{5}x + \tfrac{4}{5}.$$

The strategy of finding f^{-1} as illustrated in (1)–(4) has three phases: (i) Substitute '$f^{-1}(x)$' for 'x'; (ii) Apply (II); (iii) Solve the resulting equation for '$f^{-1}(x)$'.

The strategy just outlined has one defect: application of it will yield an inverse function even though no such inverse may exist. Consider, for instance, the function f defined by the equation:

$$f(x) = x^2 - 1. \tag{5}$$

Substituting '$f^{-1}(x)$' for 'x' and applying (II), we obtain:

$$x = f^{-1}(x)^2 - 1. \tag{6}$$

Solving (6) for '$f^{-1}(x)$' we have:

$$f^{-1}(x) = \pm\sqrt{x+1}.$$

We thus infer that

$$f^{-1}(3) = 2 \quad \text{and} \quad f^{-1}(3) = -2,$$

and hence

$$2 = -2,$$

which is absurd. The mistake here was in inferring (6) from (5). The substitution of '$f^{-1}(x)$' for 'x' is permissible only if '$f^{-1}(x)$' is a proper term. If $\breve{f}$ is not a function, the expression '$f^{-1}(x)$' is not a term, since it does not designate a unique entity.

Sometimes, mathematicians use the term 'inverse' in a slightly wider sense, meaning by 'an inverse' of a function f simply any function g whose domain of definition is the range of values of f, and which satisfies the equation

$$f[g(x)] = x,$$

for every x in the range of f (i.e., in the domain of g). In this sense, if f is the function such that, for every real number x,

$$f(x) = x^2,$$

then an inverse of f is the function g whose domain of definition is the set of non-negative real numbers, and which, for any non-negative real number x, satisfies the equation

$$g(x) = \sqrt{x}.$$

Here we have, for every x in the range of f,

$$f[g(x)] = f[\sqrt{x}\,] = (\sqrt{x}\,)^2 = x;$$

it should be noticed that here the equation

$$g[f(x)] = x$$

is not satisfied by every x in the domain of f—for we have, for example,

$$g[f(-3)] = g[(-3)^2] = g(9) = \sqrt{9} = 3 \neq -3.$$

When a function does not have a unique inverse, mathematicians sometimes pick out (rather arbitrarily) a particular inverse and refer to it as the "principal inverse"—or sometimes, less exactly, as the "principal value of the inverse". This is especially common with respect to the trigonometric functions. Thus when dealing with the function sin (i.e., with the function f such that, for every x, $f(x) = \sin x$) it is customary to introduce, as the principal inverse, the function g, whose domain of definition is the closed interval $-1 \leq x \leq +1$, and which is such that, for any number x in this interval, $g(x)$ is the numerically smallest number y such that $\sin y = x$: so that $g(\frac{1}{2}) = \pi/6$, for example, since $\sin \pi/6 = \frac{1}{2}$, and $\pi/6$ is the numerically smallest number whose sin is $\frac{1}{2}$; and similarly $g(-\frac{1}{2}) = -\pi/6$. It is customary to denote this function g by 'arcsin', or '$\sin^{-1}$'. We notice that

$$\sin [\arcsin x] = x,$$

for x any number in the domain of definition of arcsin; on the other hand, we have, for instance:

$$\arcsin \left[\sin \frac{5\pi}{6} \right] = \arcsin [\tfrac{1}{2}] = \frac{\pi}{6} \neq \frac{5\pi}{6}.$$

The relative product of two functions plays an important role in many discussions and is sometimes given a special name: the *composition* of the functions. Suppose that f and g are functions, and that x, y, and z are entities such that

$$y = f(z) \tag{7}$$

and

$$z = g(x); \tag{8}$$

then x stands to y in the relation g/f. In other terms, from equations (7) and (8) we see, that

$$y = f(g(x)).$$

Thus the relative product, or composition, of two functions is a function h such that (whenever x is in the domain of definition of g, and $g(x)$ is in the domain of definition of f)

$$h(x) = f(g(x)).$$

This operation of forming the composition of two functions is so extensively used in certain branches of mathematics that various special symbols have been used for it; we shall use a small circle '$\circ$'. Thus,

$$(f \circ g)(x) = f(g(x)).$$

We introduce this special notation instead of using the relative product notation because the order of 'f' and 'g' in '$f \circ g$' is the natural one corresponding to their order in '$f(g(x))$', whereas this order is reversed in the relative product:

$$f \circ g = g/f.$$

It is clear that composition of functions is associative but not commutative.

The problem of finding the inverse of a function is actually a special case of the problem of finding "the" function g, given the functions f and $f \circ g$. In finding the inverse, g is f^{-1}, and

$$(f \circ g)(x) = (f \circ f^{-1})(x) = f(f^{-1}(x)) = x.$$

In general the function g is not unique, and the problem is usually just to find at *least one* function g such that $f \circ g$ is identical with some given function. For example, given the functions h and f such that

$$h(x) = x + 1 \tag{9}$$

$$f(x) = x^2, \tag{10}$$

find a function g such that

$$f \circ g = h.$$

Putting '$g(x)$' for 'x' in (10), we have:

$$f(g(x)) = g(x)^2,$$

hence, by virtue of (9),

$$g(x)^2 = x + 1,$$

and

$$g(x) = \pm\sqrt{x + 1}.$$

Thus we may take either:

$$g(x) = \sqrt{x + 1}$$

or:

$$g(x) = -\sqrt{x + 1}$$

as the solution of our problem. The strategy of substituting in (10) is typical for this kind of problem. A somewhat different strategy is called for when we are given functions g and h and want to find a function f such that $f \circ g = h$. For example, given

$$h(x) = 10x^2 + 1, \tag{11}$$

$$g(x) = 5x, \tag{12}$$

find a suitable f. In this case, we first solve (12) for g^{-1}, obtaining:

$$g^{-1}(x) = \frac{x}{5}. \tag{13}$$

Then we substitute '$g^{-1}(x)$' for 'x' in (11), taking, of course, $h = f \circ g$:

$$(14) \qquad (f \circ g)(g^{-1}(x)) = 10(g^{-1}(x))^2 + 1.$$

Since composition is associative, we have:

$$((f \circ g) \circ g^{-1}) = (f \circ (g \circ g^{-1}))(x),$$

and by virtue of (II)

$$f(x) = (f \circ (g \circ g^{-1}))(x).$$

Hence we infer from (13) and (14)

$$f(x) = 10\left(\frac{x}{5}\right)^2 + 1$$

that is,

$$f(x) = \tfrac{2}{5}x^2 + 1.$$

Since this type of problem involves several steps, it may be useful to summarize the typical strategy: (i) Solve for g^{-1}. (ii) Substitute '$g^{-1}(x)$' for 'x' in the equation defining $f \circ g$. (iii) Substitute the explicit expression for g^{-1} obtained in (i) on the right-hand side of the result (ii). (iv) Solve for f.

A natural query is, What is the situation when g does not have an inverse? In such a case, find any function g_1 such that

$$g(g_1(x)) = x.$$

The procedure for finding such a function g_1 is, of course, identical with the procedure for finding g^{-1}.

Finally we remark that since functions are relations, and relations are sets, the set-theoretical operations of union, intersection, and difference apply to functions. The union of two functions is not necessarily a function, although the intersection of two functions is a function.

EXERCISES

1. In Exercise 1 of § 11.1, which of the relations are one-many, that is, of which relations are the converses functions?

2. In Exercise 4 of § 11.1, consider the relations which are functions. Of which ones do the inverses exist?

3. Find the inverses, if they exist, of the functions defined for every real number x by the following equations:

(a) $f(x) = 3x + 7$
(b) $f(x) = x^3 + 1$
(c) $f(x) = -2x - 4$
(d) $f(x) = 1$

4. Let f, g, and h be functions such that for every real number x

$$f(x) = x + 2,$$
$$g(x) = x - 2,$$
$$h(x) = 3x.$$

Find $f \circ g$, $g \circ f$, $f \circ h$, $h \circ g$, $f \circ f$, $g \circ g$, $(f \circ h) \circ g$.

5. Let f and g be functions each of which has the set of all positive integers for its domain of definition, and suppose that

$$f(n) = \begin{cases} 2n & \text{if } n < 3 \\ n & \text{if } n \geq 3 \end{cases}$$

$$g(n) = \begin{cases} 3n & \text{if } n < 3 \\ n & \text{if } n \geq 3. \end{cases}$$

Find the domain of definition and range of $f \circ f$, $f \circ g$, $g \circ f$, and $g \circ g$.

6. Let f_1 be a function whose domain of definition is A_1 and whose range is B_1; and let f_2 be a function whose domain of definition is A_2 and whose range is B_2. What is the domain of definition of $f_1 \circ f_2$?

7. Which of the following identities hold for any function f, g, and h? (When we indicate the inverse of a function, we assume that the inverse exists.)

(a) $f \circ f^{-1} = f/f^{-1}$
(b) $(f \circ g)^{-1} = (g/f)^{-1}$
(c) $(f \circ g) \circ h = (g/f)/h$
(d) $f \circ (g \circ h) = (h/g)/f$
(e) $f \circ g = g \circ f$
(f) $(f \circ g) \circ h = f \circ (g \circ h)$.

8. Let f be the function such that for every real number x,

$$f(x) = x + 2,$$

and let g be the function such that for every real number x,

$$g(x) = x - 2.$$

Which of the following are functions: $f \cup g$, $f \cap g$, $f \sim g$?

9. In each of the following, find a function g such that $h = f \circ g$.

(a) $h(x) = 3x - 5$, $f(x) = 2x$
(b) $h(x) = x^2$, $f(x) = x^2 - 1$
(c) $h(x) = x + 1$, $f(x) = x^2 + 2$
(d) $h(x) = 10x - 5$, $f(x) = \sqrt{x}$

10. In each of the following, find a function f such that $h = f \circ g$.

(a) $h(x) = 3x - 5$, $g(x) = 2x$
(b) $h(x) = 5x$, $g(x) = x - 7$
(c) $h(x) = x^2$, $g(x) = 1 - x$
(d) $h(x) = -x^2$, $g(x) = -x$
(e) $h(x) = 3 - 2x$, $g(x) = x^2$

§ **11.3 Church's Lambda Notation.*** It is extremely important to distinguish between a given function and the values of the function. For example, suppose that f is the function such that, for every real number x,

$$f(x) = x^2 + 2.$$

Then f is the set of all ordered couples of the form $\langle x, x^2 + 2\rangle$, for x a real number. On the other hand, for any particular real number x, $f(x)$ is simply the real number $x^2 + 2$. Thus it is not correct, for example, to use the phrase 'the function $f(x)$', instead of 'the function f'. To amplify the point, let

$$f_1 = \{\langle 1, \text{Washington}\rangle, \langle 8, \text{Van Buren}\rangle, \langle 13, \text{Fillmore}\rangle\}.$$

Then we easily see that f_1 is a function, and for every x in $D(f_1)$, $f_1(x)$ is a past President of the United States. It is nonsense to use the phrase 'the function $f_1(x)$', since neither Washington, Van Buren, nor Fillmore is a function.

In introducing various special functions as examples, we have been trying to make a sharp distinction between a function and its values. Our regard for this distinction has led us to modes of expression, however, which are more lengthy than those usual in mathematics (the distinction is neglected in many books on mathematics). Thus in an ordinary mathematics book one might find such a statement as 'the inverse of the function x^3 exists' but we should have to say 'the inverse of the function f such that, for every real number x

$$f(x) = x^3,$$

exists'. It is tedious to repeat phrases like 'the function f such that'.

Naturally, we want to have our cake and eat it too; we should like to be exact without being dull. A partial way of achieving this end is to introduce the *lambda notation* of Alonzo Church, which we shall now explain. The basic idea is that we prefix to a numerical expression an expression such as 'λx' to obtain an expression designating a function. Thus from the numerical term '$x + 2$', we obtain the expression '$(\lambda x)(x + 2)$' designating the function f such that for every real number x

$$f(x) = x + 2.$$

We say that we have abstracted the function $(\lambda x)(x + 2)$ from the expression '$x + 2$', and we call '(λx)' an abstraction operator. An abstraction operator has a logical status very similar to that of a quantifier. In particular, the use of abstraction operators provides a new way of *binding* variables. In the expression '$(\lambda x)(x + 2)$' both occurrences of the variable 'x'

* This section may be omitted without loss of continuity.

are bound. Let us fix upon some domain V of individuals. The general formulation of abstraction (with respect to V) is:

> *If* v *is any variable and* Φ *is any term, then* $(\lambda \mathsf{v})(\Phi)$ *designates the function whose value for* $x \in \mathrm{V}$ *is designated by the result of substituting a symbol designating* x *for* v *in* Φ.

For simplicity of formulation we have considered a fixed domain of individuals. Consequently for any expression Φ, the domain of definition of the function designated by $(\lambda \mathsf{v})\Phi$ is V. We shall be mainly interested in letting V be the set of all real numbers.

We only use lambda with variables, just as in the case of quantifiers. It is nonsense, for example, to write: $(\lambda 3)(3 + 2)$. On the other hand, it often does not matter which particular variable we use with lambda. Thus

$$(\lambda x)(x + 2) = (\lambda y)(y + 2).$$

It is also permissible to apply the lambda operator to a term which involves more than one variable; in this case, however, we do not in general obtain a definite function until a value is assigned to the variables other than the one associated with the lambda operator. Thus $(\lambda x)(ax + b)$ is a function for every number a and b. And $(\lambda x)(x^2 + y)$ is a function for every y. When there is more than one variable in Φ we must take some care in replacing the variable used with lambda by another variable. Thus if we consider the expression:

$$(\lambda x)(x^2 + z),$$

(which, for every number z, designates a function) we obtain an equivalent expression if we replace 'x' by 'y', obtaining:

$$(\lambda y)(y^2 + z);$$

but if we replace 'x' by 'z' we obtain:

$$(\lambda z)(z^2 + z);$$

and $(\lambda z)(z^2 + z)$ is a function identical with

$$(\lambda x)(x^2 + x),$$

not with

$$(\lambda x)(x^2 + z).$$

Ordinarily, the Φ to which one prefixes 'λx' will contain the variable 'x'. This is not necessarily the case, however; it is also permissible to use this notation for the so-called *constant functions*. Thus, for example, $(\lambda x)(71)$ is the function f whose domain is the set of all real numbers, and such that, for every real number x,

$$f(x) = 71.$$

The distinction between functions and their values is particularly apt to be neglected in the case of functions of constant value. But notice that whereas '71' is the name of a particular real number, '$(\lambda x)(71)$' is the name of a certain set of ordered couples: the set of all ordered couples $\langle u, v\rangle$ such that u is a real number and $v = 71$.

Since $(\lambda x)(x^2)$ is a function, then

$$(\lambda x)(x^2)(3)$$

is the value of the function for the argument 3. As further examples

$$[(\lambda x)(x^2 + 2)](5) = 5^2 + 2 = 27$$

$$[(\lambda x)(-x^3 - 17)](-2) = -(-2)^3 - 17 = -9$$

$$[(\lambda x)(x^2 + y)](2) = 2^2 + y = 4 + y.$$

Note the free variable 'y' in the last example. It should be clear how to extend the lambda notation to abstract a function from the term '$(\lambda x)(x^2 + y)$'. We simply add '(λy)':

$$(1) \qquad (\lambda y)(\lambda x)(x^2 + y).$$

We must, of course, be careful in finding values of such a function as (1) for given arguments. We use parentheses to indicate which term, say '2', "replaces" which variable. Thus,

$$(2) \qquad [(\lambda y)[(\lambda x)(x^2 + y)](2)](3) = [(\lambda y)(4 + y)](3) = 4 + 3 = 7.$$

Our procedure is to work from the inside out. We begin with the function of one variable $(\lambda x)(x^2 + y)$, and find its value for an element of its domain. The result is designated by the term '$4 + y$' in (2). The '(λy)' abstracts from this term to express a new function, and our final step is to evaluate this new function for a given argument.

It is important to note that (1) does not designate a function of two variables in the ordinary sense, for (1) designates a function whose domain is the set of all real numbers and whose range is the set of functions $(\lambda x)(x^2 + y)$. Thus

$$(\lambda y)(\lambda x)(x^2 + y) = \{\langle 1, (\lambda x)(x^2 + 1)\rangle, \langle 2, (\lambda x)(x^2 + 2)\rangle, \ldots\},$$

whereas if f is the function of two variables such that for all real numbers x and y

$$f(x, y) = x^2 + y,$$

then

$$f = \{\langle\langle 1, 1\rangle, 2\rangle, \langle\langle 1, 2\rangle, 3\rangle, \ldots\}.$$

The lambda notation is easily adapted to permit the abstraction of f, namely by reiteration of several variables with one lambda:

$$f = (\lambda xy)(x^2 + y),$$

and

$$f(\sqrt{2}, 1) = (\lambda xy)(x^2 + y)(\sqrt{2}, 1) = (\sqrt{2}\,)^2 + 1 = 3.$$

As a final example, the function g of three variables such that for all real numbers x, y, and z

$$g(x, y, z) = x^2 + xy + yz$$

may be denoted by:

$$(\lambda xyz)(x^2 + xy + yz),$$

and

$$(\lambda xyz)(x^2 + xy + yz)(2, 4, 5) = 2^2 + 2 \cdot 4 + 4 \cdot 5 = 32. \tag{3}$$

Note that the arguments of the function are listed in the same order in which the corresponding variables are listed after lambda. Thus in (3) to evaluate the function for the arguments (2, 4, 5), we replace 'x' by '2', 'y' by '4' and 'z' by '5' in the expression from which the function is abstracted.

EXERCISES

1. Use the lambda notation to describe the following functions:

   (a) the function f such that for every real number x,
   $$f(x) = x^3 + x^2 + 2;$$
   (b) the function f such that for every real number x,
   $$f(x) = \sqrt{x} + 2;$$
   (c) the function f such that for every real number x,
   $$f(x) = 0;$$
   (d) the function f such that for all real numbers x and y,
   $$f(x, y) = x + y;$$
   (e) the function f such that for all real numbers x and y,
   $$f(x, y) = x^2y + 2xy + y.$$

2. Evaluate the following:

   (a) $[(\lambda x)(x^2 + x^4 + x^{10})](1)$
   (b) $[(\lambda x)(x + x^3)](2x)$
   (c) $[(\lambda x)(1 + x)](-1)$
   (d) $[(\lambda z)(z + x)](2)$
   (e) $[(\lambda y)([(\lambda x)(x^2 + y^3)](5))](3)$
   (f) $[(\lambda y)([(\lambda x)(2xy)](3))](6)$
   (g) $[(\lambda y)[(\lambda x)(x^2)](y^2)](2)$
   (h) $[(\lambda x)[(\lambda y)(y^2)](\sqrt{x}\,)](5)$
   (i) $(\lambda xy)(x^2 - y^2)(3, 5)$
   (j) $(\lambda xyz)(x^2y - y^2z)(1, 3, 5)$

CHAPTER 12

SET-THEORETICAL FOUNDATIONS OF THE AXIOMATIC METHOD

§ 12.1 Introduction. Of all the remarkable intellectual achievements of the ancient Greeks perhaps the most outstanding is their explicit development of the axiomatic method of analysis. Euclid's *Elements*, written about 300 B.C., has probably been the most influential work in the history of science.* Every educated person knows the name of Euclid and in a rough way what he did—that he expounded geometry in a systematic manner from a set of axioms and geometrical postulates. But most of us would hardly be able to give any detailed description of Euclid's method or tell why it is considered important to develop geometry in this way. Euclid begins Book I with a list of thirty-five definitions, three postulates, and twelve axioms.† He then derives the forty-eight propositions of Book I. The derivations are of course informal rather than formal in character. There is in the *Elements* no critique of logic nor any attempt to list permissible rules of inference. On this point Euclid is in agreement with present-day standard mathematical practice, which we discussed in Chapter 7. But on other matters his approach is not in complete agreement with modern conceptions of the axiomatic method. He does not clearly see that the *axiomatic* development of geometry must begin with some ideas which are not themselves defined in terms of others. He thereby confuses formal or axiomatic questions with problems concerning the application of geometry. For example, Definition 1 of Book I asserts that a point is that which has no parts, or which has no magnitude, and Definition 2 asserts that a line is length without breadth. The fundamental, technical notions here are not those of part or breadth from which we construct the notions of point and line in a precise manner, but rather those of point and line, and it is properties of the latter notions which are stated in the postulates. What Euclid has tried to do in Definitions 1 and 2 is to explain

* The axiomatic method does not originate with Euclid. His main contribution was to systematize the results of his predecessors.

† Everyman's Library edition.

in an intuitive way the notions of point and line. These intuitive explanations are not logically necessary for the proofs of theorems, but they are suggestive in helping one to think about the formal properties of points and lines, as well as useful in learning how to apply geometry to empirical problems.

In contrast to Definitions 1 and 2, Definition 11, which says that an obtuse angle is an angle which is greater than a right angle, has a technical status. Proofs of theorems which involve the notion of obtuse angle depend in a formal way upon this definition. We thus make the following important distinction. The notions of point and line are *undefined* or *primitive* notions in Euclid's geometry, whereas the notion of obtuse angle is a *defined* notion or concept.

One of the first steps in axiomatizing a subject or theory is to list the primitive notions of the theory. It is not precisely clear what notions should be regarded as primitive in Euclid's development of geometry. However, there are a large number of modern, axiomatic treatments of Euclidean geometry which are very explicit on this point. For instance, we might take as primitives the following three notions: the notion of point; the notion of betweenness—of one point being between two others on a line; and the notion of equidistance—the distance between two given points being the same as the distance between two other given points. Using then only the apparatus of logic and set theory we can proceed to define all other geometrical notions in terms of these three notions. For example, the line generated by two points x and y is defined as the *set* of all points z which are between x and y or which are such that y is between x and z or which are such that x is between z and y.

It is to be emphasized that the primitive notions of a theory are seldom if ever uniquely determined by the intuitive content of the theory. Euclidean geometry, for instance, may be developed in terms of a wide variety of primitive notions other than the three mentioned above. For example, in the famous German mathematician Hilbert's axiomatization (1899), the five notions of point, line, plane, betweenness, and congruence are introduced as primitive. On the other hand, the Italian mathematician Pieri published in the same year an axiomatization using only the primitive notions of point and motion.

A preliminary step in fixing on the primitive notions of a theory is to become clear about what other theories are to be assumed in developing the axiomatization. For most axiomatic work in mathematics the standard development of logic and general set theory is assumed without comment. If such an assumption is not made, then a complete apparatus must be built from the ground up; that is, the theory must be constructed within a completely and explicitly formalized language. In axiomatic work in the empirical sciences, such as physics, psychology, and economics, it is customary to assume not only logic and general set theory but the standard portions of mathematics as well. This permits such concepts as those of number to be used in the axiomatizations of portions of physics or economics, say, and yet not be regarded as primitive.

After deciding on what other theories are to be assumed and fixing on the primitive notions of the theory being studied, we are in a position to state the axioms of the theory. The axioms are, of course, those statements which are basic to the theory and from which we may derive the other statements we consider true of the theory. It is desirable to have as few axioms as seems possible, and also to take as axioms statements whose meaning has a strong intuitive appeal. The only notions referred to in the axioms must be the primitive notions, notions defined in terms of the primitive notions, and the notions belonging to the theories assumed a priori. It would, for example, be improper to refer to a particular physical object such as the sun in an axiom of geometry, since it is hardly likely that any physical theory, to say nothing of one dealing specifically with the sun, would be assumed as a prerequisite to the development of geometry. A second point to be emphasized is that in deriving logical consequences of the axioms, nothing may be assumed about the primitive notions except what is stated in the axioms.

In previous chapters in various exercises and examples we have considered axiomatizations of theories like the theory of groups which need only assume first-order predicate logic with identity. Such theories are sometimes called *theories with standard formalization*. Many mathematically significant theories are susceptible of such standard formalization in a natural and simple way, and in the last two or three decades a number of interesting results have been established about various theories of this sort. In Chapter 7 we were concerned with a standard formalization of the arithmetic of real numbers.

Unfortunately, when a theory assumes more than first-order logic as already available for use in its statement and development, it is neither natural nor simple to formalize the theory in first-order logic. For example, if in axiomatizing geometry we want to define lines as certain *sets* of points, we must work within a framework that already includes the ideas of set theory. To be sure, it is theoretically possible to axiomatize simultaneously in first-order logic geometry *and* the relevant portions of set theory, but this is awkward and unduly laborious. Above all, it is repetitious, for in axiomatizing a wide variety of theories, it is necessary or at least highly expedient to make use of set theory; if formalization in first-order logic is the method used, then each such axiomatization must include appropriate axioms for set theory. Theories with more complicated structures like probability theory need to use not only general ideas of set theory but also many results concerning the real numbers and functions whose domains or ranges are sets of real numbers. Formalization of such theories in first-order logic is utterly impractical. The aim of the present chapter is to provide a general foundation for the axiomatization of such complicated theories, and to consider in some detail two substantive examples, one dealing with probability and one dealing with particle mechanics.

Although some of the general aspects of axiomatizing Euclidean geometry have been discussed above in this section, a particular axiomatization of geometry was not selected as a substantive example for two reasons. First, because a large number of axioms is needed and a fair amount of deductive development is called for to appreciate the structure of the theory, limitations of space argued against a geometrical example. In addition, since a substantial portion of Euclidean geometry is easily axiomatized in first-order logic directly, it would not provide a good illustration of axiomatic methods as applied to theories which have no natural formalization in first-order logic. The essential methodological purpose of the present chapter is to demonstrate that the same standards of clarity and precision may be achieved in axiomatizing complicated theories within set theory as are achieved by axiomatizing relatively simple theories directly in first-order logic. An appreciation of this point is necessary to comprehend the wide applicability of axiomatic methods in all domains of mathematics and in the theoretical portions of the empirical sciences.

§ 12.2 Set-Theoretical Predicates and Axiomatizations of Theories. The kernel of the procedure for axiomatizing theories within set theory may be described very briefly: to axiomatize a theory is to define a predicate in terms of notions of set theory. A predicate so defined is called a *set-theoretical* predicate. Actually, a number of such predicates have been defined in the previous three chapters, probably the most important two being the predicates 'is a binary relation' and 'is a function'.

In this chapter we shall not give a sharp definition of 'set-theoretical predicate'. Our objective is to elucidate by examples what is involved in defining such predicates. Moreover, we shall assume that the set-theoretical framework within which we operate consists not only of general set theory, as discussed in previous chapters, but also of the full apparatus of classical mathematics, that is, the real numbers, functions of real numbers, derivatives and integrals of such functions,* and the like.

At the beginning it may seem difficult to decide how much prior mathematical development it is appropriate to assume in axiomatizing a given theory. In practice, however, this question is usually easily answered. In axiomatizing physics, for example, it is natural to make use of any part of classical mathematics, but in studying the foundations of, say, the real number system, results from classical mathematics could be used only with great discretion and care if at all. In the next section when we consider

* Derivatives and integrals are needed only in the final section.

probability theory, it is appropriate to assume as already given in set theory whatever is needed concerning operations on sets or operations on the real numbers; but to make our developments depend on any results from intuitive probability theory would be incorrect.

Any ambiguities concerning what is already a part of the general set-theoretical framework we assume can in principle be completely eliminated by an axiomatic development of set theory. The properties of general operations on sets are developed first. Then the natural numbers are constructed, followed by the real numbers and the systematic development of classical mathematics. This much can be accomplished by roughly a thousand theorems and five or six hundred definitions. Definition 601, say, could define the predicate 'is a probability space'. Any use of classical mathematics in this definition or in proofs of theorems about probability spaces could then be explicitly and completely justified by reference to the appropriate preceding theorems and definitions.*

We now turn to some specific remarks about axiomatizing a theory by defining a set-theoretical predicate. Some of the remarks are minor in nature, but all of them are intended to clarify various questions which arise in axiomatic work. Our examples shall deal primarily with either the theory of quasi-orderings or the theory of groups. To begin with, we may consider a definition of quasi-orderings slightly different from that given in Chapter 10. The purely set-theoretical character of the predicate 'is a quasi-ordering', which is defined, is immediately apparent upon consideration of the terms which occur in the definiens.

DEFINITION A. *$\mathfrak{A}$ is a quasi-ordering if and only if there is a set A and a binary relation R such that $\mathfrak{A} = \langle A, R\rangle$ and*

Q1. *R is reflexive in A.*
Q2. *R is transitive in A.*

Note that it would not do to replace '$\mathfrak{A} = \langle A, R\rangle$' in the definiens of Definition A by '$\mathfrak{A} = \{A, R\}$', that is, to replace the ordered couple by an unordered set; for if A is a set of ordered couples, confusion could arise concerning which set, A or R, is meant to be the ordering relation. In the case of strict partial orderings we could even have the anomalous situation for some sets A and R that given $\{A, R\}$ we could not decide if A was meant to be a strict partial ordering of R, or vice versa. For instance, let

$$A_1 = \{\langle 1, 2\rangle\}$$

$$R_1 = \{\langle 2, 1\rangle\};$$

* A group of contemporary mathematicians writing under the collective pseudonym 'Bourbaki' are indeed pursuing such a systematic development of the whole of mathematics.

then R_1 is a strict partial ordering of A_1, and A_1 is a strict partial ordering of R_1. Difficulties of a deeper sort arise for more complicated theories when unordered sets rather than ordered n-tuples are used.

The form of Definition A satisfies the rule for proper definition of relation symbols given in Chapter 8, but it is at slight variance with the dominant style of modern mathematics. To illustrate other possibilities, we consider some alternatives.

A first alternative is to define the two-place predicate 'is a quasi-ordering of', which is what we did in Chapter 10.

> DEFINITION B. *R is a quasi-ordering of A if and only if R is a binary relation which is reflexive and transitive in A.*

In comparing Definitions A and B it might at first seem that all the advantages of simplicity lie with B, since no quantifiers or references to ordered couples are required in the definiens. However, the advantage of A is that it exhibits a uniform approach applicable to any theory, for the grammar of definitions like B becomes awkward when the predicate defined is more than two-place. Thus corresponding to B we would have for systems of particle mechanics, defined in § 12.5, something like:

> P is a system of particle mechanics with respect to T, s, m, f and g if and only if ...

A second alternative, which is very close to standard mathematical practice, is to use a conditional definition.

> DEFINITION C. *Let A be a set and R a binary relation. Then $\langle A, R\rangle$ is a quasi-ordering if and only if*
>
> Q1. *R is reflexive in A.*
> Q2. *R is transitive in A.*

In the case of C the purely set-theoretical structure of quasi-orderings is the hypothesis of the conditional definition. Such conditional definitions seem very natural, and we shall use them in the sequel, but they do promote the continual commission of a certain kind of minor error, which may be illustrated by C. Suppose someone asserts as a theorem:

> If $\langle A, R\rangle$ is a quasi-ordering then A is a subset of the field of R.

This assertion seems to be an obvious consequence of C and familiar facts about relations. However, the difficulty is that since it is not stated in the hypothesis of the theorem that A is a set and R a binary relation, we cannot use C to infer that R is reflexive in A, from which we may obtain the desired conclusion. In other words, since the conditional clause of Definition C is not satisfied we cannot significantly use C in an inference.

But in many contexts this criticism is a quibble. It is perfectly obvious that it is intended for the theorem stated to apply only to the appropriate set-theoretical entities.

In discussions dealing with entities having the same general set-theoretical structure, the problem just mentioned is often met by the device of specifying at the beginning of the discussion this general structure. For example, we might define a *simple relation structure* as an ordered couple consisting of a set and a binary relation. Then all our definitions of ordering relations are for such structures. Thus

DEFINITION D. *A simple relation structure* $\langle A, R \rangle$ *is a quasi-ordering if and only if* R *is reflexive and transitive in* A.

In considering groups and related algebraic structures, we may consider *algebras* consisting of a non-empty set A and a binary operation $\circ$ from $A \times A$ to A, that is, $\circ$ is a function whose domain is $A \times A$ and whose range is a subset of A. We then define:

DEFINITION E. *An algebra* $\langle A, \circ \rangle$ *is a group if and only if for every* x, y, *and* z *in* A

A1. $x \circ (y \circ z) = (x \circ y) \circ z$.
A2. *There is a* w *in* A *such that*

$$x = y \circ w.$$

A3. *There is a* w *in* A *such that*

$$x = w \circ y.$$

(Here we have used the formulation of the axioms for groups already stated in Exercise 5 of § 5.2.) Now Definitions C, D, and E all violate the rules for conditional definitions given in Chapter 8. For instance, in the case of C, the definiendum uses the term '$\langle A, R \rangle$' where a single variable should be used. To rectify this mistake and then point out how it is conveniently met in practice we may consider Definition E. We may reformulate it.

DEFINITION F. If $\mathfrak{A} = \langle A, \circ \rangle$ *and* $\langle A, \circ \rangle$ *is an algebra, then* $\mathfrak{A}$ *is a group if and only if* ...

(The German letter '$\mathfrak{A}$' is used in deference to a common usage in the literature. The reason for the usage is this: A denotes the basic set of the group, whereas $\mathfrak{A}$ is the basic set together with the operation.) The lengthy conditional clause of F is tedious to repeat continually. The standard abbreviation is to write instead:

DEFINITION G. *An algebra* $\mathfrak{A} = \langle A, \circ \rangle$ *is a group if and only if* ...

If for any reason an exact formulation is wanted, the phrase 'An algebra

$\mathfrak{A} = \langle A, \circ \rangle$' of Definition G can always be expanded in the style of Definition F.

In connection with the definition of groups it should be remarked that many mathematicians prefer a definition like:

> DEFINITION H. *A set A is a group with respect to the binary operation $\circ$ if and only if ...*

But the predicate here is so lengthy that it is customary to refer to the set A alone as a group, which is literally false and may lead to mistakes. Definition G has simultaneously the virtues of brevity and explicitness.

In ordinary mathematical contexts definitions are frequently formulated in a metamathematical fashion, but this metamathematical formulation does not involve any real metamathematical commitments, that is, commitments to prove assertions about expressions of some given, fixed language.* Definition G, for instance, might be formulated:

> DEFINITION I. *An algebra $\mathfrak{A} = \langle A, \circ \rangle$ is a group if and only if the following three axioms are satisfied ...*

Mathematical proofs about groups would not use in any explicit or deep fashion properties of the metamathematical notion of satisfaction, which is used in Definition I.

When a theory is axiomatized by defining a set-theoretical predicate, by a *model* for the theory we mean simply an entity which satisfies the predicate. For the theory of quasi-orderings we could put the point trivially as follows. If $\langle A, R \rangle$ is a quasi-ordering, then $\langle A, R \rangle$ is a model for the theory of quasi-ordering. Correspondingly, if an algebra $\langle A, \circ \rangle$ is a group then $\langle A, \circ \rangle$ is a model for the theory of groups. When the theory of groups is axiomatized directly in first-order logic, the notion of model is defined so that the same set-theoretical entities are models for the theory thus formulated, and similarly for other theories which may be axiomatized either directly in first-order logic or by defining a set-theoretical predicate: the two axiomatizations have the same entities as models.

During the last two decades the phrases 'model' and 'mathematical model' have been widely used, particularly in the behavioral sciences. These phrases seem to be used in at least three distinct senses. The sense of the phrase in logic has just been described. A second meaning of 'model' for mathematical economists is closely related: *the* model for a theory is the set of all models for the theory in the logicians' sense. What the logicians call a model is labeled a *structure*. In this terminology, if an algebra $\langle A, \circ \rangle$ is a group then it is a structure for the theory of groups.

* Roughly speaking, metamathematics is that branch of mathematics which investigates the structure of formalized languages or theories and their relation to other mathematical entities. Many philosophers tend to call the study of formalized languages *semantics* and *logical syntax* rather than metamathematics.

The third meaning of 'model', the one most popular with empirical scientists, is what we have meant by 'theory' in preceding pages. In this sense, to give a mathematical model for some branch of empirical science is to state an exact mathematical theory. In such empirical contexts the word 'theory' is often reserved for non-mathematical, relatively inexact statements about the fundamental ideas of a given domain of science. The important difference between the first two senses of model and the third is that only in the third sense are models linguistic entities.

When theories are formalized in first-order logic, theorems relating different models for the theory are necessarily metamathematical in their statement and proof.* In contrast, theorems comparing various models for a theory may be stated in direct mathematical fashion, when the theory is axiomatized by defining a set-theoretical predicate. For example, consider the following theorem about quasi-orderings.

THEOREM 1. *If* $\langle A, R_1 \rangle$ *and* $\langle A, R_2 \rangle$ *are quasi-orderings then* $\langle A, R_1 \cap R_2 \rangle$ *is a quasi-ordering.*

PROOF. We need to show that $R_1 \cap R_2$ is reflexive and transitive in A. Let x be an arbitrary element of A. By hypothesis of the theorem, we have:

$$x\, R_1\, x \;\&\; x\, R_2\, x,$$

whence

$$x\, R_1 \cap R_2\, x,$$

which proves $R_1 \cap R_2$ is reflexive in A.

Now suppose we have for any elements x, y, and z in A:

$$(1) \qquad x\, R_1 \cap R_2\, y \;\&\; y\, R_1 \cap R_2\, z,$$

which is equivalent to:

$$(2) \qquad xR_1y \;\&\; yR_1z \;\&\; xR_2y \;\&\; yR_2z.$$

From (2) and the fact that by hypothesis R_1 and R_2 are transitive in A, we infer:

$$(3) \qquad xR_1z \;\&\; xR_2z,$$

which is equivalent to:

$$x\, R_1 \cap R_2\, z,$$

and thus $R_1 \cap R_2$ is transitive in A. Q.E.D.

* Generally speaking, metamathematical methods are not used by mathematicians when they can be avoided, for their exact application requires the often tedious and difficult task of working with a completely specified and formalized language. This remark is not meant to devalue in any way the significance of metamathematics. Many important results can be established only by metamathematical methods.

If the theory of quasi-orderings were formalized in first-order logic, we would then need to formulate Theorem 1 in a metamathematical fashion.

THEOREM 1′. *If* $\langle A, R_1 \rangle$ *and* $\langle A, R_2 \rangle$ *are models satisfying the theory of quasi-orderings, then* $\langle A, R_1 \cap R_2 \rangle$ *is also such a model.*

Essentially, the proof of Theorem 1′ is just like that of Theorem 1 and would require the same set-theoretical framework. The advantage of Theorem 1 over Theorem 1′ is that the proof of Theorem 1 requires no shifting back and forth from a theory formalized in first-order predicate logic to a metamathematical framework which includes all the apparatus of set theory.

In the informal proof of Theorem 1 we have used in an intuitive and somewhat casual fashion properties of sets familiar from Chapters 9–11. It is important to reiterate the remark made in the first section of this chapter: in axiomatizing particular theories *within* set theory it is possible for almost all mathematical purposes to proceed without an axiomatization of set theory itself explicitly at hand. Familiar properties of sets are used in proofs without explicit appeal to theorems derived from some given axioms of set theory. This practice will be held to in subsequent sections of this chapter and should be adopted in working various of the exercises.

When a theory is axiomatized by defining a set-theoretical predicate the independence of axioms or primitive notions is established by the kind of methods previously described and used. The axioms are listed in the definiens of the definition of the given predicate; and a given axiom is shown to be independent by exhibiting a set-theoretical entity which satisfies the predicate defined by the original definiens minus the given axiom, but does not satisfy the full definiens. Intuitively this just amounts to finding a model satisfying all but the given axiom. It needs to be noted that the purely set-theoretical structure of entities satisfying a predicate is usually not characterized in the axioms proper, but rather is stated in one of the following three places.

(i) The running text of the definiens immediately preceding the axioms (see Definition A).
(ii) The hypothesis of the definition if the definition is conditional in form (see Definition C).
(iii) Informally in the discussion preceding the definition (see Definitions D and E).

From the standpoint of the systematic theory of definition, (i) is superior; but from the standpoint of brevity and elegance, (iii) is to be preferred. Whichever alternative is adopted, the set-theoretical structure of models for a theory is not stated in the sentences labeled *axioms.** This has the desirable consequence that the proofs of independence of axioms of a given theory all make use of the same general kind of set-theoretical entities. For instance, each of the models proving the independence of one of the axioms for groups of Definition E must be an algebra, that is, an ordered couple whose first member is a set A and whose second member is a binary operation from the Cartesian product $A \times A$ to A.

Since the definitions favored in the above discussion of alternatives define one-place predicates, some clarification about the status of primitive notions is needed. Intuitively, the primitive notions of a theory are just the sets, relations, and operations which are members of an ordered n-tuple satisfying the given predicate. Thus the theory of quasi-orderings and the theory of groups as formulated in this section are each based on two primitive notions.† Unfortunately a certain ambiguity surrounds the use of the term 'primitive notion'. For example, consider the predicate 'is a group'. An infinity of ordered couples $\langle A, \circ \rangle$ satisfy this predicate, but we do not want to infer from this that there is an infinity of sets and binary operations which are primitive notions for group theory. Rather we want to say that (with reference to Definition E) there are exactly two primitive notions. We can make the primitive notions of a theory definite mathematical objects by relativizing them to a model for the theory. Thus we say that any model for the theory of groups has exactly two primitive notions, namely a set and a binary operation. The proof of independence of primitive notions then amounts to showing that the primitive notions of an arbitrary model for the theory cannot be defined in terms of each other. When no confusion is possible, we shall refer simply to the primitive notions of a *theory*, but in all cases it will be perfectly obvious how such language may be replaced by the more correct phrase 'primitive notions of a model for the theory'.

If the formalization of a theory in first-order predicate logic is compared with its axiomatization by defining a set-theoretical predicate, it will be noticed that almost always the latter requires one more primitive notion than the former does primitive symbols. Namely, there is added the primitive notion of a set corresponding to the domain of a model of the formalized theory. For example, the set A in a group $\langle A, \circ \rangle$ corresponds to no primitive symbol in the corresponding axiomatization given in § 5.2. The reasons for the additional primitive notion grow clear if we reflect that

* Each axiom should say something intuitively significant about the theory. The assertion, for instance, that a given primitive notion is a binary relation hardly satisfies this requirement and thus should not be labeled an axiom.

† It is customary to refer to the *primitive notions* rather than the *primitive symbols* of a theory axiomatized by defining a set-theoretical predicate, for (literally speaking) in defining the appropriate predicate no new primitive symbols are added to the language of set theory.

when the axioms for groups are formalized in first-order predicate logic, the variables are interpreted to range over the elements of the group; but in the set-theoretical framework of this section the variables range over all individuals and sets, and the new primitive notion is needed to "relativize" the variables to range over the elements of the group. Thus in the formalized theory, we have simply:

$$x \circ (y \circ z) = (x \circ y) \circ z, \tag{1}$$

but here (1) is replaced by:

$$x \in A \;\&\; y \in A \;\&\; z \in A \rightarrow x \circ (y \circ z) = (x \circ y) \circ z,$$

where $\langle A, \circ \rangle$ is a group.

The exact way in which the basic set of elements is required to be related to the other primitive notions affects both its independence as a primitive notion and the truth or falsity of certain simple statements. For example, on the basis of Definition E, for any group $\langle A, \circ \rangle$ the set A may be defined as the set whose Cartesian product is the domain of definition of the binary operation $\circ$. By a slight reformulation of the general set-theoretical requirements, the basic set A may be made independent of the other primitive notion. This may be illustrated by reformulating Definition E for groups. The two essential changes are to drop the requirement that $\circ$ be a binary operation from $A \times A$ to A, and to add the *closure axiom* that if $x, y \in A$ then $x \circ y \in A$. An *algebra* $\langle A, \circ \rangle$ is now defined to be simply a non-empty set A and a binary operation $\circ$, with no restriction on the relation between A and $\circ$, and we define groups by:

Definition J. *An algebra* $\mathfrak{A} = \langle A, \circ \rangle$ *is a group if and only if for every* x, y, *and* z *in* A

A1. $x \circ y \in A$.
A2. $x \circ (y \circ z) = (x \circ y) \circ z$.
A3. *There is a* w *in* A *such that*

$$x = y \circ w.$$

A4. *There is a* w *in* A *such that*

$$x = w \circ y.$$

The following two groups satisfying Definition J show that A is independent. Let $+$ be the set of ordered triples of integers $\langle x, y, z \rangle$ such that $x + y = z$. Let A_1 be the set of integers and let A_2 be the set of even integers including zero. Then $\langle A_1, + \rangle$ and $\langle A_2, + \rangle$ are both groups in the sense of Definition J and by application of Padoa's principle (see Chapter 8) it is clear that these two groups establish the independence of A as a primitive notion. Yet the obtaining of independence of primitive notions is not sufficient to justify use of Definition J rather than Definition E. In fact, many algebraists prefer E. Some fundamental reasons will be forthcoming in the next section. On the other hand, one advantage of J is mentioned in the exercises. Note that we have defined quasi-orderings in a manner analogous to J rather than E.

In many contexts—one of them is exemplified in the next section—it is convenient to have a group be an ordered quadruple rather than an ordered couple. And we introduce algebras which are ordered quadruples $\langle A, \circ, {}^{-1}, e\rangle$, where $\circ$ is a binary operation from $A \times A$ to A, ${}^{-1}$ is a unary operation from A to A, and e is an element of A. Corresponding to the axioms for groups given in the text of § 5.1, we then have:

Definition K. *An algebra* $\mathfrak{A} = \langle A, \circ, {}^{-1}, e\rangle$ *is a group if and only if for every* x, y, *and* z *in* A

A1. $x \circ (y \circ z) = (x \circ y) \circ z$.
A2. $x \circ e = x$.
A3. $x \circ x^{-1} = e$.

On first reflection it might seem that the addition of Definition K introduces an ambiguity in our use of the predicate 'is a group'. However, notice that both Definition E and Definition K are disguised conditional definitions, a point made explicit by Definition F. And the appropriate hypotheses of the two definitions are mutually exclusive, so that any set-theoretical entity satisfying the one cannot satisfy the other. In particular, the hypothesis of Definition E is that $\mathfrak{A}$ is an ordered couple $\langle A, \circ\rangle$, and the hypothesis of Definition K that $\mathfrak{A}$ is an ordered quadruple $\langle A, \circ, {}^{-1}, e\rangle$. The following theorem relates the two definitions in an exact way, which we make use of in the next section. The proof is left as an exercise. (The word 'algebra' in the hypothesis of the theorem is used in the sense defined just before Definition E.)

Theorem 2. *If* $\langle A, \circ\rangle$ *is an algebra, then* $\langle A, \circ\rangle$ *is a group (in the sense of Definition E) if and only if there is a unary operation* ${}^{-1}$ *from* A *to* A *and an element* e *of* A *such that* $\langle A, \circ, {}^{-1}, e\rangle$ *is a group (in the sense of Definition K).*

In § 12.4 and § 12.5 we turn to two examples of axiomatizations which follow the ideas laid down in the present section. The first example is concerned with probability, and the second with mechanics. Neither of the two has a simple and natural formalization in first-order predicate logic. Consequently each is intended to exemplify the relative power and flexibility of the axiomatic approach which consists of defining an appropriate set-theoretical predicate. Before turning to these two examples, the important notion of isomorphism of models for a theory is introduced in the next section.

EXERCISES

1. Axiomatize the theory of Boolean algebras discussed in § 9.9 (see particularly Exercise 9) by defining the appropriate set-theoretical predicate

(a) in the style of Definition A;
(b) in the style of Definition E;
(c) in the style of Definition J (be sure to include closure axioms in this case).

2. Axiomatize the theory of the measurement of mass discussed in Exercise 9 of § 4.5 by defining the appropriate set-theoretical predicate.

3. Using Definition A decide which of the following are true. If any is true, prove it. If false, give a counterexample.

(a) If $\langle A, R_1 \rangle$ and $\langle A, R_2 \rangle$ are quasi-orderings then $\langle A, R_1 \cup R_2 \rangle$ is a quasi-ordering.
(b) If $\langle A, R_1 \rangle$ and $\langle A, R_2 \rangle$ are quasi-orderings then $\langle A, R_1 \sim R_2 \rangle$ is a quasi-ordering.
(c) If $\langle A, R_1 \rangle$ and $\langle A, R_2 \rangle$ are quasi-orderings then $\langle A, R_1/R_2 \rangle$ is a quasi-ordering.
(d) If $\langle A_1, R \rangle$ and $\langle A_2, R \rangle$ are quasi-orderings then $\langle A_1 \cap A_2, R \rangle$ is a quasi-ordering.
(e) If $\langle A_1, R \rangle$ and $\langle A_2, R \rangle$ are quasi-orderings, then $\langle A_1 \cup A_2, R \rangle$ is a quasi-ordering.
(f) If $\langle A_1, R_1 \rangle$ and $\langle A_2, R_2 \rangle$ are quasi-orderings then $\langle A_1 \cap A_2, R_1 \cap R_2 \rangle$ is a quasi-ordering.
(g) If $\langle A, R_1 \rangle$ and $\langle A, R_2 \rangle$ are quasi-orderings then $\langle A \times A, R_1 \times R_2 \rangle$ is a quasi-ordering.

4. One of the fundamental problems of welfare economics and political theory is the optimal method of aggregation of individual preferences to determine preferences for the social unit. For both individuals and social units we define:* *$\mathfrak{A}$ is a weak preference pattern if and only if there is a set A, and binary relations P and I such that $\mathfrak{A} = \langle A, P, I \rangle$ and*

AXIOM 1. *P is transitive in A.*
AXIOM 2. *I is transitive in A.*
AXIOM 3. *For any x and y in A, exactly one of the following: xPy, yPx, xIy.*

(For an intuitive interpretation see Exercise 6 of § 10.5.)

Let us now consider a social unit consisting of just two individuals with weak preference patterns $\langle A, P_1, I_1 \rangle$ and $\langle A, P_2, I_2 \rangle$ (for some fixed set A of alternatives facing the unit).

(a) Is $\langle A, P_1 \cap P_2, I_1 \cap I_2 \rangle$ a weak preference pattern (for the unit), i.e., are the three axioms satisfied?

* We use the adjective 'weak' because the axioms do not say very much about the theory of preference. For an extensive discussion of this problem in the style of this exercise, see K. J. Arrow, *Social Choice and Individual Values*, New York, 1951.

(b) Define for $x, y \in A$:

$$x\,P\,y \leftrightarrow [x\,P_1\,y \,\&\, (x\,P_2\,y \vee x\,I_2\,y)] \vee [x\,P_2\,y \,\&\, (x\,P_1\,y \vee x\,I_1\,y)],$$

$$x\,I\,y \leftrightarrow -(x\,P\,y) \,\&\, -(y\,P\,x).$$

Is $\langle A, P, I\rangle$ a weak preference pattern for the unit?

(c) Formally describe, in the manner of (b), the method of aggregation which you think most closely corresponds to the method of majority vote for a social unit of three individuals.

5. Following the developments of Exercise 5 of § 5.2, it is not difficult to justify the addition of the following two definitional equivalences to the axioms for groups of Definition J:

$$e = y \leftrightarrow (x)(x \in A \rightarrow x \circ y = x \,\&\, y \circ x = x),$$

$$x^{-1} = y \leftrightarrow x \circ y = e.$$

(In other words, we introduce by these definitions the identity element and the inverse operation of a group.) We now define: *If* $\mathfrak{A} = \langle A, \circ\rangle$ *is a group and* $B \subseteq A$ *then* $\langle B, \circ\rangle$ *is a subgroup of* $\mathfrak{A}$ *if and only if for every* x *and* y *in* B

(i) $x \circ y \in B$
(ii) $e \in B$
(iii) $x^{-1} \in B$.

(Note that the definition of subgroup must be complicated if Definition E for groups is used, since the operation $\circ$ is not an operation from $B \times B$ to B.)

(a) Prove that if $B \neq \Lambda$ and $B \subseteq A$ then $\langle B, \circ\rangle$ is a subgroup of $\mathfrak{A}$ if and only if for every x and y in B, $x \circ y \in B$ and $x^{-1} \in B$.

(b) Prove that if $\langle B_1, \circ\rangle$ and $\langle B_2, \circ\rangle$ are subgroups of $\mathfrak{A}$, then $\langle B_1 \cap B_2, \circ\rangle$ is a subgroup of $\mathfrak{A}$.

(c) If $\langle B_1, \circ\rangle$ and $\langle B_2, \circ\rangle$ are subgroups of $\mathfrak{A}$, is $\langle B_1 \cup B_2, \circ\rangle$ a subgroup of $\mathfrak{A}$?

(d) For $B_1 \subseteq A$ and $B_2 \subseteq A$, define:

$$B_1 \circ B_2 = C \leftrightarrow (z)(z \in C \leftrightarrow (\exists x)(\exists y)(x \in B_1 \,\&\, y \in B_2 \,\&\, x \circ y = z).$$

Prove that if $\langle B_1, \circ\rangle$ and $\langle B_2, \circ\rangle$ are subgroups of $\mathfrak{A}$ then $\langle B_1 \circ B_2, \circ\rangle$ is a subgroup of $\mathfrak{A}$ if and only if $B_1 \circ B_2 = B_2 \circ B_1$.

6. Referring to the preceding exercise, define the notion of subgroup when Definition E is used for groups (and thus the notion of an algebra corresponds to that defined immediately prior to Definition E). Also define the notion of subgroup for groups $\langle A, \circ, {}^{-1}, e\rangle$ satisfying Definition K.

7. Prove Theorem 2.

§ 12.3 Isomorphism of Models for a Theory. The separation of the purely set-theoretical characterization of the structure of models for a theory from the axioms proper is significant in defining certain important notions concerning models for a theory. For example, the notion of two groups being *isomorphic* is often said to be axiom-free, since the definition of isomorphism for groups depends on none of the axioms. In fact, the definition is really for isomorphic algebras and applies to algebras which are not groups. A cursory inspection of the definition, which we now state, verifies these remarks. (In the definition, an algebra is understood to be an ordered couple $\langle A, \circ\rangle$ such that $\circ$ is a binary operation from $A \times A$ to A, which is the first of the senses of 'algebra' introduced in the preceding section.)

DEFINITION 1. *An algebra* $\mathfrak{A} = \langle A, \circ\rangle$ *is isomorphic to an algebra* $\mathfrak{A}' = \langle A', \circ'\rangle$ *if and only if there is a function f such that*

(i) $D(f) = A$ & $\mathcal{R}(f) = A'$,
(ii) *f is a one-one function,*
(iii) *if* $x, y \in A$ *then*

$$f(x \circ y) = f(x) \circ' f(y).$$

The intuitive idea is, of course, that two algebras are isomorphic just when they have the same structure. The properties required of the function f make precise the idea of *same structure*. The relation of isomorphism between algebras is an equivalence relation, i.e., it is reflexive, symmetric, and transitive. (We leave showing this as an exercise.) An example of two distinct groups which are isomorphic is afforded by the following. Let A be the set of integers, let A' be the set of even integers and, let $+'$ be addition of even integers (where zero is counted as an even integer). Then $\langle A, +\rangle$ is isomorphic to $\langle A', +'\rangle$, for it is easy to find a function f with properties (i)–(iii). Let f be the function such that for any integer n

$$f(n) = 2n.$$

Then we verify at once that $D(f) = A$, $\mathcal{R}(f) = A'$ and f is a one-one function. Furthermore, for any two integers m and n

$$f(m + n) = 2(m + n) = 2m +' 2n = f(m) +' f(n),$$

which verifies (iii).

If an algebra is defined as an ordered couple $\langle A, \circ\rangle$ where the operation $\circ$ need have no connection with the set A, Definition 1 must be modified, since $x \circ y$ may not be in the domain of f. Emphasizing that the operations $\circ$ and $\circ'$ are sets of ordered triples, we replace (iii) of Definition 1 by:

(iii′) *If* $x, y, z \in A$ *then*

$$\langle x, y, z\rangle \in \circ \leftrightarrow \langle f(x), f(y), f(z)\rangle \in \circ'.$$

We may rewrite (iii′):

(iii″) *If* $x, y, z \in A$ *then*

$$x \circ y = z \leftrightarrow f(x) \circ' f(y) = f(z).$$

When algebras in this second sense are considered, care must be exercised in finding out when performing the operation $\circ$ on two elements of the set A results in an element of A. Moreover, only under the most trivial circumstances can we determine if two algebras are identical, since the binary operations of any two algebras can have any sort of structure outside of the set $A \times A \times A$. In the exercises at the end of this section we consider only algebras $\langle A, \circ \rangle$ in the first sense, that is, where the operation $\circ$ is from $A \times A$ to A.

A satisfactory general definition of isomorphism for two set-theoretical entities of any kind is difficult if not impossible to formulate. The standard mathematical practice is to formulate a separate definition for each general kind of ordered n-tuple. When the n-tuples are complicated as in the case of models for the theory of particle mechanics, it is sometimes difficult to decide exactly what is to be meant by two isomorphic models; but for algebras or simple relation structures the choice of the appropriate conditions is clearer. To illustrate the condition on binary relations we state the definition of isomorphism for simple relation structures.

DEFINITION 2. *A simple relation structure* $\langle A, R \rangle$ *is isomorphic to a simple relation structure* $\langle B, S \rangle$ *if, and only if, there is a function f such that*

(i) $D(f) = A \mathbin{\&} \mathscr{R}(f) = B$,
(ii) *f is a one-one function,*
(iii) *If $x, y \in A$ then $xRy \leftrightarrow f(x)\, S\, f(y)$.*

(In place of '$\langle B, S \rangle$' we could as well have written '$\langle A', R' \rangle$' with corresponding changes in (i)–(iii).)

Using this definition the relation structure

$$\langle \{1, 2\}, \{\langle 1, 2 \rangle, \langle 2, 2 \rangle\} \rangle$$

is isomorphic to the relation structure

$$\langle \{3, 4\}, \{\langle 4, 3 \rangle, \langle 3, 3 \rangle\} \rangle,$$

for we may take as an appropriate function f:

$$f(1) = 4$$

$$f(2) = 3.$$

The notion of isomorphism has important applications in all domains of modern mathematics. When the special situation obtains that any two models for a theory are isomorphic, then the theory is said to be *categorical.** Simple counterexamples may be found to show that neither the theory of quasi-orderings nor the theory of groups is categorical. For

* This notion originates (1904) with the American mathematician Oswald Veblen. It may be of some interest to philosophers to know that the word 'categorical' was suggested to Veblen by John Dewey (see *Transactions of the American Mathematical Society* Vol. 5 (1904) p. 346).

example, in the case of quasi-orderings, we may take two quasi-orderings $\langle A, R\rangle$ and $\langle B, S\rangle$ such that the set A has two elements and the set B three elements. Then there is no one-one function whose domain is A and range is B. This is trivial. It is a little more interesting to exhibit two quasi-orderings $\langle A, R\rangle$ and $\langle B, S\rangle$ such that A and B have the same number of elements and yet $\langle A, R\rangle$ is not isomorphic to $\langle B, S\rangle$. Let

$$A = \{1, 2\}$$

$$R = \{\langle 1, 1\rangle, \langle 2, 2\rangle, \langle 1, 2\rangle\}$$

$$B = \{3, 4\}$$

$$S = \{\langle 3, 3\rangle, \langle 4, 4\rangle\}.$$

Then if the appropriate f for isomorphism of $\langle A, R\rangle$ and $\langle B, S\rangle$ existed, from

$$1\, R\, 2$$

we would infer

$$f(1)\, S\, f(2).$$

Since f is a one-one function, $f(1) \neq f(2)$, but the relation S does not hold between any two distinct elements. Consequently there can be no such function f.

Some simple examples of categorical theories are given in the exercises.* When a theory is not categorical, an important problem is to discover if an interesting subset of models for the theory may be found such that any model for the theory is isomorphic to some member of this subset. To find such a distinguished subset of models for a theory and show that it has the property indicated is to prove a *representation theorem* for the theory. Such a theorem may be proved for groups; namely, every group is isomorphic to a *group of transformations*. Roughly speaking, a group of transformations may be described as follows. Let M be a non-empty set, and let B be a set of one-one functions whose domains and ranges are M (such a function is called a *transformation* on M). Then if B is appropriately chosen $\langle B, \circ\rangle$ will be a *group* of transformations, where $\circ$ is composition of functions. Thus if $M = \{1, 2\}$, and $B =$ the set of all transformations on M, then there are two functions f_1 and f_2 in B:

$$f_1(1) = 1 \;\&\; f_1(2) = 2,$$

$$f_2(1) = 2 \;\&\; f_2(2) = 1.$$

The function f_1 is the *identity* transformation, since it maps each element

* The theory of the real numbers and the theory of the positive integers are probably the two most important examples of categorical theories.

of M into itself. We also note that both f_1 and f_2 are identical with their inverses, that is,

$$f_1^{-1} = f_1$$

$$f_2^{-1} = f_2.$$

We may now check that $\langle B, \circ \rangle$ is a *group* of transformations (where $\circ$ is composition of functions). In making such a check it is convenient to use Theorem 2 of § 12.2; that is, we show that the algebra $\mathfrak{B} = \langle B, \circ \rangle$ is a group by showing that there is an inverse operation $^{-1}$ and an identity element e such that $\langle B, \circ, ^{-1}, e \rangle$ is a group. We see at once that the identity element of $\mathfrak{B}$ is the identity transformation f_1, since

$$f_1 \circ f_2 = f_2 \circ f_1 = f_2.$$

And the group inverse element of a function in B is just the function inverse, whence immediately

$$f_1 \circ f_1^{-1} = f_1^{-1} \circ f_1 = f_1$$

$$f_2 \circ f_2^{-1} = f_2^{-1} \circ f_2 = f_1.$$

We now prove the representation theorem mentioned.*

THEOREM 2. *Every group is isomorphic to a group of transformations.*

PROOF. Let $\mathfrak{A} = \langle A, \circ \rangle$ be an arbitrary group. For each x in A we define a function f_x which maps A into A as follows: for every y in A

$$f_x(y) = x \circ y. \tag{1}$$

From the left-hand cancellation law for groups (§ 5.2, Exercise 4, Theorem 5) it follows that f_x is a one-one function. For suppose it were not. Then there would be elements y and y' in A such that $y \neq y'$ and

$$f_x(y) = f_x(y');$$

but then by (1)

$$x \circ y = x \circ y',$$

and by the cancellation law

$$y = y',$$

which is absurd.

From Axiom A2 of Definition E it follows that for any z in A there is a y in A such that

$$f_x(y) = z.$$

Thus the range of f_x is A, whence f_x is a transformation.

* Readers interested in a further development of group theory and related topics of modern algebra will find useful G. Birkhoff and S. MacLane, *A Survey of Modern Algebra*, rev. ed., New York, 1953.

Now to show that the set of all such transformations f_x form a group with respect to the operation of composition of functions. (To avoid confusion between the operation of composing functions and the group operation of $\mathfrak{A}$, we denote composition of functions by juxtaposition in this proof.) We use Theorem 2 of § 12.2. Let e be the identity element of $\mathfrak{A}$; then f_e is the identity transformation on A, since

$$f_e(x) = e \circ x = x.$$

Whence,

$$f_x f_e = f_e f_x = f_x.$$

Moreover, since for any x in A, f_x is 1 to 1, the inverse function f_x^{-1} exists and we have at once

$$f_x f_x^{-1} = f_x^{-1} f_x = f_e.$$

Finally, to complete the proof of our theorem we need to show that the group of transformations on A that we have defined is isomorphic to $\mathfrak{A}$. For this, we first need to observe that for any x and y in A

$$f_x f_y = f_{x \circ y},$$

since for any z in A

$$(2) \qquad f_x f_y(z) = f_x(y \circ z) = x \circ (y \circ z) = (x \circ y) \circ z = f_{x \circ y}(z).$$

Now for the function demonstrating the isomorphism of the two groups, we use, as would be expected, the function φ such that for any x in A

$$\varphi(x) = f_x.$$

Obviously the domain and range of φ are what they should be. To see that φ is a one-one function, we notice that if $x \neq y$ then

$$f_x(e) \neq f_y(e),$$

for if $f_x = f_y$ when $x \neq y$ then

$$x = x \circ e = f_x(e) = f_y(e) = y \circ e = y.$$

To establish (iii) of Definition 1 we use (2) to infer that

$$\varphi(x \circ y) = f_{x \circ y} = f_x f_y = \varphi(x)\varphi(y). \qquad \text{Q.E.D.}$$

The notion of a representation theorem, and thereby the notion of isomorphism, has important applications in the philosophy of science. The primary aim of the theory of measurement, for instance, is to show in a precise fashion how to pass from qualitative observations ('This rod is longer than that one,' 'the left pan of the balance is higher than the right one') to the quantitative assertions needed in empirical science ('The length of this rod is 7.2 centimeters,' 'the mass of this chemical sample is 5.4 grams'). In other words, the theory of measurement should provide an exact analysis of how we may infer quantitative assertions from fundamentally qualitative observations. Such an analysis is provided by axiomatizing appropriate algebras of experimentally realizable operations and relations. A partial example for the measurement of mass constituted Exercise 9 of § 4.5. Given an axiomatized theory of measurement of some empirical quantity, the mathematical task is to prove a representation theorem for models for the theory which establishes, roughly speaking, that any model is isomorphic to some numerical model for the theory. The existence of this isomorphism justifies the application of numbers to things. We cannot literally take a number in our hands and "apply" it to a physical object, say. What we can do is show that the structure of a set of phenomena under certain empirical operations and relations is the *same as* the structure of some set of numbers under certain arithmetical operations and relations. The definition of isomorphism in the given context makes the intuitive idea of *same structure* precise, as has already been remarked. The great significance of finding such an isomorphism of structures is that we may then use all our familiar knowledge of computational methods, as applied to the arithmetical structure, to infer facts about the isomorphic empirical structure.

Unfortunately most of the proofs of representation theorems in the theory of measurement are too long to include in either the text or exercises. To illustrate the methods, we shall consider a very simple set of axioms for measuring perceived or felt differences in various classes of phenomena, like differences in pitch or loudness of a set of sounds, differences in intensity of pain of a set of stimuli, differences in visual brightness of a set of color stimuli, and differences in value (or utility) of a set of economic goods. The axioms are based on three primitive notions: the set A of objects or stimuli, the binary relation P which represents the ordering of the stimuli, and the quaternary relation E which represents equality of difference of pairs of stimuli.

Suppose, for example, we wanted to develop a numerical scale for pitch. Then A would be a certain set of sounds. For two sounds x and y in A, xPy if and only if the pitch of x is judged lower than y. For four sounds x, y, u, and v in A, $x, y\ E\ u, v$ if and only if the (algebraic) difference in pitch between x and y is judged to be the same as that between u and v. We say *algebraic* difference, since E takes account of ordering; that is, if x, y, u, and v were numbers then we would have:

$$x, y\ E\ u, v \leftrightarrow x - y = u - v.$$

Generally speaking, our objective is to state axioms on A, P, and E such that we can prove there is a real-valued function φ defined on A such that

$$xPy \leftrightarrow \varphi(x) < \varphi(y),$$

$$x, y\,E\,u, v \leftrightarrow \varphi(x) - \varphi(y) = \varphi(u) - \varphi(v).$$

We now systematize our ideas. With applications in psychology in mind, we call any ordered triple $\langle A, P, E\rangle$, where A is a set, P a binary relation and E a quaternary relation, a *difference structure.* To emphasize that the appropriate definition of isomorphism is axiom-free, we may state it prior to considering the axioms of our theory of measurement.

DEFINITION 3. *A difference structure $\langle A, P, E\rangle$ is isomorphic to a difference structure $\langle A', P', E'\rangle$ if and only if there is a function f such that*

(i) $D(f) = A$ & $\mathscr{R}(f) = A'$,
(ii) *f is a one-one function,*
(iii) *If $x, y \in A$ then $xPy \leftrightarrow f(x)\,P'\,f(y)$,*
(iv) *If $x, y, u, v \in A$ then $x, y\,E\,u, v \leftrightarrow f(x), f(y)\,E'\,f(u), f(v)$.*

This definition is simply an extension in a natural way of Definition 2. We now turn to the axioms which a difference structure must satisfy in order to be an *equal* difference structure. The intuitive idea of the axiomatization is to require that we order the stimuli according to pitch, intensity of visual brightness, or some other characteristic, and then demand that any two stimuli adjacent in the ordering have the same difference in intensity as any two other such adjacent stimuli. Moreover, we restrict ourselves to *finite* sets of stimuli. A set of five stimuli would have to be arranged like

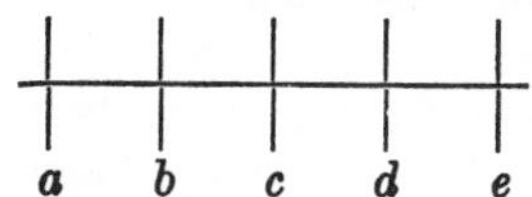

to be an equal difference structure. Obviously we may assign the number 1 to a, 2 to b, etc. Our axioms state sufficient conditions to guarantee that such an assignment is always possible.

Before stating the axioms, a defined notion needs to be introduced which it is convenient to use in both axioms and theorems. Moreover, the introduction of this defined notion may serve to illustrate some pertinent points regarding the proper form of definition for such defined notions.

DEFINITION 4A. *If $x, y \in A$, then*

$$xJy \leftrightarrow [xPy \;\&\; (z)(z \in A \;\&\; xPz \rightarrow y = z \vee yPz)].$$

The intuitive interpretation of the binary relation J is that xJy if and only if y is the *unique immediate successor* of x with respect to the relation

P. Obviously Definition 4A is not a proper definition in general set theory for the definiens contains the free variables 'A' and 'P' which do not occur in the definiendum. The definition would be proper only if 'A' and 'P' were constants rather than variables. A proper definition is:

DEFINITION 4B. *If $\langle A, P, E\rangle$ is a difference structure and if $x, y \in A$, then*

$$xJ(A, P)y \leftrightarrow [xPy \;\&\; (z)(z \in A \;\&\; xPz \rightarrow y = z \vee yPz)].$$

Also, since the variable 'E' does not occur in the consequent of 4B, we can as well write:

DEFINITION 4C. *If there is an E such that $\langle A, P, E\rangle$ is a difference structure and if $x, y \in A$ then*

$$xJ(A, P)y \leftrightarrow [xPy \;\&\; (z)(z \in A \;\&\; xPz \rightarrow y = z \vee yPz)].$$

But since for any set A and binary relation P there is an E such that $\langle A, P, E\rangle$ is a difference structure, we may as well simplify the definition and refer only to simple relation structures $\langle A, P\rangle$.

DEFINITION 4D. *If $\langle A, P\rangle$ is a simple relation structure and if $x, y \in A$, then*

$$xJ(A, P)y \leftrightarrow [xPy \;\&\; (z)(z \in A \;\&\; xPz \rightarrow y = z \vee yPz)].$$

For 4B, 4C, and 4D the rule for conditional definitions of relation symbols is satisfied, and 'J' is a defined quaternary relation symbol of set theory. But this result seems slightly artificial. We expect 'J' to have the same status as 'A' and 'P', which are variables. For this reason, the universal practice is to use 4A and to explain away its shortcomings in the following manner. Basically the definition of the J relation is the sort of definition we could add as a definitional axiom to whatever list of axioms we impose on difference structures. If our theory of measurement were axiomatized in first-order logic, then 'A' and 'P' would be primitive symbols and 'J' would be a defined binary relation symbol properly introduced by 4A. Such definitions (i.e., 4A) are labeled 'elementary' with respect to the theory being considered, and are used constantly to avoid the troublesome notation required by definitions in general set theory like 4B, 4C, and 4D. When considering some fixed theory, we shall label definitions like 4A with a prefix 'E' to indicate their elementary character.* The same labeling is

* All of these remarks apply to the definition of the identity element and inverse operation for groups $\langle A, \circ\rangle$. Thus when a definition in general set theory is used the identity element e is denoted by a binary operation symbol '$e(A, \circ)$': If an algebra $\langle A, \circ\rangle$ is a group then

$$e(A, \circ) = y \leftrightarrow (x)(x \in A \rightarrow x \circ y = x).$$

used for elementary theorems, which are theorems derivable from the axioms of the given theory and which hold for any model for the theory. (This point will be returned to.)

We return to our main task, development of the theory of measurement for finite equal difference structures. The relation J is referred to in the final axiom of the basic definition which we now state.*

DEFINITION 5. *A difference structure* $\langle A, P, E\rangle$ *is a finite equal difference structure if and only if* A *is a finite set and for every* $x, y, z, u, v,$ *and* w *in* A *the following nine axioms are satisfied:*

AXIOM D1. *If* xPy *then not* yPx.
AXIOM D2. *If* xPy & yPz *then* xPz.
AXIOM D3. *If* $x \neq y$ *then* xPy *or* yPx.
AXIOM D4. $x, x\ E\ y, y$.
AXIOM D5. *If* $x, y\ E\ u, v$ *then* $u, v\ E\ x, y$.
AXIOM D6. *If* $x, y\ E\ u, v$ *then* $x, u\ E\ y, v$.
AXIOM D7. *If* $x, y\ E\ u, v$ & $u, v\ E\ z, w$ *then* $x, y\ E\ z, w$.
AXIOM D8. *If* xPy & $x, y\ E\ u, v$ *then* uPv.
AXIOM D9. *If* xJy & uJv *then* $x, y\ E\ u, v$.

Remembering the empirical interpretation discussed above, we can grasp the intuitive interpretation of each axiom. For example, the first three axioms just require that P be a strict simple ordering of A. The next four axioms state properties of E only. For example, Axiom D4 says that the difference in pitch between any sound x and itself is the same as the difference in pitch between any other sound y and itself. On the basis of the whole set of axioms we may prove that this difference is zero as would be expected. The numerical interpretation of D4 is just that

$$x - x = y - y,$$

a trivial arithmetical truth for any numbers x and y. The final two axioms relate P and E. It is Axiom D9 that imposes the equal difference spacing on stimuli. Regarding experimental application of these axioms, it should be noticed that D9 has a different status from Axioms D1–D8. We might expect a careful subject approximately to satisfy D1–D8 for any set of stimuli with respect to some characteristic such as pitch or visual brightness or value. But it would be surprising indeed to find D9 satisfied by an arbitrarily selected set of stimuli. The appropriate set of stimuli must be

* A related set of axioms is to be found in D. Davidson and P. Suppes, "A Finitistic Axiomatization of Subjective Probability and Utility," *Econometrica*, Vol. 24 (1956) pp. 264–275. A set of axioms for infinite difference structures is given in P. Suppes and M. Winet, "An Axiomatization of Utility Based on the Notion of Utility Differences," *Management Science*, Vol. 1 (1955) pp. 259–270.

carefully selected by the experimenter in order to find a model satisfying D9.*

The elementary theorems which we need to prove our representation theorem for finite equal difference structures are listed in the exercises at the end of this section. From the preceding remarks on elementary definitions, it should be clear how one may convert any elementary theorem about finite equal difference structures into a theorem of general set theory. For instance, the first theorem is:

THEOREM E1. *If $x, y, z \in A$ and if $x, y \; E \; x, z$, then $y = z$.*

This may be formulated as a theorem of general set theory as follows.

THEOREM. *If $\langle A, P, E \rangle$ is a finite equal difference structure, if $x, y, z \in A$, and if $x, y \; E \; x, z$, then $y = z$.*

Regarding the elementary theorems, it is common practice to omit the requirement in the hypothesis that $x, y, z \in A$. Thus we come to the simple formulation:

THEOREM E1′. *If $x, y \; E \; x, z$, then $y = z$.*

which is the style followed in the exercises; but it should be understood that literal correctness demands an addition to the hypothesis—that the elements considered belong to A.

To state the representation theorem, we need to define *numerical* equal difference structures.

DEFINITION 6. *Let N be a finite set of numbers such that differences between numbers adjacent under the natural ordering $<$ are equal. Let W be the quaternary relation such that for any numbers x, y, u, and v,*

$$x, y \; W \; u, v \leftrightarrow x - y = u - v.$$

Then $\langle N, <, W \rangle$ is a finite numerical equal difference structure.

THEOREM 3 (REPRESENTATION THEOREM). *If a difference structure is a finite equal difference structure, then it is isomorphic to a finite numerical equal difference structure.*

Some hints concerning the proof of this theorem are given in the exercises.

For a theory of measurement we demand not only a numerical representation theorem but also a theorem concerning the uniqueness of the representation. For our example we have the following.

* The two major weaknesses of the axioms from the standpoint of the kind of judgments or responses individuals actually make regarding felt or perceived differences of some characteristic of stimuli are the following: (i) judgments of order and of differences are not perfectly transitive; (ii) at different times different responses will be given to the same stimulus presentation. Methods for changing the axioms to accommodate either of these phenomena are too complicated to discuss here.

THEOREM 4 (UNIQUENESS THEOREM). *Any two finite numerical equal difference structures isomorphic to a given finite equal difference structure are related by a linear transformation. That is, if* $\langle N, <, W\rangle$ *and* $\langle N', <, W\rangle$ *are the two numerical structures, then there is a positive number* α *and a number* β *such that for any* x *in* N, *there is an unique* x' *in* N' *such that*

$$x' = \alpha x + \beta.*$$

The second major application in the philosophy of science of the notion of a representation theorem is to the problem of showing that one branch of science may be *reduced* to another. For example, it is known that a large portion of classical thermodynamics may be derived from statistical mechanics, in the sense that many laws of thermodynamics may be derived from various fundamental laws of statistical mechanics. To show in a sharp sense that thermodynamics may be reduced to statistical mechanics, we would need to axiomatize both disciplines by defining appropriate set-theoretical predicates, and then show that given any model T of thermodynamics we may find a model of statistical mechanics on the basis of which we may construct a model isomorphic to T. Substantive examples of such a reduction are too complicated to include in the text.†

EXERCISES

1. Using Definition 1 prove:

(a) Any algebra is isomorphic to itself.

(b) If an algebra $\mathfrak{A} = \langle A, \circ\rangle$ is isomorphic to an algebra $\mathfrak{A}' = \langle A', \circ'\rangle$, then $\mathfrak{A}'$ is isomorphic to $\mathfrak{A}$.

(c) If an algebra $\mathfrak{A} = \langle A, \circ\rangle$ is isomorphic to an algebra $\mathfrak{A}' = \langle A', \circ'\rangle$ and if $\mathfrak{A}'$ is isomorphic to an algebra $\mathfrak{A}'' = \langle A'', \circ''\rangle$, then $\mathfrak{A}$ is isomorphic to $\mathfrak{A}''$.

2. Let $\mathfrak{A} = \langle A, \circ\rangle$ be an algebra which is a group, and let $\mathfrak{A}$ be isomorphic to an algebra $\mathfrak{A}' = \langle A', \circ'\rangle$. Prove in detail that $\mathfrak{A}'$ is a group.

3. Give an example of two groups which are not isomorphic, and prove that they are not.

4. Let $\langle A, R\rangle$ and $\langle B, S\rangle$ be simple relation structures. Prove any of the following statements which are true. For those which are false, give counterexamples.

(a) $\langle A, \breve{R}\rangle$ is isomorphic to $\langle A, R\rangle$.

(b) $\langle A \times B, R \times S\rangle$ is isomorphic to $\langle B \times A, S \times R\rangle$.

* Naturally, different sorts of uniqueness are obtained for different kinds of measurement. For instance, in the theory of measurement of mass we get uniqueness up to a similarity transformation; that is, from one of two numerical models isomorphic to a given model for the theory, one can obtain the other by multiplication by a positive number (which intuitively corresponds to a change in the unit of mass measurement). With respect to uniqueness the kind of measurement we get for finite equal difference scales is like that obtained for longitude or ordinary temperature measurements. An arbitrary unit *and* origin are selected.

† A detailed and exact analysis of the reduction of rigid-body mechanics to particle mechanics is to be found in Ernest Adams, *Axiomatic Foundations of Rigid Body Mechanics*, dissertation, Stanford University, 1955.

(c) If $R = S$ then $\langle A, R\rangle$ is isomorphic to $\langle B, S\rangle$.
(d) If $R \neq S$ then $\langle A, R\rangle$ is not isomorphic to $\langle B, S\rangle$.
(e) If $A \subset B$ then $\langle A, R\rangle$ is not isomorphic to $\langle B, S\rangle$.
(f) If $\langle A, R\rangle$ is isomorphic to $\langle B, S\rangle$ then $\langle A, R/S\rangle$ is isomorphic to $\langle B, S/R\rangle$.
(g) If $\langle A, R\rangle$ is isomorphic to $\langle B, S\rangle$ then $\langle A, R/R\rangle$ is isomorphic to $\langle B, S/S\rangle$.

5. We define: *An algebra* $\mathfrak{A} = \langle A, \circ\rangle$ *is utterly trivial if and only if for every* x *and* y *in* A, $x = y$. Prove that the theory of utter triviality is categorical. (Remember that if $\langle A, \circ\rangle$ is an algebra, then A is a non-empty set.)

6. We define: *A simple relation structure* $\langle A, R\rangle$ *is a simple ordering of order* n *if and only if the following five axioms are satisfied:*

AXIOM 1. *The set A has exactly n elements.*
AXIOM 2. *R is reflexive in A.*
AXIOM 3. *R is antisymmetric in A.*
AXIOM 4. *R is transitive in A.*
AXIOM 5. *R is connected in A.*

Prove that the theory of simple orderings of order n is categorical.

7. Let M be a non-empty set. Prove that the set of all transformations on M is a group with respect to composition of functions.

8. Prove the following elementary theorems for finite equal difference structures. Those proofs which require use of the principle of mathematical induction are indicated.* The inductions are in fact always on powers of J, which we now define. (This recursive definition is actually the appropriate one for powers of any binary relation.)

$$xJ^1y \leftrightarrow xJy$$

$$xJ^{n+1}y \leftrightarrow (\exists z)(xJ^nz \ \&\ zJy), \tag{1}$$

or in terms of relative product:

$$xJ^1y \leftrightarrow xJy$$

$$xJ^{n+1}y \leftrightarrow xJ^n/Jy.$$

One set-theoretical point needs to be emphasized. The theorems are actually concerned with powers of J restricted to A, so (1) needs to be replaced by:

$$xJ^{n+1}y \leftrightarrow (\exists z)(z \in A \ \&\ xJ^nz \ \&\ zJy).$$

Remember that it is tacitly understood that all elements referred to in the elementary theorems belong to A.

THEOREM E1. *If $x, y\ E\ x, z$, then $y = z$.* (HINT: Use particularly Axioms D5, D6, and D8 and give an indirect proof.)

THEOREM E2. *If $x, x\ E\ u, v$, then $u = v$.*

THEOREM E3. *If $x, y\ E\ u, v$, then $y, x\ E\ v, u$.*

THEOREM E4. *If $x, y\ E\ u, v\ \&\ y, z\ E\ v, w$, then $x, z\ E\ u, w$.*

* This fundamental principle may be formulated symbolically:

$$\varphi(1) \ \&\ (n)(\varphi(n) \rightarrow \varphi(n+1)) \rightarrow (n)\varphi(n). \tag{I}$$

To prove that a formula φ holds for all n, we need only establish the hypothesis of (I), namely that φ holds for 1, and that *if* φ holds for n then it holds for $n + 1$.

THEOREM E5. *If xJ^ny, then xPy.* (HINT: Use induction on n.) *

THEOREM E6. *If xPy, then there is a (positive integer) n such that xJ^ny.*

THEOREM E7. *If xJ^ny & xJ^nz, then $y = z$.* (HINT: Use induction on n.)

THEOREM E8. *If xJ^my & yJ^nz, then $xJ^{m+n}z$.* (HINT: Use induction on n.)

THEOREM E9. *If xJ^my & $xJ^{m+n}z$, then yJ^nz.* (HINT: Use induction on n.)

THEOREM E10. *If $xJ^{m+n}y$, then there is a z (in A) such that xJ^mz.* (HINT: Use induction on m.)

Note that Theorems E5–E10 depend only on Axioms D1–D3.

THEOREM E11. *If xJ^ny & uJ^nv, then $x, y\ E\ u, v$.* (HINT: Use induction on n.)

THEOREM E12. $x, y\ E\ x, y$.

Note that Theorem E12 and Axioms D5 and D7 together assert that E is an equivalence relation on $A \times A$.

THEOREM E13. *If $x, y\ E\ u, v$, then either $\exists n$ such that xJ^ny & uJ^nv, or $\exists n$ such that yJ^nx & vJ^nu, or $x = y$ & $u = v$.*

9. Prove the representation theorem for finite equal difference structures (Theorem 3). HINT: Let z^* be the first element of A with respect to the ordering P. Define the numerical function φ on A as follows for every x in A:

$$\varphi(x) = \begin{cases} 1 & \text{if} \quad x = z^* \\ n + 1 & \text{if} \quad z^*J^nx. \end{cases}$$

Then using the elementary theorems prove:

(i) $xPy \leftrightarrow \varphi(x) < \varphi(y)$
(ii) $x, y\ E\ u, v \leftrightarrow \varphi(x) - \varphi(y) = \varphi(u) - \varphi(v)$.

Prove (ii) by considering the three cases listed in Elementary Theorem E13.

10. Prove the uniqueness theorem for numerical representations of finite equal difference structures (Theorem 4). Let $\mathfrak{N}_1 = \langle N_1, <, W\rangle$ and $\mathfrak{N}_2 = \langle N_2, <, W\rangle$ be two finite numerical difference structures isomorphic to a given finite equal difference structure $\langle A, P, E\rangle$ and let φ_1 and φ_2 be appropriate isomorphism functions for $\mathfrak{N}_1$ and $\mathfrak{N}_2$ respectively, that is, $\mathcal{D}(\varphi_1) = \mathcal{D}(\varphi_2) = A$, $\mathcal{R}(\varphi_1) = N_1$, $\mathcal{R}(\varphi_2) = N_2$, etc. Now define for every x in A two functions h_1 and h_2.

$$h_1(x) = \frac{\varphi_1(x) - \varphi_1(z^*)}{\varphi_1(z^{**}) - \varphi_1(z^*)}$$

$$h_2(x) = \frac{\varphi_2(x) - \varphi_2(z^*)}{\varphi_2(z^{**}) - \varphi_2(z^*)},$$

where z^* is the first element of A under the ordering P and z^{**} the second element. Show then (i) h_1 is a linear transformation of φ_1 and h_2 is a linear transformation of

* The proofs by induction needed for these elementary theorems are all extremely easy.

φ_2, and (ii) $h_1 = h_2$. It is then easy to prove that φ_1 is a linear transformation of φ_2, i.e., there are numbers α, β with $\alpha > 0$ such that for every x in A

$$\varphi_1(x) = \alpha\varphi_2(x) + \beta.$$

The proof of the identity of h_1 and h_2 requires use of induction.

§ 12.4 Example: Probability. In this section we want to exemplify the axiomatic methods described in the preceding sections. We have chosen the elementary theory of probability spaces because development of this theory calls for numerous direct and simple applications of set operations and relations. Additional reasons for this choice are the importance of probability theory in all areas of empirical science and the historically close connections between the foundations of probability and logic. It is not possible within the limitations of this section to discuss the general foundations of probability, but some attempt will be made to relate the formal developments to foundational problems.

Our axiomatization of probability proceeds by defining the set-theoretical predicate 'is a finitely additive probability space'. The axioms are based on three primitive notions: a non-empty set X of possible outcomes, a family $\mathfrak{F}$ of subsets of X representing possible events, and a real-valued function P on $\mathfrak{F}$; for $E \in \mathfrak{F}$, $P(E)$ is interpreted as the probability of E. These three notions may be illustrated by a simple example. Let X be the set of all possible outcomes of two flips of a coin. Then

$$X = \{\langle H, H\rangle, \langle H, T\rangle, \langle T, H\rangle, \langle T, T\rangle\}.$$

Let $\mathfrak{F}$ be the family of all subsets of X. The *event* of getting at least one head is the set

$$A = \{\langle H, H\rangle, \langle H, T\rangle, \langle T, H\rangle\}.$$

The event of getting exactly one head is the set

$$E = \{\langle H, T\rangle, \langle T, H\rangle\},$$

and so on. The important basic idea is that any event which could occur as the result of flipping the coin twice may be represented by a subset of X such that the subset of X has as elements those possible outcomes, the occurrence of any one of which would imply the occurrence of the event. Thus the event of getting at least one head occurs if both flips are actually heads, the first is head and the second tail, or the first tail and the second head.

If the coin is fair, then for each x in X

$$P(\{x\}) = 1/4.$$

And the probability of any other event may be obtained simply by adding up the number of elements of X in the subset. Thus the probability of getting at least one head in two flips of the coin is $P(A)$ and

$$(1)\qquad \begin{cases} P(A) = P(\{\langle H, H\rangle, \langle H, T\rangle, \langle T, H\rangle\}) \\ \qquad = P(\{\langle H, H\rangle\}) + P(\{\langle H, T\rangle\}) + P(\{\langle T, H\rangle\}) \\ \qquad = 1/4 + 1/4 + 1/4 = 3/4. \end{cases}$$

The additive property exemplified by (1) is one of the fundamental properties of probability which we postulate as an axiom: the probability of either one of two mutually exclusive events is equal to the sum of their individual probabilities, that is, if $A \cap B = \Lambda$ then

$$P(A \cup B) = P(A) + P(B).$$

Since throughout this section we apply probability language to sets which are interpreted as events, it will be useful to have a table relating the set-theoretical notation and probability terms. Thus, the assertion that $A \in \mathfrak{F}$ corresponds to asserting that A is an event; the assertion that $A \cap B = \Lambda$ to the assertion that events A and B are incompatible. In this table and subsequently we use a compact notation for the intersection or union of a finite number of sets:

$$\bigcap_{i=1}^{n} A_i = A_1 \cap A_2 \cap \ldots \cap A_n$$

$$\bigcup_{i=1}^{n} A_i = A_1 \cup A_2 \cup \ldots \cup A_n.$$

Correspondingly, later we shall use sums:

$$\sum_{i=1}^{n} P(A_i) = P(A_1) + P(A_2) + \ldots + P(A_n)$$

or more generally for any numbers a_i, $i = 1, \ldots, n$

$$\sum_{i=1}^{n} a_i = a_1 + a_2 + \ldots + a_n.$$

Also if X is a finite set, and f is a numerical function defined on X, then

$$\sum_{x \in X} f(x)$$

is just the sum of the values of f for all elements in X. Thus if $X = \{a, b, c\}$

$$\sum_{x \in X} f(x) = f(a) + f(b) + f(c).$$

SET-THEORETICAL NOTATION AND PROBABILITY TERMS

SET THEORY	PROBABILITY THEORY *
(a) $A \in \mathfrak{F}$	(a) A is an event.
(b) $A \cap B = \Lambda$	(b) Events A and B are incompatible.
(c) $A_i \cap A_j = \Lambda$ for $i \neq j$, $1 \leq i, j \leq n$	(c) Events $A_1, A_2, \ldots, A_n$ are pairwise incompatible.
(d) $\bigcap_{i=1}^{n} A_i = B$	(d) B is the event which occurs when events $A_1, A_2, \ldots, A_n$ occur all together.
(e) $\bigcup_{i=1}^{n} A_i = B$	(e) B is the event which occurs when at least one of the events $A_1, \ldots, A_n$ occurs.
(f) $\sim A$	(f) The event which occurs when A does not.
(g) $A = \Lambda$	(g) Event A is impossible.
(h) $A = X$	(h) Event A is inevitable or certain.
(i) $B \subseteq A$	(i) If event B occurs, then A must occur.

In (f), complementation is, of course, with respect to X. Thus if A is the event of getting at least one head, in our example $\sim A$ is the event of getting no heads, and

$$\sim A = X \sim A = \{\langle T, T \rangle\}.$$

The relation expressed by (i) is important. Set-theoretical inclusion of events corresponds to the occurrence of one event implying the occurrence of another. Thus in our example the occurrence of the event of getting exactly one head implies the occurrence of the event of getting at least one head, and we easily see that $E \subseteq A$.

Some further intuitive remarks justifying the particular primitive notions selected are interspersed with the formal developments. Our first axiom on probability spaces will be that $\mathfrak{F}$ is a field of sets on X, which notion we now define.†

DEFINITION 1. *$\mathfrak{F}$ is a field of sets on X if and only if $\mathfrak{F}$ is a non-empty family of subsets of X and for every A and B in $\mathfrak{F}$*

F1. $A \cup B \in \mathfrak{F}$;
F2. $\sim A \in \mathfrak{F}$.

In other words, a field of sets on X is a non-empty family of subsets closed under union and complementation. The proof of the following theorem about elementary properties of fields of sets is left as an exercise.

* This table and much of the material in this section is derived from Chapter 1 of A. N. Kolmogorov's classic work, *Foundations of the Theory of Probability*, English edition, New York, 1950.

† We say that $\mathfrak{F}$ is a field of sets on X rather than that $\langle X, \mathfrak{F} \rangle$ is a field of sets in deference to the usage in the literature.

THEOREM 1. *If $\mathfrak{F}$ is a field of sets on X then*

(i) $X \in \mathfrak{F}$,
(ii) $\Lambda \in \mathfrak{F}$,
(iii) *If* $A \in \mathfrak{F}$ & $B \in \mathfrak{F}$, *then* $A \cap B \in \mathfrak{F}$,
(iv) *If* $A \in \mathfrak{F}$ & $B \in \mathfrak{F}$, *then* $A \sim B \in \mathfrak{F}$.

As we see from Theorem 1, fields of sets are also closed under intersection and difference of sets. Furthermore, the impossible event Λ and the certain event X both belong to the field. It should be clear that for ordinary applications of probability theory an arbitrary family of events would be difficult to work with and something like a field is necessary. For example, it would be extremely awkward not to be able to treat the intersection of two events as an event, or the complement of an event as an event. When X is finite, we usually consider the field which is the family of all subsets of X. This is, of course, the largest field on X. The smallest field on X is the trivial one which has as members just X and the empty set.

We now define the basic concept of this section. The reason for the adjectival phrase 'finitely additive' will shortly be given. Note that in the definition we assume the set-theoretical structure of X, $\mathfrak{F}$, and P already mentioned: X is a non-empty set, $\mathfrak{F}$ is a family of subsets of X, and P is a real-valued function on $\mathfrak{F}$, i.e., a function whose range is a set of real numbers. Thus the definition applies to ordered triples $\langle X, \mathfrak{F}, P\rangle$, which we shall call *set function structures*, since P is a function on a family of sets.

DEFINITION 2. *A set function structure* $\mathfrak{X} = \langle X, \mathfrak{F}, P\rangle$ *is a finitely additive probability space if and only if for every A and B in $\mathfrak{F}$*

P1. *$\mathfrak{F}$ is a field of sets on X.*
P2. $P(A) \geq 0$.
P3. $P(X) = 1$.
P4. *If* $A \cap B = \Lambda$, *then* $P(A \cup B) = P(A) + P(B)$.

X is often called the *sample space* and P the *probability measure*. The first axiom, P1, has already been discussed. The second simply requires that the probability of any event be a non-negative number, and the third that the probability of the certain event be one. The final axiom states the additivity assumption already made. On the basis of P4 we may prove that the probability measure P is finitely additive; that is, we may prove that for $A_1, \ldots, A_n$ pairwise incompatible events

$$P\left(\bigcup_{i=1}^{n} A_i\right) = \sum_{i=1}^{n} P(A_i).$$

However, in the general theory of probability for several fundamental reasons it is desirable to have that if $\langle A_1, A_2, \ldots, A_n \rangle$ is a sequence * of pairwise incompatible events then

$$P\left(\bigcup_{i=1}^{\infty} A_i\right) = \sum_{i=1}^{\infty} P(A_i). \tag{2}$$

A probability measure satisfying (2) is said to be *countably additive.* Equation (2) cannot be derived from Definition 2; this is the reason for the restrictive phrase 'finitely additive'. However, if $\mathfrak{F}$ is finite, (2) holds. In the applications considered here we shall in fact restrict ourselves to finite sets X. Some further remarks concerning countable additivity follow Theorem 3.

Some elementary consequences of Definition 2 are listed in the following theorem. Since the basic entities under discussion in this section are finitely additive probability spaces, subsequent theorems and definitions referring to an arbitrary finitely additive probability space are indicated by a letter 'E' and no explicit reference in these elementary theorems and definitions will be made to the basic space under consideration. As was remarked in the preceding section, such an elementary theorem or definition may be converted into a general theorem or definition of set theory by adding the hypothesis: *if* $\langle X, P, A \rangle$ *is a finitely additive probability space,* and making certain other appropriate minor changes.

In formulating elementary theorems we use probability language when possible rather than set-theoretical language. For instance, we say 'for every event A' rather than 'for every set A in $\mathfrak{F}$'.

THEOREM E2.

(i) $P(\Lambda) = 0.$
(ii) *For every event* A, $P(A) + P(\sim A) = 1.$
(iii) *If the events* A_i *are pairwise incompatible for* $1 \leq i \leq n$, *then*

$$P\left(\bigcup_{i=1}^{n} A_i\right) = \sum_{i=1}^{n} P(A_i). \tag{1}$$

PROOF. We only prove (iii), which requires use of mathematical induction.† For $n = 1$, clearly

$$P(A_1) = P(A_1).$$

Our inductive hypothesis for n is that

$$P\left(\bigcup_{i=1}^{n} A_i\right) = \sum_{i=1}^{n} P(A_i).$$

* A sequence is just a function whose domain is the set of positive integers.

† See footnote to Exercise 8 of § 12.3 for a formulation of the principle of mathematical induction (p. 272).

Now consider

$$P\left(\bigcup_{i=1}^{n+1} A_i\right).$$

Since

$$(2) \qquad \bigcup_{i=1}^{n+1} A_i = \bigcup_{i=1}^{n} A_i \cup A_{n+1},$$

we have:

$$\begin{aligned}
P\left(\bigcup_{i=1}^{n+1} A_i\right) &= P\left(\bigcup_{i=1}^{n} A_i \cup A_{n+1}\right) && \text{By (2)} \\
&= P\left(\bigcup_{i=1}^{n} A_i\right) + P(A_{n+1}) && \text{By P4 and hypothesis of the theorem} \\
&= \sum_{i=1}^{n} P(A_i) + P(A_{n+1}) && \text{By inductive hypothesis} \\
&= \sum_{i=1}^{n+1} P(A_i) && \text{Q.E.D.}
\end{aligned}$$

The above proof is not entirely satisfactory, since a prior discussion of set-theoretical identities like (2) has not been given. However, it is intuitively clear that (2) is true. In the present section all needed elementary properties of set and arithmetical operations are assumed known.

We now define the notion of a probability distribution on a finite set. The notion can be extended to infinite sets under certain restrictions which will not be discussed here.

DEFINITION 3. *If X is a finite, non-empty set, then a probability distribution on X is a non-negative real-valued function p defined on X such that*

$$\sum_{x \in X} p(x) = 1.$$

For example, consider the coin tossing above in which $X = \{\langle H, H\rangle, \langle H, T\rangle, \langle T, H\rangle, \langle T, T\rangle\}$. Then for any $x \in X$, $p(x) = 1/4$, assuming a fair coin.

Probability distributions on finite sets and finitely additive probability spaces are related by the following theorem. (Note that this theorem has no prefix 'E' before its number, since it is not a theorem about an arbitrary finitely additive probability space.)

THEOREM 3. *Let X be a non-empty, finite set, let p be a probability distribution on X, and let $\mathfrak{F}$ be a family of subsets of X such that $\mathfrak{F}$ is a field on X. Let P be a real-valued function defined on $\mathfrak{F}$ such that if*

$A \in \mathfrak{F}$ then $P(A) = \sum_{x \in A} p(x)$. Then $\langle X, \mathfrak{F}, P\rangle$ is a finitely additive probability space.

We leave the proof as an exercise.

Since probability distributions appear to be so much simpler to define than finitely additive probability spaces, it might be thought that they should be the basic entities discussed in this section. Fields of sets could be dispensed with entirely, and the probability of *any* set would be obtained by summing over the probabilities of its individual elements. Unfortunately this simplicity disappears when infinite sets are considered, and many problems arise. It is beyond the scope of this section on elementary probability theory to discuss the mathematical methods actually used to deal with infinite sets. Suffice it to say that it consists of a natural and relatively simple extension of our concept of a finitely additive probability space to that of a countably additive probability space where the underlying field of sets is also countably additive in the sense that if each member of a sequence of sets $\langle A_1, A_2, \ldots\rangle$ is in $\mathfrak{F}$ then

$$\bigcup_{i=1}^{\infty} A_i \in \mathfrak{F}.$$

The countable operations are required to permit various important limiting operations to be well-defined.

We now define the important notion of *conditional probability.*

DEFINITION E4. *If A and B are events and $P(A) > 0$, then*

$$P_A(B) = \frac{P(A \cap B)}{P(A)}.$$

The symbol '$P_A(B)$' is read 'the conditional probability of the occurrence of event B given the occurrence of event A', or often, for brevity, 'the conditional probability of B given A'. A second notation for conditional probability is also useful, and we shall use either interchangeably.

DEFINITION E5. *If A and B are events, then*

$$P(B|A) = P_A(B).$$

As an example, let us consider the situation in which a fair coin is flipped three times.

$$X = \{\langle H, H, H\rangle, \langle H, H, T\rangle, \langle H, T, H\rangle, \langle H, T, T\rangle, \langle T, H, H\rangle, \langle T, H, T\rangle, \langle T, T, H\rangle, \langle T, T, T\rangle\}$$

$\mathfrak{F}$ = *set of all subsets of X*

$P(\{x\}) = 1/8$ *for all x in X.*

Let A be the event of at least one head in the last two tosses, and let B be the event of exactly two tails in the three tosses. Then

$$A = \{\langle H, H, H\rangle, \langle H, H, T\rangle, \langle H, T, H\rangle, \langle T, H, H\rangle, \langle T, H, T\rangle, \langle T, T, H\rangle\}$$

$$B = \{\langle H, T, T\rangle, \langle T, H, T\rangle, \langle T, T, H\rangle\}$$

$$A \cap B = \{\langle T, H, T\rangle, \langle T, T, H\rangle\}$$

$$P(A) = 3/4$$

$$P(B) = 3/8$$

$$P(A \cap B) = 1/4,$$

and therefore,

$$P(B|A) = \frac{P(A \cap B)}{P(A)} = \frac{1/4}{3/4} = 1/3.$$

$$P(A|B) = \frac{P(A \cap B)}{P(B)} = \frac{1/4}{3/8} = 2/3.$$

Thus if we know that in three flips of a fair coin at least one head occurred in the last two tosses, then the probability of exactly two tails having occurred is $1/3$. The intuitive idea is simply that two of the six outcomes in A are also in $A \cap B$. On the other hand if we know that exactly two tails occurred in the three flips, the probability of at least one head in the last two tosses is $2/3$. As this example shows, in general

$$P(B|A) \neq P(A|B).$$

The elementary properties of conditional probabilities are summarized in the following theorem, the proof of which is left as an exercise.

THEOREM E4. *If A, B, and C are events and $P(A) > 0$, then:*

(i) $P_A(B) \geq 0$.
(ii) $P_A(X) = 1$.
(iii) $P_A(A) = 1$.
(iv) *If* $B \cap C = \Lambda$, *then* $P_A(B \cup C) = P_A(B) + P_A(C)$.
(v) *If* $P(B) > 0$, *then* $P_B(A) = P(A)P_A(B)/P(B)$.
(vi) *If* $A \cap B = \Lambda$, *then* $P_A(B) = 0$.
(vii) *If* $A \subseteq B$, *then* $P_A(B) = 1$.

We may use the various parts of Theorem E4 to prove that given the knowledge of an event A with positive probability, we may generate a new finitely additive probability space from the given one.

THEOREM E5. *If A is an event and $P(A) > 0$, then $\langle X, \mathfrak{F}, P_A \rangle$ is a finitely additive probability space.*

PROOF. We need to show that $\langle X, \mathfrak{F}, P_A \rangle$ satisfies the four axioms of Definition 2. It is already given that $\mathfrak{F}$ is a field of sets on X, since $\langle X, \mathfrak{F}, P \rangle$ is a finitely additive probability space and thus Axiom P1 is satisfied. From part (i) of Theorem E4 it follows that Axiom P2 is satisfied, from part (ii) that Axiom P3 is satisfied, and from part (iv) that Axiom P4 is satisfied. Q.E.D.

The following theorem is extremely useful in applications; it expresses the probability of an event as a sum of certain products. It is often called the *Theorem on Total Probability.*

THEOREM E6. *If $X = \bigcup_{i=1}^{n} A_i$ and if the events A_i are pairwise incompatible and $P(A_i) > 0$ for $1 \le i \le n$, then for every event B*

$$P(B) = \sum_{i=1}^{n} P(B|A_i)P(A_i).$$

PROOF. Using the hypothesis of the theorem and the fact that intersection of sets is distributive with respect to union of a finite number of sets, we have:

$$B = B \cap X = B \cap \bigcup_{i=1}^{n} A_i = \bigcup_{i=1}^{n} B \cap A_i. \tag{1}$$

Hence

$$P(B) = P\left(\bigcup_{i=1}^{n} B \cap A_i\right). \tag{2}$$

Since by hypothesis $A_i \cap A_j = \Lambda$ for $i \neq j$, we may apply our previous result on finite additivity part (iii) of Theorem E2, to (2) and obtain:

$$P(B) = \sum_{i=1}^{n} P(B \cap A_i). \tag{3}$$

But by Definition E5

$$P(B \cap A_i) = P(B|A_i)P(A_i) \tag{4}$$

and our theorem follows immediately from (3) and (4). Q.E.D.

An example of the application of the theorem on total probability within a scientific theory may be described in qualitative terms by considering statistical learning theory as developed in the last few years by W. K. Estes, R. R. Bush, F. Mosteller, and others. We are concerned with an experimental situation in which on any trial a subject may make one of two responses. For instance, a left or right turn in a T-maze by a rat, or the prediction by a human subject that a left or right light will flash. Furthermore, according to some pattern fixed upon by the experimenter the subject is rewarded for making certain responses on certain trials. The general problem of the theory is to predict what response a subject will make on any trial, the rough idea being that a subject learns to make those responses which are more often rewarded. The axioms of the theory make precise this rough idea by stating how the response probabilities change: (i) given a rewarded response; (ii) given an unrewarded response with subject observing that the other response would have been rewarded; (iii) given an unrewarded response and subject observing that no response would have been rewarded. There is an axiom for each of the three cases, exactly one of which must occur on any trial. Thus we may use the learning axioms and the theorem on total probability to evaluate the probability of a left response, say, on trial n. (A more exact characterization of statistical learning theory would constitute too lengthy a digression.)

We next state and prove *Bayes' theorem*, the applications of which have been the source of endless controversies. The proof is relatively trivial.

THEOREM E7. *If*

$$X = \bigcup_{i=1}^{n} H_i,$$

and if the events H_i are pairwise incompatible and $P(H_i) > 0$ for $1 \leq i \leq n$, and if B is an event such that $P(B) > 0$, then

$$P(H_i|B) = \frac{P(B|H_i)P(H_i)}{\sum_{i=1}^{n} P(B|H_i)P(H_i)}.$$

PROOF. By part (v) of Theorem E4

$$P(H_i|B) = \frac{P(B|H_i)P(H_i)}{P(B)}. \tag{1}$$

But by Theorem E6

$$P(B) = \sum_{i=1}^{n} P(B|H_i)P(H_i). \tag{2}$$

The desired conclusion follows at once from (1) and (2). Q.E.D.

The terminology ordinarily used in connection with Bayes' theorem should be noted. The events H_i are called *hypotheses*, which are, of course, pairwise incompatible and exhaustive. One of the hypotheses should offer a "best explanation" of the event B. The probability $P(H_i)$ is called the *a priori probability* of the hypothesis H_i. These a priori probabilities are the center of most of the controversy concerning applications of the theorem. The conditional probability $P(B|H_i)$ is called the *likelihood* of H_i when B is observed.* The conditional probability $P(H_i|B)$ is called the *a posteriori probability* of the hypothesis H_i given the observed event B.

A simple example will illustrate some of the problems which arise in applications of probability, particularly in application of Bayes' theorem. Suppose there is an urn containing four balls about which we know that either one of the following may be true:

H_1: All four balls are white.
H_2: Two balls are white and two balls are black.

Suppose further, that a ball is drawn at random and is seen to be white. Our concern is to compute the a posteriori probabilities of H_1 and H_2. Three problems have to be met in an exact analysis of the situation:

(i) What is the correct sample space X?
(ii) What is to be said about the a priori probabilities of H_1 and H_2?
(iii) What is to be said about the likelihoods of H_1 and H_2?

Turning to the first problem, we see at once that the preceding definitions and theorems in the theory of probability yield no exact rule for constructing the sample space. A little reflection indicates that it can be constructed in such a manner that four consecutive selections without replacement of balls from the urn will necessarily determine H_1 or H_2 as true and thus having probability 1. Let us designate this particular sample space by 'X_1'. Knowing that either H_1 or H_2 is true, X_1 is simply the following set of ordered quadruples:

$$X_1 = \{\langle W, W, W, W\rangle, \langle W, W, B, B\rangle, \langle B, B, W, W\rangle, \langle W, B, B, W\rangle, \langle B, W, W, B\rangle, \langle W, B, W, B\rangle, \langle B, W, B, W\rangle\}.$$

It will then be observed that H_1 and H_2 are the following subsets of X.

$$H_1 = \{\langle W. W, W, W\rangle\}$$

$$H_2 = \{\langle W, W, B, B\rangle, \langle B, B, W, W\rangle, \langle W, B, B, W\rangle, \langle B, W, W, B\rangle, \langle W, B, W, B\rangle, \langle B, W, B, W\rangle\} = \sim H_1.$$

If E is the event of getting a white ball on the first draw, then

$$E = \{\langle W, W, W, W\rangle, \langle W, B, W, B\rangle, \langle W, W, B, B\rangle, \langle W, B, B, W\rangle\}.$$

* This *likelihood* terminology, common in modern statistics, reverses the natural conditional probability language: the probability of B given H_i. The reasons for this inversion should become clearer in the sequel.

The set X_1 is not, however, the only correct sample space for the analysis of our urn problem. It is clear that which hypothesis is correct could be completely decided on the basis of only three draws without replacement. Thus, we might use:

$$X_2 = \{\langle W, W, W\rangle, \langle W, W, B\rangle, \langle W, B, W\rangle, \langle B, W, W\rangle, \langle B, B, W\rangle,$$
$$\langle B, W, B\rangle, \langle W, B, B\rangle\}.$$

Observe that the number of elements in X_2 is equal to that in X_1. That is, their cardinality is the same. If X_2 is the basic sample space then H_1, H_2, and E are the following sets:

$$H_1 = \{\langle W, W, W\rangle\}$$
$$H_2 = \{\langle W, W, B\rangle, \langle W, B, W\rangle, \langle B, W, W\rangle, \langle B, B, W\rangle, \langle B, W, B\rangle,$$
$$\langle W, B, B\rangle\} = {\sim}H_1$$
$$E = \{\langle W, W, W\rangle, \langle W, W, B\rangle, \langle W, B, W\rangle, \langle W, B, B\rangle\}.$$

On the other hand, if unordered sets are used as elements of the sample space, then difficulties arise. For instance, if we used:

$$X_3 = \{\{W_1, W_2, W_3, W_4\}, \{W_1, W_2, B_1, B_2\}\}.$$

Then

$$H_1 = \{\{W_1, W_2, W_3, W_4\}\}$$
$$H_2 = \{\{W_1, W_2, B_1, B_2\}\},$$

but there is no appropriate subset of X_3 to represent the event of drawing a white ball. Yet X_1 and X_2 do not constitute the only legitimate possibilities. For several reasons many statisticians would prefer the following sample space:

$$X_4 = \{\langle 4, W\rangle, \langle 2, W\rangle, \langle 2, B\rangle\},$$

and thus

$$H_1 = \{\langle 4, W\rangle\}$$
$$H_2 = \{\langle 2, W\rangle, \langle 2, B\rangle\}$$
$$E = \{\langle 4, W\rangle, \langle 2, W\rangle\}.$$

The number which is the first member of each ordered couple indicates the total number of white balls in the urn on the given hypothesis (H_1 or H_2). The second member indicates a possible outcome of drawing one ball. An advantage of X_4 over X_1 and X_2 is that X_4 has just three elements. The reduction of the number of elements is a consequence of the fact that X_4 expresses the possible outcomes of exactly one drawing, whereas X_1 indicates the possible outcomes of four samples without replacement, and X_2

of three samples without replacement. The sample space X_4 is tailored to our particular urn problem in a way in which X_1 and X_2 are not.

Fixing upon X_4 as our sample space, we may turn to our second problem, how to determine the a priori probabilities $P(H_1)$ and $P(H_2)$. It is a happy characteristic of textbook problems (as opposed to real-life problems) that they may be artificially and so completely prescribed. In the description of our urn problem, absolutely nothing was said about H_1 and H_2 except that exactly one of them is true. To a situation of this character we may apply what is without doubt the most controversial rule of application in the field of probability, namely, the *principle of indifference:* * If there is no evidence for regarding one event as more probable than a second, assign them equal probabilities. This principle yields that

$$P(H_1) = P(H_2) = 1/2.$$

The closely related but more palatable principle of maximum likelihood is discussed below.

Our third problem concerns the likelihoods $P(E|H_1)$ and $P(E|H_2)$. Although a rule of application is needed to evaluate these likelihoods, it is not nearly so much disputed as the various rules proposed for evaluating a priori probabilities. The rule we have in mind may be called the *principle of (simple) random sampling*. Roughly speaking, the idea of the principle is that we use a procedure of sampling such that if H is the true hypothesis, any possible outcome of sampling compatible with H is as probable as any other such outcome. Thus here

$$P(E|H_2) = 1/2$$

$$P(\sim E|H_2) = 1/2.$$

Of course, independent of any sampling principle, in the urn problem being considered:

$$P(E|H_1) = 1.$$

The computation of the a posteriori probabilities $P(H_1|E)$ and $P(H_2|E)$ is now straightforward by use of Bayes' theorem:

$$\begin{aligned} P(H_1|E) &= \frac{P(E|H_1)P(H_1)}{P(E|H_1)P(H_1) + P(E|H_2)P(H_2)} \\ &= \frac{1\cdot 1/2}{1\cdot 1/2 + 1/2\cdot 1/2} \\ &= 2/3, \end{aligned}$$

and similarly, we find that

$$P(H_2|E) = 1/3.$$

* Usually credited to James Bernouilli, *Ars Conjectandi*, Basel, 1713; but its prominence in the history of probability is due more to the use made of it by Pierre Laplace

Obviously anyone using the principle of indifference would on the basis of the sample E favor hypothesis H_1.

It is important to realize that use of the principle of random sampling is sensitive to the particular sample space fixed upon for the analysis of a problem. For instance, to change our problem slightly, let H_3 be the hypothesis that three of the four balls in the urn are white, and the fourth is black. Suppose further, for the moment, that H_1 and H_3 are exhaustive hypotheses. Analogous to X_4 we would have a sample space

$$X_5 = \{\langle 4, W\rangle, \langle 3, W\rangle, \langle 3, B\rangle\}$$

and

$$H_1 = \{\langle 4, W\rangle\}$$

$$H_3 = \{\langle 3, W\rangle, \langle 3, B\rangle\}$$

$$E = \{\langle 4, W\rangle, \langle 3, W\rangle\},$$

but in this case we should have the following likelihood value:

$$P(E|H_3) = 3/4,$$

since if H_3 is true, of the four balls three are white, and thus the chance of drawing a white ball is $3/4$.

The unsatisfactory character of the relatively vague methods for applying the principle of random sampling can be considerably reduced by using a sample space similar in structure to X_4 and X_5 but more concrete. The idea is to assign names to the individual balls (as we tacitly did for the unsatisfactory space X_3). If H_1 is true, let W_1, W_2, W_3, W_4 be the four balls. If H_3 is true, let W_1, W_2, W_3, B_1 be the four balls. Our sample space is then:

$$X_6 = \{\langle 4, W_1\rangle, \langle 4, W_2\rangle, \langle 4, W_3\rangle, \langle 4, W_4\rangle, \langle 3, W_1\rangle, \langle 3, W_2\rangle, \langle 3, W_3\rangle, \langle 3, B_1\rangle\},$$

and it is obvious which subsets of X_6 are H_1, H_3, and E. By an *atomic event* we mean a unit subset of X. Thus $\{\langle 3, W_2\rangle\}$ is an atomic event, for instance. The principle of random sampling may now be expressed very simply. Let there be n atomic events included in H. Then the likelihood given H of an atomic event A included in H is just:

$$P(A|H) = \frac{1}{n}.$$

in his great treatise *Théorie analytique des probabilités*, Paris, 1812; 2d ed., 1814. The principle of indifference is also called the principle of non-sufficient reason. It is not possible here to discuss the relative soundness of the principle nor to analyze any of the paradoxes which it has given rise to.

And for X_6 we then have on this basis:

$$P(E|H_3) = \sum_{i=1}^{4} P(\{\langle 4, W_i\rangle\}|H_3) + \sum_{i=1}^{3} P(\{\langle 3, W_i\rangle\}|H_3)$$
$$= 0 + 1/4 + 1/4 + 1/4$$
$$= 3/4.$$

It will be of some interest to consider a slightly more complicated urn problem, which we analyze without so many diversions. Suppose again that the urn contains four balls and either H_2 or H_3 is true, that is, either the urn has two white balls and two black ones, or three white ones and one black. Suppose further that a random sample is drawn consisting of one ball without replacement, and then two balls with replacement, and that the first ball is observed to be white, and the next two black. We could take as elements of our sample space ordered quadruples of the form $\langle 2, W_1, B_1, B_2\rangle$, but since on the principle of random sampling which we assume, the order of the balls in the double sample with replacement does not matter, we may reduce the number of elements in the sample space by using ordered triples of the form $\langle 2, W_1, \{B_1, B_2\}\rangle$, where the last member of each triple is a set, namely, the set of balls drawn in the double sample with replacement. Considering all possible combinations of outcomes for both H_2 and H_3, we see that the sample space consists of forty-eight elements:

$$X_7 = \{\langle 2, W_1, \{W_2, B_1\}\rangle, \langle 2, W_1, \{W_2, B_2\}\rangle, \langle 2, W_1, \{B_1, B_2\}\rangle,$$
$$\langle 2, W_1, \{W_2\}\rangle, \langle 2, W_1, \{B_1\}\rangle, \langle 2, W_1, \{B_2\}\rangle, \ldots,$$
$$\langle 2, B_2, \{B_1\}\rangle, \langle 3, W_1, \{W_2, W_3\}\rangle, \ldots, \langle 3, B_1, \{W_3\}\rangle\}.$$

Note that an element like $\langle 2, W_1, \{W_2\}\rangle$ represents the fact that ball W_2 was drawn twice; since

$$\{W_2, W_2\} = \{W_2\},$$

we use the unit set.

H_2 is simply the set of elements whose first members are 2, and H_3 those which are 3. The sample F consisting of a white ball without replacement and then two black balls with replacement is the set

$$F = \{\langle 2, W_1, \{B_1, B_2\}\rangle, \langle 2, W_1, \{B_1\}\rangle, \langle 2, W_1, \{B_2\}\rangle,$$
$$\langle 2, W_2, \{B_1, B_2\}\rangle, \langle 2, W_2, \{B_1\}\rangle, \langle 2, W_2, \{B_2\}\rangle,$$
$$\langle 3, W_1, \{B_1\}\rangle, \langle 3, W_2, \{B_1\}\rangle, \langle 3, W_3, \{B_1\}\rangle\}.$$

Assuming again the principle of indifference,

$$P(H_2) = P(H_3) = 1/2.$$

Moreover, using the principle of random sampling, we easily compute the likelihoods:

$$P(F|H_2) = 6/24 = 1/4,$$

$$P(F|H_3) = 3/24 = 1/8.$$

We then have for the a posteriori probabilities:

$$\begin{aligned} P(H_2|F) &= \frac{P(F|H_2)P(H_2)}{P(F|H_2)P(H_2) + P(F|H_3)P(H_3)} \\ &= \frac{1/4 \cdot 1/2}{1/4 \cdot 1/2 + 1/8 \cdot 1/2} \\ &= 2/3, \end{aligned}$$

and

$$P(H_3|F) = 1 - P(H_2|F) = 1/3.$$

Assuming the principle of indifference the evidence F clearly favors hypothesis H_2.

It has not been possible to present here a critique of the principle of indifference, but we may mention R. A. Fisher's *principle of maximum likelihood* which circumvents the necessity of fixing upon any a priori probabilities. In the case of our urn problems, the principle says that one should accept that hypothesis whose likelihood is largest, that is, maximum for the observed sample. For the urn problems considered it is also immediately obvious which hypothesis has the maximum likelihood. Applications of the principle to estimating some quantity, like the number of fish in a lake, on the basis of a sample are more complicated.

In discussing these simple urn problems the construction of an appropriate sample space has been emphasized, and the selection of a field of sets has been ignored, for it has been tacitly assumed that the field of sets is the set of all subsets of the sample space, which choice is almost always appropriate when the sample space has a finite number of elements. Similarly, the probability measure P has not been completely specified, but only for those arguments relevant to the problem at hand. In the problems considered it is clear that the principle of indifference and the principle of random sampling completely determine P, but this is often not the case. For example, if the principle of maximum likelihood is used, and the principle of indifference is rejected, in the urn problems given above it is not possible to determine P completely from the data given. For this reason even when the sample space X is fixed there are frequently many distinct finitely additive probability spaces adequate for analysis of a given problem. This situation is not peculiar to probability theory, but usually prevails in application of any exactly formulated theory to problems described in an intuitive, informal manner.

In this section we have been able to consider only a fragment of probability theory, which is one of the richest and deepest subjects in modern mathematics.* Just as lamentable from a philosophical standpoint is the omission of any detailed discussion of the foundations of probability, which has been one of the most important topics in systematic philosophy since the eighteenth century. Here we can discuss neither of the two most important schools of thought: those who hold to the objective view that probabilities of events are measured by relative frequencies of their occurrences, and those who hold to the subjective view that probabilities of events are first and foremost measures of degrees of reasonable belief that the events will occur.†

EXERCISES

1. Prove Theorem 1.
2. Rewrite Definition 2 in the style of Definition A of Section 2.
3. Prove (i) and (ii) of Theorem E2.
4. Prove Theorem 3.
5. Prove Theorem E4.
6. Consider the following situation. We have an urn containing five balls. We know that one of the following two hypotheses is true:

H_1: Two balls are white and three are black.
H_2: Three balls are white and two are black.

For each of the following construct an appropriate sample space and assuming the principles of indifference and random sampling, compute the a posteriori probabilities of the two hypotheses.

(a) We draw a random sample with replacement consisting of two white and one black ball.
(b) We draw a random sample of the following sort: a black ball without replacement, and then a white ball.
(c) We draw a random sample of the following sort: a sample of two without replacement consisting of a black and white ball, and then a single sample consisting of a black ball.

* The developments begun in this chapter are carried far in W. Feller, *An Introduction to Probability Theory and Its Applications*, Vol. I, New York, 1950. From this book the reader may go to advanced treatises on the subject of probability.

† For the relative-frequency view, see R. von Mises, *Probability, Statistics and Truth*, English translation, New York, 1939; E. Nagel, *Principles of the Theory of Probability*, International Encyclopedia of Unified Science, Vol. I, No. 6, Chicago, 1939; H. Reichenbach, *The Theory of Probability*, English translation, Los Angeles, 1949. For the subjective view, see H. Jeffreys, *Theory of Probability*, 2d ed., Oxford, 1948; L. J. Savage, *Foundation of Statistics*, New York, 1954. For a variant of the subjective theory closely connected with logic, see R. Carnap, *Logical Foundations of Probability*, Chicago, 1950.

7. Given the situation of Exercise 6, and permission to draw a random sample of two balls, which method of sampling would you recommend—drawing the first ball with replacement, or drawing it without replacement? Justify your answer.

8. Assuming that the probability that a baby is male is $\frac{1}{2}$, construct the appropriate finitely additive probability space for a family of five children. What is the probability that

(a) All children will be boys?
(b) There will be two boys and three girls?
(c) There will be at least one boy?
(d) There will be at least one girl, given that there are at least three boys?
(e) There will be exactly two boys, given that there are at least two girls?

9. A classical problem of induction is the following. A man draws at random m balls with replacement from an urn containing N black and white balls, possibly with either no blacks or no whites. Suppose he draws m black balls. What is the probability that on the $(m + 1)$th draw he will draw a black ball?

(a) Let $N = 4$ and $m = 2$, and construct the appropriate finitely additive probability space, assuming the principles of indifference and random sampling. Note that there are five mutually exclusive and exhaustive hypotheses, since there are five possibilities (from zero to four) for the number of black balls.
(b) Show that the general answer is:

$$\frac{\sum_{i=0}^{N} \left(\frac{i}{N}\right)^{m+1}}{\sum_{i=0}^{N} \left(\frac{i}{N}\right)^{m}}.$$

(It may be shown that as N increases without limit the probability of a black ball on the $(m + 1)$th draw becomes $(m + 1)/(m + 2)$.)

§ 12.5 Example: Mechanics. One of the most exact and well-established branches of empirical science is the classical theory of particle mechanics. Indeed, for a period of almost two hundred years from the beginning of the eighteenth century until the end of the nineteenth century it was widely held that particle mechanics was *the* fundamental theory of the universe. In this section we want briefly to consider an axiomatization of this theory for the purpose of demonstrating that even a relatively complicated theory can be given a clear and exact formulation within set theory. Unfortunately our development of mechanics must use certain mathematical notions which it is not expedient to explain *ab initio* here.* At this point we will define the more important notions needed. An *interval* of real numbers is the set of all real numbers between two numbers, greater than some one number or less than some one number. More exactly, if a and b are real numbers, $[a, b]$ is the set of all real numbers x such that $a \leq x \leq b$. The interval $[a, b]$ is *closed* since it includes its end points a

* This is the only section with respect to which this book is not mathematically self-contained.

and b. Square brackets are customarily used to designate closed intervals, round brackets to designate *open* intervals. Thus (a, b) is the open interval which is the set of all numbers x such that $a < x < b$. As would be expected $(a, b]$ is a half-open, half-closed interval, etc. $(-\infty, a)$ is the set of all real numbers x such that $x < a$, and the interval $[a, \infty)$ is the set of all real numbers x such that $x \geq a$. Finally, the interval $(-\infty, \infty)$ is the set of all real numbers.

A (*three-dimensional*) *vector* is an ordered triple of real numbers. Thus $\langle 4, 7, 3\rangle$ is a vector. The usual physical interpretation of vectors is that they represent the magnitude and direction of some physical quantity like velocity or force with respect to three mutually perpendicular spatial coordinate axes. Binary operations on vectors, or vectors and real numbers, are defined in a natural way. Thus if $x = \langle x_1, x_2, x_3\rangle$ is a vector and a is a real number,

$$ax = a\langle x_1, x_2, x_3\rangle = \langle ax_1, ax_2, ax_3\rangle = xa.$$

If $x = \langle x_1, x_2, x_3\rangle$ and $y = \langle y_1, y_2, y_3\rangle$ are vectors then

$$x + y = \langle x_1, x_2, x_3\rangle + \langle y_1, y_2, y_3\rangle = \langle x_1 + y_1, x_2 + y_2, x_3 + y_3\rangle$$

and

$$x - y = \langle x_1, x_2, x_3\rangle - \langle y_1, y_2, y_3\rangle = \langle x_1 - y_1, x_2 - y_2, x_3 - y_3\rangle.$$

Moreover,

$$-x = -\langle x_1, x_2, x_3\rangle = \langle -x_1, -x_2, -x_3\rangle.$$

The *inner* or *scalar product* operation on vectors x and y yields a real number and is designated by a dot. Thus,

$$x \cdot y = \langle x_1, x_2, x_3\rangle \cdot \langle y_1, y_2, y_3\rangle = x_1y_1 + x_2y_2 + x_3y_3.$$

The *vector* or *cross product* operation on vectors x and y is a vector and is designated by '$\times$' (not to be confused with the Cartesian product operation for sets). Thus

$$\begin{aligned} x \times y &= \langle x_1, x_2, x_3\rangle \times \langle y_1, y_2, y_3\rangle \\ &= \langle x_2y_3 - x_3y_2, x_3y_1 - x_1y_3, x_1y_2 - x_2y_1\rangle. \end{aligned}$$

If $x = \langle x_1, x_2, x_3\rangle$ is a vector then

$$|x| = \sqrt{x_1^2 + x_2^2 + x_3^2}.$$

The number $|x|$ is the *absolute value* of the vector x. Formal properties of these various operations are stated in the exercises.

Some intuitive remarks about the last three operations may be helpful. When, for instance, x is a velocity vector, then $|x|$ is the magnitude of the velocity (independent of direction). The scalar product has the following geometric interpretation:

$$x \cdot y = |x|\,|y| \cos(x, y),$$

where, $\cos(x, y)$ is the cosine of the angle between the two vectors. The magnitude of the vector product is given by the equation:

$$|x \times y| = |x|\,|y|\,|\sin(x, y)|.$$

The direction of the vector product is perpendicular to the plane formed by the vectors x and y. If f is a real or vector-valued function then Df is the *derivative* of f. A few remarks on derivatives will not be amiss. Let f be a function which is defined on an interval T of real numbers and whose range is a set of real numbers. Then Df is a function whose domain of definition is a subset of T. If $a \in T$ then a is in the domain of Df if and only if there is a number l such that for every $\epsilon > 0$ there is a δ such that if $x \in T$, $x \neq a$ and $|x - a| < \delta$ then

$$\left|\frac{f(x) - f(a)}{x - a} - l\right| < \epsilon.*$$

Moreover, when a is in the domain of Df, the function f is said to be differentiable at a. If f is differentiable at every point † of T it is said to be *differentiable on* T. We have explicitly introduced these definitions to avoid the usual phrases like 'the derivative of f exists at a'. Although this language of existence is perfectly clear to the experienced reader, from a logical standpoint it is not a happy idiom, and we shall avoid it. The second derivative of a function f is simply the function $D(Df)$, which we shall designate: D^2f. If the interval T is the domain of definition of a function f then f is *twice differentiable* at a in T if a is in the domain of D^2f, etc. One of our axioms for mechanics is that the position function of a particle is twice differentiable on the time interval of the mechanical system.

We now turn to our six primitive notions for particle mechanics. P and T are sets, s is a binary function, m a unary function, f a ternary function, and g a binary function.

The intended physical interpretation of P is that it is the set of *particles*. T is interpreted physically as a set of real numbers measuring elapsed times. If $p \in P$ and $t \in T$ then $s(p, t)$ is a vector which is physically interpreted as the *position* of particle p at time t. For each p in P it will be convenient to introduce its position function s_p.‡ Note that the primitive

* This definition is easily extended to vector-valued functions. If a is in the domain of Df then $Df(a) = l$.

† In these contexts 'point' is often used as a synonym for 'number'.

‡ Using the lambda notation of § 11.3, we could write: $(\lambda t)s(p, t)$ instead of: s_p.

s fixes the choice of a coordinate system. These first three primitive notions are the *kinematical* notions of particle mechanics.

If $p \in P$, then $m(p)$ is interpreted as the numerical value of the *mass* of the particle p. If $p, q \in P$, and $t \in T$, then $f(p, q, t)$ is the *force* which particle q exerts on particle p at time t. Such forces are called the *internal* forces of the mechanical system. If $p \in P$ and $t \in T$, then $g(p, t)$ is the *resultant external force* acting on particle p at time t. For instance, consider the two particle systems consisting of the planets Earth and Venus. Then the internal force on Earth is the gravitational attraction of Venus, and the resultant external force on Earth is the vector sum of the gravitational attractive forces of all other bodies in the solar system (and the rest of the universe for that matter). There are good arguments for insisting that the *given* external forces on a particle should be primitive rather than the resultant force, particularly since the latter is definable in terms of the former but not vice versa. However, it is formally simpler to treat only the resultant external force on a particle.* The notions of mass, internal force, and external force are the primitive *dynamical* notions of particle mechanics.

We are now ready to state our kinematical and dynamical axioms for systems of particle mechanics.

DEFINITION 1. *A system* $\mathfrak{P} = \langle P, T, s, m, f, g\rangle$ *is a system of particle mechanics if and only if the following seven axioms are satisfied:*

KINEMATICAL AXIOMS

AXIOM P1. *The set* P *is finite and non-empty.*

AXIOM P2. *The set* T *is an interval of real numbers.*

AXIOM P3. *For* p *in* P, s_p *is twice differentiable on* T.

DYNAMICAL AXIOMS

AXIOM P4. *For* p *in* P, $m(p)$ *is a positive real number.*

AXIOM P5. *For* p *and* q *in* P *and* t *in* T,

$$f(p, q, t) = -f(q, p, t).$$

AXIOM P6. *For* p *and* q *in* P *and* t *in* T,

$$s(p, t) \times f(p, q, t) = -s(q, t) \times f(q, p, t).$$

AXIOM P7. *For* p *in* P *and* t *in* T,

$$m(p)D^2s_p(t) = \sum_{q \in P} f(p, q, t) + g(p, t).$$

* For a discussion of this and other such points see J. C. C. McKinsey, A. C. Sugar, and Patrick Suppes, "Axiomatic foundations of classical particle mechanics," *Journal of Rational Mechanics and Analysis*, Vol. 2 (1953) pp. 253–272. The axiomatization

Although a complete discussion of the individual axioms and their physical interpretation is not possible here, some clarifying remarks can be made.

The condition in the first axiom that P be finite is needed to guarantee that the mass and kinetic energy of the whole system be well-defined. Since there is no interest in having P empty, we have followed the tradition in algebra and made P non-empty. (Remember that in § 12.2 we required that for $\langle A, \circ \rangle$ to be an algebra, A must be non-empty.)

Regarding the second axiom it might be thought simpler to take the interval of time as the set of all real numbers, but in numerous applications this is not convenient; for, as Aristotle might put it, particles are continually coming to be and passing away. Moreover, our axioms are not adequate to deal with impacts, but we can treat the motion of particles prior to the instant of impact by selecting the appropriate time interval.

The philosophical doctrine of conventionalism,* that is, the doctrine that certain assumptions are made in empirical science for purposes of convenience only, is illustrated by the second axiom. From the point of view of empirical testability this axiom could be replaced by the requirement that T be the set of rational numbers in some interval. Certainly it is impossible to build a clock for which the measurement of elapsed times cannot be represented satisfactorily by rational numbers. On the other hand, the assumption that T is a set of rational numbers is mathematically inconvenient, for it blocks direct application of the standard methods of mathematical analysis—in particular, the methods of the differential and integral calculus.

The requirement of the third axiom that the position function of a particle be twice differentiable on T is also an idealization, for any set of empirical data on the trajectory of a particle can be approximated arbitrarily closely by a function which is twice differentiable at no point in T. A function of the latter sort would be utterly intractable mathematically; if such a function were used to represent the motion of a particle, many of the simplest problems of mechanics would be rendered extremely difficult if not impossible to solve. On the other hand, there are workable conditions weaker than that of twice differentiability which may be used.† We have characterized the first three axioms as the kinematical axioms of me-

given in this paper is closely related to the one given here. For an axiomatization of relativistic particle mechanics, see H. Rubin and P. Suppes, "Transformations of systems of relativistic particle mechanics," *Pacific Journal of Mathematics*, Vol. 4 (1954) pp. 563–601.

* The famous French mathematician and philosopher Henri Poincaré (1854–1912) is prominently associated with this doctrine.

† See, for example, G. Hamel, "Die Axiome der Mechanik," *Handbuch der Physik*, Vol. 5, pp. 1–42. However, Hamel's axiomatization does not satisfy the set-theoretical criteria developed in this chapter.

chanics. Maxwell gives the following succinct characterization of kinematics: *

> We begin with kinematics, or the science of pure motion. In this division of the subject the ideas brought before us are those of space and time. The only attribute of matter which comes before us is its continuity of existence in space and time—the fact, namely, that every particle of matter, at any instant of time, is in one place and in one only, and that its change of place during any interval of time is accomplished by moving along a continuous path.
>
> Neither the force which affects the motion of the body nor the mass of the body, on which the amount of force required to produce the motion depends, come under our notice in the pure science of motion.

Maxwell's continuous-path requirement is not strong enough, for there are functions continuous everywhere but nowhere differentiable. If such a function were used to represent the motion of a particle, the instantaneous velocity of the particle would be nowhere defined, since the velocity function of a particle is the first derivative of its position function.

Axiom P4 requires that the mass of a particle always be positive. If the mass of some particle were zero then by use of the final axiom, P7, the acceleration of the particle could not be determined from a knowledge of the forces acting on the particle. In the theory of rockets, for instance, it is desirable to permit the mass to vary with time (corresponding to the consumption of fuel), but by making mass independent of time we have stayed within the traditional framework of classical mechanics.

Axioms P5 and P6 provide an exact formulation of Newton's third law of motion: † To every action there is always opposed an equal reaction: or, the mutual actions of two bodies upon each other are always equal, and directed to contrary parts. The fifth axiom corresponds to what Hamel calls the *first complete reaction principle*, and the sixth axiom to the *second complete reaction principle*.‡ The intuitive content of the two axioms should be clear. The fifth axiom requires that the force exerted by q on p be equal and opposite to that exerted by p on q. The sixth axiom requires that the direction of the two internal forces be along the line connecting the position of the two particles. That this is expressed by the vector product equation of P6 may be seen by the following con-

* J. Clerk Maxwell, *Matter and Motion*, 1920 ed., London, p. 79.

† Isaac Newton, *Principia*, Cajori translation, Berkeley, Calif., 1934, p. 13. The *Principia* was first published in 1687. Next to Euclid's *Elements* it has probably been the most influential work in the history of science. Regarding the third law, Newton goes on to remark, "Whatever draws or presses another is as much drawn or pressed by that other. If you press a stone with your finger, the finger is also pressed by the stone. If a horse draws a stone tied to a rope, the horse (if I may say so) will be equally drawn back towards the stone; for the distended rope, by the same endeavor to relax or unbend itself, will draw the horse as much towards the stone as it does towards the horse, and will obstruct the progress of the one as much as it advances that of the other."

‡ Hamel, *op. cit.*, p. 25.

siderations. Two vectors x and y are parallel if and only if there is a real number a such that

$$ax = y.$$

(Note the vector $\langle 0, 0, 0 \rangle$ is parallel to every vector *.) Consequently, when two vectors x and y are parallel, the vector product $x \times y$ is the zero vector $\langle 0, 0, 0 \rangle$, for

$$\langle ax_1, ax_2, ax_3 \rangle = \langle y_1, y_2, y_3 \rangle,$$

that is,

$$ax_i = y_i \qquad \text{for} \quad i = 1, 2, 3,$$

whence

$$x_2y_3 - x_3y_2 = ax_2x_3 - ax_3x_2 = 0,$$

and similarly for the other two components of the vector product. Now the axiom asserts that

$$s(p, t) \times f(p, q, t) = -s(q, t) \times f(q, p, t). \tag{1}$$

We add to both sides of (1) $-s(q, t) \times f(p, q, t)$, and use the fact that the vector product operation is distributive with respect to addition to obtain:

$$(s(p, t) - s(q, t)) \times f(p, q, t) = -s(q, t) \times (f(q, p, t) + f(p, q, t)). \tag{2}$$

By Axiom P5

$$f(q, p, t) + f(p, q, t) = \langle 0, 0, 0 \rangle,$$

whence we infer from (2):

$$(s(p, t) - s(q, t)) \times f(p, q, t) = \langle 0, 0, 0 \rangle,$$

that is, $f(p, q, t)$ is parallel to $s(p, t) - s(q, t)$, which means that the line of action of $f(p, q, t)$ is along the line connecting the two particles. A similar argument shows that $f(q, p, t)$ is parallel to $s(p, t) - s(q, t)$. It is a consequence of this result that every internal force is either a force of attraction or of repulsion.

The final axiom, P7, formulates Newton's second law of motion, which he originally phrased: † The change of motion is proportional to the motive force impressed; and is made in the direction of the right line in which that force is impressed. Some authors ‡ have proposed that we convert the second law, that is, P7, into a definition of the total force acting on a particle. There seem to be two serious objections to this procedure. It prohibits within the axiomatic framework any analysis of the

* Some authors define parallelism only for non-zero vectors.

† *Op. cit.*, p. 13.

‡ For instance G. Kirchhoff in his classical work *Vorlesungen über mathematische Physik*, I: *Mechanik*, 1878.

internal and external forces acting on a particle. That is, if all notions of force are eliminated as primitive and P7 is used as a definition, then the notions of internal and external force are not definable within the given axiomatic framework. Secondly, it converts one of the most important laws of mechanics into a definition, which is something conventionalists find attractive, but scarcely jibes with an empirically oriented philosophy of science. With regard to the axiomatization given here, it may be shown by Padoa's principle that the notions of mass and internal force are each independent of the remaining primitive notions. (We leave this as an exercise.) On the other hand, the notion of (resultant) external force is definable, since the equation of Axiom P7 may be written:

$$g(p, t) = m(p)D^2s_p(t) - \sum_{q \in P} f(p, q, t).$$

However, if the notion of *given* external force had been taken as primitive, it could easily have been shown to be independent. The resultant of the given external forces was used here to avoid the complication of introducing the notion of an infinite series (corresponding to an infinite number of given external forces).

Those readers familiar with the literature on the foundations of mechanics may find strange the claim that the notions of mass and internal force are independent of the other primitives, since definitions of these notions are so widely discussed. However, these much-discussed definitions, like Mach's "definition" of the relative mass of two bodies as the inverse ratio of their "mutually induced" accelerations when they are isolated from other bodies,* are only definitions in some Pickwickian sense that has little connection with the theory of definition discussed in Chapter 8.

We turn now to some theorems which illuminate various facets of particle mechanics. The first theorem concerns the center of mass of a system of particle mechanics (formally defined below) and it is ordinarily formulated in the subjunctive mood: the center of mass of a system moves *as if* all the mass *were concentrated* there, and the resultant of all the forces *acted* there. In recent years some philosophers have claimed that use of contrary-to-fact conditionals and thus use of the non-truth-functional subjunctive mood are necessary to properly formulate empirical laws (though this necessity is not usually claimed for the example given here). However, I know of no systematic example like the present one which cannot be satisfactorily reformulated in the indicative mood, and certainly I know of no theorem in any branch of empirical science whose proof hinges upon some peculiar non-truth-functional property of contrary-to-fact condi-

* E. Mach, *Science of Mechanics*, 5th American ed., LaSalle, Ill., 1942, pp. 264–277. From a formal standpoint Mach's book is a mass of confusions.

tionals.* An elementary formal definition of the center of mass function for a system of particles is straightforward.

DEFINITION E2. *If $t \in T$, then*

$$c(t) = \frac{\sum_{p \in P} m(p)s(p, t)}{\sum_{p \in P} m(p)}.$$

And the appropriate theorem is:

THEOREM E1. *If $t \in T$, then*

$$\sum_{p \in P} m(p)D^2c(t) = \sum_{p \in P} g(p, t).$$

Note that the motion depends only on the external forces, since the internal forces cancel each other, from which observation together with Axiom P7 and the above definition we immediately infer the theorem.

In the third section of this chapter the notion of two models of a theory being isomorphic was discussed in some detail, and it was pointed out that one of the most interesting applications of the idea of isomorphism is to the establishing of a representation theorem for a theory. Sometimes it is not possible to prove an interesting representation theorem. When this situation obtains it is natural to ask if an *embedding* theorem can be proved for the theory, that is, to prove that there is an interesting class K of models such that every model for the theory is isomorphic to a submodel of a model belonging to K. The exact definition of submodel varies from one theory to another. If $\mathfrak{A} = \langle A, \circ \rangle$ is an algebra, for instance, that is, A is a set and $\circ$ is a binary operation from $A \times A$ to A, then an algebra $\langle A', \circ' \rangle$ is a *subalgebra* of $\mathfrak{A}$ if A' is a subset of A and $\circ' = \circ \cap (A' \times A' \times A')$ that is, $\circ'$ is the operation $\circ$ restricted to A'. In the case of the theory of particle mechanics, the relative complexity of the primitive notions entails that several alternative definitions of submodel are natural. For instance, in a subsystem (i.e., submodel) of a system of particle mechanics, should we permit the subsystem to have a smaller time interval? The rather arbitrary decision made here is to keep the time interval the same in the subsystem. The crucial part of the definition of subsystems peculiar to mechanics is the handling of internal forces. The internal forces on a particle due to particles not in the subsystem are added to the resultant external force on the particle, a procedure which seems to agree with the intuitive idea in physics of a subsystem.

* This is not to deny that the logic of the subjunctive mood is of considerable general philosophical interest.

DEFINITION 3. *Let* $\mathfrak{P} = \langle P, T, s, m, f, g\rangle$ *be a system of particle mechanics, let* P' *be a non-empty subset of* P, *let* s' *and* m' *be the functions* s *and* m *with their first arguments restricted to* P', *let* f' *be the function* f *with its first two arguments restricted to* P', *and let* g' *be the function such that for every* p *in* P' *and* t *in* T

$$g'(p, t) = g(p, t) + \sum_{q \in P \sim P'} f(p, q, t).$$

Then $\langle P', T, s', m', f', g'\rangle$ *is a subsystem of* $\mathfrak{P}$.

This definition could have been put in more set-theoretical language. We could have written:

$$P' \subseteq P$$

$$P' \neq \Lambda,$$

$$s' = s|(P' \times T)$$

$$m' = m|P'$$

$$f' = f|(P' \times P' \times T),$$

where in general if f is a function $f|A$ is the function h such that

$$h = f \cap (A \times \mathcal{R}(f)),$$

that is, $f|A$ is the function f with its domain restricted to A.

We leave as an exercise proof of the theorem that

THEOREM 2. *Every subsystem of a system of particle mechanics is itself a system of particle mechanics.*

Of course, Theorem 2 does not constitute an embedding theorem, and this is the problem to which we now turn. A philosophically and physically interesting class of systems of particle mechanics is the class of *isolated* systems, that is, those for which the resultant external force on every particle is zero (more precisely, the null vector $\langle 0, 0, 0\rangle$).

DEFINITION 4. *A system of particle mechanics* $\mathfrak{P} = \langle P, T, s, m, f, g\rangle$ *is isolated if and only if for every* p *in* P *and* t *in* T

$$g(p, t) = \langle 0, 0, 0\rangle.$$

The notion of an isolated system is connected with the physicists' notion of a *closed* or *independent* mechanical system, which, roughly speaking, is an isolated system with the internal force of any one particle on another expressible as a function only of the distance between the two particles. The total mechanical energy of closed systems is constant in time; more important, when the internal forces do not explicitly depend on time, we may regard the causal analysis of the motions of the particles as in one sense complete. Unfortunately the problem of embedding any system of mechanics in a closed system is too complicated to discuss here, and we shall content ourselves with a weaker embedding result concerning isolated systems. We first define the notion of two systems of mechanics being *equivalent;* the intuitive idea is that the systems are identical except for the individual forces acting on the particles, the resultant force on a particle being the same in both systems.

DEFINITION 5. *Two systems of particle mechanics* $\mathfrak{P} = \langle P, T, s, m, f, g\rangle$ *and* $\mathfrak{P}' = \langle P', T', s', m', f', g'\rangle$ *are equivalent if and only if*

$$P = P'$$

$$T = T'$$

$$s = s'$$

$$m = m'.$$

Notice that this notion of equivalence is one of several concepts which are both weaker and stronger than the "natural" notion of isomorphism; * it is weaker in that two equivalent systems do not have the same structure of individual forces, but it is stronger in that two equivalent systems must be kinematically identical and identical in their mass functions.

We leave proofs of the two following theorems as an exercise; they justify use of the term 'equivalence' in Definition 5.

THEOREM 3. *The relation of equivalence between systems of particle mechanics is reflexive, symmetric, and transitive in the set of all systems of particle mechanics.*

The next theorem precisely formulates the idea that the resultant forces are identical in two equivalent systems.

THEOREM 4. *If* $\mathfrak{P} = \langle P, T, m, s, f, g\rangle$ *and* $\mathfrak{P}' = \langle P', T', m', s', f', g'\rangle$ *are two equivalent systems of particle mechanics, then for every p in P and t in T*

$$\sum_{q\in P} f(p, q, t) + g(p, t) = \sum_{q\in P'} f'(p, q, t) + g'(p, t).$$

(Note that since $\mathfrak{P}$ and $\mathfrak{P}'$ are equivalent, both sides of the equation are well-defined.)

The embedding theorem which we want now to prove is that every system of particle mechanics is equivalent to a subsystem of an isolated system. This theorem is closely related to some historically famous positions concerning the foundations of physics. For instance, Roger Joseph Bosco-

* Exactly what *the* natural definition is of isomorphism for systems of particle mechanics is a rather complicated question which we shall not try to discuss here.

vich, the prominent eighteenth-century Jesuit physicist, maintained that the matter of the universe is composed of a finite number of non-extended points, and the only forces in the universe are attractive and repulsive forces acting between the points and satisfying our axioms P5 and P6. All the observed phenomena of nature are to be explained solely in terms of the distribution and motion of these points and the forces between them.* Certainly Boscovich would regard our embedding theorem as a definite step toward establishing his thesis concerning the nature of the universe. Naturally we are making no such claims here, for we do not want to assert that the new particles, added to make up the isolated system in which a given system is embedded, have any real physical existence.

Our embedding theorem is closest to the analysis of the foundations of mechanics provided by the great nineteenth-century physicist Heinrich Hertz, whose *Principles of Mechanics* was first published in Germany in 1894.† The basic idea of his approach is that any complicated mechanical system is to be explained by assuming that it is part of a larger system, the positions of whose particles account for the complicated character of the original system in a natural and simple way. It is to be remarked that Hertz would only consider the embedding of a system in a closed, not merely an isolated, system as a fully satisfactory result. We turn now to our theorem.

THEOREM 5. *Every system of particle mechanics is equivalent to a subsystem of an isolated system of particle mechanics.*‡

PROOF. For simplicity we consider only a two-dimensional system, that is, a system for which the third component of all position and force vectors is zero. Let $\mathfrak{P} = \langle P, T, s, m, f, g \rangle$ be an arbitrary such system of mechanics. The idea of the proof is to *introduce* for each p in P three new particles such that the four particles together are placed symmetrically with respect to the two coordinate axes. We then resolve the total resultant force on p into internal forces between the given particle p and two of the new particles. The remaining new particle is needed to provide an appropriate balance of internal forces on the first two new particles.

We now turn to the formal proof. It is convenient to introduce two functions which single out the first two components of a vector. If $\langle x_1, x_2, x_3 \rangle$ is any vector, then

$$\varphi_1(\langle x_1, x_2, x_3 \rangle) = x_1,$$

$$\varphi_2(\langle x_1, x_2, x_3 \rangle) = x_2.$$

* Boscovich's main treatise is his *Theoria Philosophiae Naturalis*, first published in Vienna in 1758.

† English translation reprinted in 1956 by Dover Publications, New York.

‡ This theorem is closely related to Theorem 8 of the paper by McKinsey, Sugar, and Suppes previously cited.

For simplicity of notation and without any essential loss of generality we may suppose that P has exactly two particles, say,

$$P = \{p, q\}.$$

We now use the system $\mathfrak{P} = \langle P, T, s, m, f, g\rangle$ to define a new system consisting of eight distinct particles:

$$P' = \{p, q, p_1, p_2, p_3, q_1, q_2, q_3\}$$
$$T' = T$$
$$m'(p_i) = m'(p) = m(p) \qquad \text{for} \quad i = 1, 2, 3$$
$$m'(q_i) = m'(q) = m(q) \qquad \text{for} \quad i = 1, 2, 3,$$

and for every t in T'

$$s'(p, t) = s(p, t)$$
$$s'(p_1, t) = \langle -\varphi_1[s(p, t)], \varphi_2[s(p, t)], 0\rangle$$
$$s'(p_2, t) = \langle \varphi_1[s(p, t)], -\varphi_2[s(p, t)], 0\rangle$$
$$s'(p_3, t) = \langle -\varphi_1[s(p, t)], -\varphi_2[s(p, t)], 0\rangle$$
$$s'(q, t) = s(q, t)$$
$$s'(q_1, t) = \langle -\varphi_1[s(q, t)], \varphi_2[s(q, t)], 0\rangle$$
$$s'(q_2, t) = \langle \varphi_1[s(q, t)], -\varphi_2[s(q, t)], 0\rangle$$
$$s'(q_3, t) = \langle -\varphi_1[s(q, t)], -\varphi_2[s(q, t)], 0\rangle$$

and

$$f'(p, q, t) = f'(q, p, t) = \langle 0, 0, 0\rangle.$$

Let

$$F(p, t) = g(p, t) + f(p, q, t)$$
$$F(q, t) = g(q, t) + f(q, p, t),$$

that is, $F(p, t)$ is the total resultant force on particle p at time t in the system $\mathfrak{P}$. Then we set:

$$f'(p, p_1, t) = \langle \varphi_1[F(p, t)], 0, 0\rangle$$
$$f'(p_1, p, t) = -f'(p, p_1, t)$$
$$f'(p, p_2, t) = \langle 0, \varphi_2[F(p, t)], 0\rangle$$
$$f'(p_2, p, t) = -f'(p, p_2, t)$$
$$f'(p, p_3, t) = f'(p_3, p, t) = \langle 0, 0, 0\rangle$$
$$f'(p_1, p_2, t) = f'(p_2, p_1, t) = \langle 0, 0, 0\rangle$$
$$f'(p_1, p_3, t) = \langle 0, \varphi_2[F(p, t)], 0\rangle$$
$$f'(p_3, p_1, t) = -f'(p_1, p_3, t)$$
$$f'(p_2, p_3, t) = \langle \varphi_1[F(p, t)], 0, 0\rangle$$
$$f'(p_3, p_2, t) = -f'(p_2, p_3, t).$$

Replacing 'p' by 'q', 'p_1' by 'q_1' etc., in the immediately preceding ten lines we obtain the new force function f' for the particles q, q_1, q_2, q_3. (There is no need to write down these additional ten lines.) Finally, we define the external force function g' so that for all particles in P' and all t in T' the external force is the null vector $\langle 0, 0, 0\rangle$.

To complete the proof of the theorem two things need to be verified (which we leave as an exercise):

(i) The system $\mathfrak{P}' = \langle P', T', s', m', f', g'\rangle$ is a system of particle mechanics, that is, satisfies Axioms P1–P7 (if $\mathfrak{P}'$ is a system of particle mechanics, it follows at once from the definition of g' that it is isolated).

(ii) $\mathfrak{P}$ is equivalent to a subsystem of $\mathfrak{P}'$. Q.E.D.

The theory of mechanics is like the theory of probability in that many of the most interesting and difficult problems are concerned with special classes of models of the theory. Probably the most famous example in mechanics is the three body problem, that is, the problem of determining the motion of three bodies when the only forces acting on them are the mutual attractive gravitational forces varying inversely as the square of their distances apart. The complete solution of this problem is still not known. Newton's solution of the two body problem constituted a derivation of Kepler's three kinematical laws of motion from the dynamical law of gravitation. Unfortunately the confines of the present section are too limited to permit a further systematic development of mechanics. A certain amount of additional material is included in the exercises.

EXERCISES

1. Using the definitions given at the beginning of this section and familiar facts about real numbers, such as the axioms and theorems of Chapter 7, prove:

(a) Vector addition is commutative and associative.

(b) Multiplication of a vector by a real number is distributive with respect to vector addition, i.e., if a is a real number and x and y are vectors, then

$$a(x + y) = ax + ay.$$

(c) Multiplication of a vector by a real number is distributive with respect to addition of real numbers, i.e., if a and b are real numbers and x is a vector, then

$$(a + b)x = ax + bx.$$

(d) Multiplication of a vector by a real number is associative, i.e., if a and b are real numbers and x is a vector then

$$a(bx) = (ab)x.$$

(e) The scalar product operation is commutative.

(f) The scalar product operation is distributive with respect to vector addition, i.e., if x, y, and z are vectors, then

$$x \cdot (y + z) = x \cdot y + x \cdot z.$$

(g) The vector product operation is distributive with respect to vector addition, i.e., if x, y, and z are vectors, then

$$x \times (y + z) = x \times y + x \times z.$$

(h) The vector product operation is anticommutative, i.e., if x and y are vectors, then

$$x \times y = -y \times x.$$

(i) (SCHWARZ'S INEQUALITY) If x and y are vectors, then

$$|x \cdot y| \leq |x|\,|y|.$$

2. The set of all vectors is a group with respect to which of the following operations (if so prove it, if not, give a counterexample):

(a) Vector addition.
(b) Vector subtraction.
(c) Scalar product operation.
(d) Vector product operation.

3. Rewrite Definition 1 in the style of Definition A of § 12.2.
4. Prove Theorem 2.
5. Prove Theorem 3.
6. Prove Theorem 4.
7. Complete the proof of Theorem 5.
8. For general three-dimensional systems of mechanics how many new particles must be added when embedding the system in an isolated system? Prove Theorem 5 for general three-dimensional systems.
9. Using Padoa's principle prove that the notions of mass and internal force are each independent of the remaining primitive notions of particle mechanics.
10. Prove by the method of interpretation that each of the dynamical axioms of Definition 1 are independent of the remaining axioms.
11. Define the moment of a force about a fixed point (points and vectors are the same entities, namely, ordered triples of real numbers) and prove that the total moment of the internal forces of a system of particle mechanics is the null vector.*
12. Define the angular momentum of a system of particles about a fixed point, and prove that the rate of change of the angular momentum of a system about a point is equal to the total moment of the external forces about this point.
13. Exactly formulate and derive from the axioms of Definition 1 a form of Lagrange's equations for systems of particle mechanics.
14. Derive Kepler's three laws of motion from the Newtonian hypothesis concerning gravitational forces.

* This and the remaining exercises require some intuitive knowledge of particle mechanics, but the usual definitions should be reformulated as elementary definitions satisfying the rules of Chapter 8.

INDEX

A CATALOG OF SELECTED

DOVER BOOKS

IN SCIENCE AND MATHEMATICS

A CATALOG OF SELECTED
DOVER BOOKS
IN SCIENCE AND MATHEMATICS

QUALITATIVE THEORY OF DIFFERENTIAL EQUATIONS, V.V. Nemytskii and V.V. Stepanov. Classic graduate-level text by two prominent Soviet mathematicians covers classical differential equations as well as topological dynamics and ergodic theory. Bibliographies. 523pp. 5⅜ x 8½. 65954-2 Pa. $14.95

MATRICES AND LINEAR ALGEBRA, Hans Schneider and George Phillip Barker. Basic textbook covers theory of matrices and its applications to systems of linear equations and related topics such as determinants, eigenvalues and differential equations. Numerous exercises. 432pp. 5⅜ x 8½. 66014-1 Pa. $12.95

QUANTUM THEORY, David Bohm. This advanced undergraduate-level text presents the quantum theory in terms of qualitative and imaginative concepts, followed by specific applications worked out in mathematical detail. Preface. Index. 655pp. 5⅜ x 8½. 65969-0 Pa. $15.95

ATOMIC PHYSICS (8th edition), Max Born. Nobel laureate's lucid treatment of kinetic theory of gases, elementary particles, nuclear atom, wave-corpuscles, atomic structure and spectral lines, much more. Over 40 appendices, bibliography. 495pp. 5⅜ x 8½. 65984-4 Pa. $13.95

ELECTRONIC STRUCTURE AND THE PROPERTIES OF SOLIDS: The Physics of the Chemical Bond, Walter A. Harrison. Innovative text offers basic understanding of the electronic structure of covalent and ionic solids, simple metals, transition metals and their compounds. Problems. 1980 edition. 582pp. 6⅛ x 9¼. 66021-4 Pa. $19.95

BOUNDARY VALUE PROBLEMS OF HEAT CONDUCTION, M. Necati Özisik. Systematic, comprehensive treatment of modern mathematical methods of solving problems in heat conduction and diffusion. Numerous examples and problems. Selected references. Appendices. 505pp. 5⅜ x 8½. 65990-9 Pa. $12.95

A SHORT HISTORY OF CHEMISTRY (3rd edition), J.R. Partington. Classic exposition explores origins of chemistry, alchemy, early medical chemistry, nature of atmosphere, theory of valency, laws and structure of atomic theory, much more. 428pp. 5⅜ x 8½. (Available in U.S. only) 65977-1 Pa. $12.95

A HISTORY OF ASTRONOMY, A. Pannekoek. Well-balanced, carefully reasoned study covers such topics as Ptolemaic theory, work of Copernicus, Kepler, Newton, Eddington's work on stars, much more. Illustrated. References. 521pp. 5⅜ x 8½. 65994-1 Pa. $15.95

PRINCIPLES OF METEOROLOGICAL ANALYSIS, Walter J. Saucier. Highly respected, abundantly illustrated classic reviews atmospheric variables, hydrostatics, static stability, various analyses (scalar, cross-section, isobaric, isentropic, more). For intermediate meteorology students. 454pp. 6⅛ x 9¼. 65979-8 Pa. $14.95

SPECIAL FUNCTIONS, N.N. Lebedev. Translated by Richard Silverman. Famous Russian work treating more important special functions, with applications to specific problems of physics and engineering. 38 figures. 308pp. 5⅜ x 8½. 60624-4 Pa. $9.95

THE EXTRATERRESTRIAL LIFE DEBATE, 1750–1900, Michael J. Crowe. First detailed, scholarly study in English of the many ideas that developed between 1750 and 1900 regarding the existence of intelligent extraterrestrial life. Examines ideas of Kant, Herschel, Voltaire, Percival Lowell, many other scientists and thinkers. 16 illustrations. 704pp. 5⅜ x 8½. 40675-X Pa. $19.95

INTEGRAL EQUATIONS, F.G. Tricomi. Authoritative, well-written treatment of extremely useful mathematical tool with wide applications. Volterra Equations, Fredholm Equations, much more. Advanced undergraduate to graduate level. Exercises. Bibliography. 238pp. 5⅜ x 8½. 64828-1 Pa. $8.95

POPULAR LECTURES ON MATHEMATICAL LOGIC, Hao Wang. Noted logician's lucid treatment of historical developments, set theory, model theory, recursion theory and constructivism, proof theory, more. 3 appendixes. Bibliography. 1981 edition. ix + 283pp. 5⅜ x 8½. 67632-3 Pa. $10.95

MODERN NONLINEAR EQUATIONS, Thomas L. Saaty. Emphasizes practical solution of problems; covers seven types of equations. ". . . a welcome contribution to the existing literature...."–*Math Reviews*. 490pp. 5⅜ x 8½. 64232-1 Pa. $13.95

FUNDAMENTALS OF ASTRODYNAMICS, Roger Bate et al. Modern approach developed by U.S. Air Force Academy. Designed as a first course. Problems, exercises. Numerous illustrations. 455pp. 5⅜ x 8½. 60061-0 Pa. $12.95

INTRODUCTION TO LINEAR ALGEBRA AND DIFFERENTIAL EQUATIONS, John W. Dettman. Excellent text covers complex numbers, determinants, orthonormal bases, Laplace transforms, much more. Exercises with solutions. Undergraduate level. 416pp. 5⅜ x 8½. 65191-6 Pa. $11.95

INCOMPRESSIBLE AERODYNAMICS, edited by Bryan Thwaites. Covers theoretical and experimental treatment of the uniform flow of air and viscous fluids past two-dimensional aerofoils and three-dimensional wings; many other topics. 654pp. 5⅜ x 8½. 65465-6 Pa. $16.95

INTRODUCTION TO DIFFERENCE EQUATIONS, Samuel Goldberg. Exceptionally clear exposition of important discipline with applications to sociology, psychology, economics. Many illustrative examples; over 250 problems. 260pp. 5⅜ x 8½. 65084-7 Pa. $10.95

THREE PEARLS OF NUMBER THEORY, A. Y. Khinchin. Three compelling puzzles require proof of a basic law governing the world of numbers. Challenges concern van der Waerden's theorem, the Landau-Schnirelmann hypothesis and Mann's theorem, and a solution to Waring's problem. Solutions included. 64pp. 5⅜ x 8½. 40026-3 Pa. $4.95

LECTURES ON CLASSICAL DIFFERENTIAL GEOMETRY, Second Edition, Dirk J. Struik. Excellent brief introduction covers curves, theory of surfaces, fundamental equations, geometry on a surface, conformal mapping, other topics. Problems. 240pp. 5⅜ x 8½. 65609-8 Pa. $9.95

NUMERICAL METHODS FOR SCIENTISTS AND ENGINEERS, Richard Hamming. Classic text stresses frequency approach in coverage of algorithms, polynomial approximation, Fourier approximation, exponential approximation, other topics. Revised and enlarged 2nd edition. 721pp. 5⅜ x 8½. 65241-6 Pa. $16.95

THEORETICAL SOLID STATE PHYSICS, Vol. 1: Perfect Lattices in Equilibrium; Vol. II: Non-Equilibrium and Disorder, William Jones and Norman H. March. Monumental reference work covers fundamental theory of equilibrium properties of perfect crystalline solids, non-equilibrium properties, defects and disordered systems. Appendices. Problems. Preface. Diagrams. Index. Bibliography. Total of 1,301pp. 5⅜ x 8½. Two volumes. Vol. I: 65015-4 Pa. $16.95
Vol. II: 65016-2 Pa. $16.95

OPTIMIZATION THEORY WITH APPLICATIONS, Donald A. Pierre. Broad spectrum approach to important topic. Classical theory of minima and maxima, calculus of variations, simplex technique and linear programming, more. Many problems, examples. 640pp. 5⅜ x 8½. 65205-X Pa. $17.95

THE CONTINUUM: A Critical Examination of the Foundation of Analysis, Hermann Weyl. Classic of 20th-century foundational research deals with the conceptual problem posed by the continuum. 156pp. 5⅜ x 8½. 67982-9 Pa. $8.95

ESSAYS ON THE THEORY OF NUMBERS, Richard Dedekind. Two classic essays by great German mathematician: on the theory of irrational numbers; and on transfinite numbers and properties of natural numbers. 115pp. 5⅜ x 8½.
21010-3 Pa. $6.95

THE FUNCTIONS OF MATHEMATICAL PHYSICS, Harry Hochstadt. Comprehensive treatment of orthogonal polynomials, hypergeometric functions, Hill's equation, much more. Bibliography. Index. 322pp. 5⅜ x 8½. 65214-9 Pa. $12.95

NUMBER THEORY AND ITS HISTORY, Oystein Ore. Unusually clear, accessible introduction covers counting, properties of numbers, prime numbers, much more. Bibliography. 380pp. 5⅜ x 8½. 65620-9 Pa. $10.95

THE VARIATIONAL PRINCIPLES OF MECHANICS, Cornelius Lanczos. Graduate level coverage of calculus of variations, equations of motion, relativistic mechanics, more. First inexpensive paperbound edition of classic treatise. Index. Bibliography. 418pp. 5⅜ x 8½. 65067-7 Pa. $14.95

COMBINATORIAL TOPOLOGY, P. S. Alexandrov. Clearly written, well-organized, three-part text begins by dealing with certain classic problems without using the formal techniques of homology theory and advances to the central concept, the Betti groups. Numerous detailed examples. 654pp. 5⅜ x 8½. 40179-0 Pa. $18.95

THEORETICAL PHYSICS, Georg Joos, with Ira M. Freeman. Classic overview covers essential math, mechanics, electromagnetic theory, thermodynamics, quantum mechanics, nuclear physics, other topics. First paperback edition. xxiii + 885pp. 5⅜ x 8½. 65227-0 Pa. $21.95